Detection Theory: A User's Guide

(2nd edition)

NEIL A. MACMILLAN
University of Massachusetts

and

C. DOUGLAS CREELMAN
University of Toronto

LEA
2005

LAWRENCE ERLBAUM ASSOCIATES, PUBLISHERS
Mahwah, New Jersey   London

Copyright © 2005 by Lawrence Erlbaum Associates, Inc.

Lawrence Erlbaum Associates, Inc., Publishers
10 Industrial Avenue
Mahwah, New Jersey 07430

Cover design by Kathryn Houghtaling Lacey

Library of Congress Cataloging-in-Publication Data

Macmillan, Neil A.
Detection theory : a user's guide / Neil A. Macmillan, C. Douglas Creelman.
—2nd ed.

    p. cm.

Includes bibliographical references and index.
ISBN 0-8058-4230-6 (cloth : alk. paper)
ISBN 0-8058-4231-4 (pbk. : alk. paper)
1. Signal detection (Psychology).   I. Creelman, C. Douglas.   II. Title.
BF237.M25   2004
152.8—dc22   2004043261
                                              CIP

Books published by Lawrence Erlbaum Associates are printed on acid-free paper, and their bindings are chosen for strength and durability.

Printed in the United States of America
10   9   8   7   6   5   4   3   2

Detection Theory:
A User's Guide

(2nd edition)

To

David M. Green, R. Duncan Luce, John A. Swets,
and the memory of Wilson P. Tanner, Jr.

Contents

Preface

Detection theory entered psychology as a way to explain detection experiments, in which weak visual or auditory signals must be distinguished from a "noisy" background. In *Signal Detection Theory and Psychophysics* (1966), David Green and John Swets portrayed observers as decision makers trying to optimize performance in the face of unpredictable variability, and they prescribed experimental methods and data analyses for separating decision factors from sensory ones.

Since Green and Swets' classic was published, both the content of detection theory and the way it is used have changed. The theory has deepened to include alternative theoretical assumptions and has been used to analyze many experimental tasks. The range of substantive problems to which the theory has been applied has broadened greatly. The contemporary user of detection theory may be a sensory psychologist, but more typically is interested in memory, cognition, or systems for medical or nonmedical diagnosis. In this book, we draw heavily on the work of Green, Swets, and other pioneers, but aim for a seamless meshing of historical beginnings and current perspective. In recognition that these methods are often used in situations far from the original problem of finding a "signal" in background noise, we have omitted the word *signal* from the title and usually refer to these methods simply as *detection theory*.

We are writing with two types of readers in mind: those learning detection theory, and those applying it. For those encountering detection theory for the first time, this book is a textbook. It could be the basic text in a one-semester graduate or upper level undergraduate course, or it could be a supplementary text in a broader course on psychophysics, methodology, or a substantive topic. We imagine a student who has survived one semester of "behavioral" statistics at the undergraduate level, and have tried to make the book accessible to such a person in several ways. First, we provide appendixes on probability and statistics (Appendix 1) and logarithms (Appendix 2). Second, there are a large number of problems, some with answers. Third, to the extent possible, the more complex mathematical derivations have been placed in "Computational Appendixes" at the ends of chapters. Finally, some conceptually advanced but essential ideas, especially from multidimensional detection theory, are presented in tutorial detail.

For researchers who use detection theory, this book is a handbook. As far as possible, the material needed to apply the described techniques is complete in the book. A road map to most methods is provided by the flowcharts of Appendix 3, which direct the user to appropriate equations (Appendix 4) and tables (Appendix 5). The software appendix (Appendix 6) provides a listing of a program for finding the most common detection theory statistics, and directions to standard software and Web sites for a wide range of calculations.

An important difference between this second edition and its predecessor is the prominence of multidimensional detection theory, to which the five chapters of Part II are devoted. This topic was covered in a single chapter of the first edition, and the increase is due to two factors. First, there has been an explosion of multidimensional applications in the past decade or so. Second, one essential area of detection theory—the analysis of different discrimination paradigms—requires multidimensional methods that were introduced in passing in the first edition, but are now integrated into a systematic presentation of these methods. Someone concerned only with analyzing specific paradigms will be most interested in chapters 1 to 3, 5, 7, 9, and 10. The intervening chapters provide greater theoretical depth (chaps. 4 and 8) as well as a careful introduction to multidimensional analysis (chap. 6).

The flowcharts (Appendix 3) are inspired by similar charts in *Behavioral Statistics* by R. B. Darlington and P. M. Carlson (1987). We thank Pat Carlson for persuasive discussions of the value of this tool and for helping us use it to best advantage.

We are grateful to many people who helped us complete this project. We taught courses based on preliminary drafts at Brooklyn College and the University of Massachusetts. Colleagues used parts of the book in courses at Purdue University (Hong Tan), the University of California at San Diego (John Wixted), and the University of Florida (Bob Sorkin). We thank these instructors and their students for providing us with feedback. We owe a debt to many other colleagues who commented on one or more chapters in preliminary drafts, and we particularly wish to thank Danny Algom, Michael Hautus, John Irwin, Marjorie Leek, Todd Maddox, Dawn Morales, Jeff

Miller, and Dick Pastore. Caren Rotello's comments, covering almost the entire book, were consistently both telling and supportive.

Our warmest appreciation and thanks go to our wives, Judy Mullins (Macmillan) and Lynne Beal (Creelman), for their generous support and patience with a project that —like the first edition—provided serious competition for their company.

We also thank Bill Webber, our editor, and Lawrence Erlbaum Associates for adopting this project and making it their own.

Finally, we continue to feel a great debt to the parents of detection theory. Among many who contributed to the theory in its early days, our thinking owes the most to four people. We dedicate this book to David M. Green, R. Duncan Luce, and John A. Swets, and to the memory of Wilson P. Tanner, Jr. Without them there would be no users for us to guide.

Introduction

Detection theory is a general psychophysical approach to measuring performance. Its scope includes the everyday experimentation of many psychologists, social and medical scientists, and students of decision processes. Among the problems to which it can be applied are these:

- assessing a person's ability to recognize whether a photograph is of someone previously seen or someone new,
- measuring the skill of a medical diagnostician in distinguishing X-rays displaying tumors from those showing healthy tissue,
- finding the intensity of a sound that can be heard 80% of the time, and
- determining whether a person can identify which of several words has been presented on a screen, and whether identification is still possible if the person reports that a word has not appeared at all.

In each of these situations, the person whose performance we are studying encounters stimuli of different types and must assign distinct responses to them. There is a *correspondence*[1] between the stimuli and the responses so that each response belongs with one of the stimulus classes. The viewer of photographs, for example, is presented with some photos of Old,[2] previously seen faces, as well as some that are New, and must respond "old" to the Old faces and "new" to the New. Accurate performance consists of using the corresponding responses as defined by the experimenter.

A *correspondence experiment* is one in which each possible stimulus is assigned a correct response from a finite set. In *complete correspondence experiments*, which include all the designs in chapters 1, 2, 4, 6, 7, 9, 10, and 11, this partition is rigidly set by the experimenter. In *incomplete corre-*

---

[1]Most italicized words are defined in the Glossary.
[2]Throughout the book, we capitalize the names of stimuli and stimulus classes.

spondence experiments (such as the rating design described in chap. 3 and the classification tasks of chap. 5), there is a class of possible correspondences, each describing ideal performance.

Correspondence provides an objective standard or expectation against which to evaluate performance. Detection theory measures the discrepancy between the two and may therefore be viewed as a technique for understanding error. Errors are assumed to arise from inevitable variability, either in the stimulus input or within the observer. If this noise does not appreciably affect performance, responses correspond perfectly to stimuli, and their correctness provides no useful information. Response time is often the dependent variable in such situations, and models for interpreting this performance measure are well developed (Luce, 1986).

The possibility of error generally brings with it the possibility of different kinds of errors—*misses* and *false alarms*. Medical diagnosticians can miss the shadow of a tumor on an X-ray or raise a false alarm by reporting the presence of one that is not there. A previously encountered face may be forgotten or a new one may be falsely recognized as familiar. The two types of error typically have different consequences, as these examples make clear: If the viewer of photographs is in fact an eyewitness to a crime, a miss will result in the guilty going free, a false alarm in the innocent being accused. A reasonable goal of a training program for X-ray readers would be to encourage an appropriate balance between misses and false alarms (in particular, to keep the number of misses very small).

Detection theory, then, provides a method for measuring people's accuracy (and understanding their errors) in correspondence experiments. This is not a definition—we offer a tentative one at the end of chapter 1—but may suggest the directions in which a discussion of the theory must lead.

Organization of the Book

This book is divided into four parts. Part I describes the measurement of sensitivity and response bias in situations that are experimentally and theoretically the simplest. One stimulus is presented on each trial, and the representation of the stimuli is one dimensional. In Part II, multidimensional representations are used, allowing the analysis of a variety of classification and identification experiments. Common but complex discrimination designs in which two or more stimuli are presented on each trial are a special case. In Part III, we consider two important topics in which stimulus characteristics are central. Chapter 11 discusses adaptive techniques for the estimation of thresholds. Chapter 12 describes ways in which detection theory can be used to relate sensitivity to stimulus parameters and partition sensitivity into its components. Part IV (chap.13) offers some statistical procedures for evaluating correspondence data.

Organization of Each Chapter

Each chapter is organized around one or more examples modeled on experiments that have been reported in the behavioral literature. (We do not attempt to reanalyze actual experiments, which are always more complicated than the pedagogical uses to which we might put them.) For each design, we present one or more appropriate methods for analyzing the illustrative data. The examples make our points concrete and suggest the breadth of application of detection theory, but they are not prescriptive: The use of a recognition memory task to illustrate the two-alternative forced-choice paradigm (chap. 7) does not mean, for instance, that we believe this design to be the only or even the best tool for studying recognition memory. The appropriate design for studying a particular topic should always be dictated by practical and theoretical aspects of the content area.

The book as a whole represents our opinions about how best to apply detection theory. For the most part, our recommendations are not controversial, but in some places we have occasion to be speculative, argumentative, or curmudgeonly. Sections in which we take a broader, narrower, or more peculiar view than usual are labeled *essays* as a warning to the reader.

I

Basic Detection Theory
and One-Interval Designs

Part I introduces the *one-interval design*, in which a single stimulus is presented on each trial. The simplest and most important example is a correspondence experiment in which the stimulus is drawn from one of two stimulus classes and the observer tries to say from which class it is drawn. In auditory experiments, for example, the two stimuli might be a weak tone and no sound, tone sequences that may be slow or fast, or passages from the works of Mozart and Beethoven.

We begin by describing the use of one-interval designs to measure *discrimination*, the ability to tell two stimuli apart. Two types of such experiments may be distinguished. If one of the two stimulus classes contains only the null stimulus, as in the tone-versus-background experiment, the task is called *detection*. (This historically important application is responsible for the use of the term *detection theory* to refer to these methods.) If neither stimulus is null, the experiment is called *recognition*, as in the other examples. The methods for analyzing detection and recognition are the same, and we make no distinction between them (until chap. 10, where we consider experiments in which the two tasks are combined).

In chapters 1 and 2, we focus on designs with two possible responses as well as two stimulus classes. Because the possible responses in some applications (e.g., the tone detection experiment) are "yes" and "no," the paradigm with two stimuli, one interval, and two responses is sometimes termed *yes-no* even when the actual responses are, say, "slow" and "fast." Performance can be analyzed into two distinct elements: the degree to which the observer's responses mirror the stimuli (chap. 1) and the degree to which they display bias (chap. 2). Measuring these two elements requires a theory; we use the most common, normal-distribution variant of detection theory to accomplish this end. Chapter 4 broadens the perspective on yes-no sensitivity and bias to include three classes of alternatives to this model: threshold theory, choice theory, and "nonparametric" techniques.

One-interval experiments may involve more than two responses or more than two possible stimuli. As an example of a larger response set, listeners could rate the likelihood that a passage was composed by Mozart rather than Beethoven on a 6-point scale. One-interval rating designs are discussed in chapter 3. As an example of a larger stimulus set, listeners could hear sequences presented at one of several different rates. If the requirement is to assign a different response to each stimulus, the task is called *identification*; if the stimuli are to be sorted into a smaller number of classes (perhaps slow, medium, and fast), it is *classification*. Chapter 5 applies detection-theory tools to identification and classification tasks, but only those in which elements of the stimulus sets differ in a single characteristic such as tempo. Identification and classification of more heterogeneous stimulus sets are considered in Part II.

1

The Yes-No Experiment: Sensitivity

In this book, we analyze experiments that measure the ability to distinguish between stimuli. An important characteristic of such experiments is that observers can be more or less *accurate*. For example, a radiologist's goal is to identify accurately those X-rays that display abnormalities, and participants in a recognition memory study are accurate to the degree that they can tell previously presented stimuli from novel ones. Measures of performance in these kinds of tasks are also called *sensitivity measures: High sensitivity* refers to good ability to discriminate, *low sensitivity* to poor ability. This is a natural term in detection studies—a sensitive listener hears things an insensitive one does not—but it applies as well to the radiology and memory examples.

Understanding Yes-No Data

Example 1: Face Recognition

We begin with a memory experiment. In a task relevant to understanding eyewitness testimony in the courtroom, participants are presented with a series of slides portraying people's faces, perhaps with the instruction to remember them. After a period of time (and perhaps some unrelated activity), recognition is tested by presenting the same participants with a second series that includes some of the same pictures, shuffled to a new random order, along with a number of "lures"—faces that were not in the original set. Memory is good if the person doing the remembering properly recognizes the Old faces, but not New ones. We wish to measure the ability to distinguish between these two classes of slides. Experiments of this sort have been performed to compare memory for faces of different races, orientations (upright vs. inverted), and many other variables (for a review, see Shapiro & Penrod, 1986).

3

Let us look at some (hypothetical) data from such a task. We are interested in just one characteristic of each picture: whether it is an Old face (one
presented earlier) or a New face. Because the experiment concerns two
kinds of faces and two possible responses, "yes" (I've seen this person before in this experiment) and "no" (I haven't), any of four types of events can
occur on a single experimental trial. The number of trials of each type can be
tabulated in a stimulus-response matrix like the following.

Stimulus Class	*Response*		
	"Yes"	"No"	Total
Old	20	5	25
New	10	15	25

The purpose of this yes-no task is to determine the participant's sensitivity to the Old/New difference. High sensitivity is indicated by a concentra-
tion of trials counted in the upper left and lower right of the matrix ("yes"
responses to Old stimuli, "no" responses to New).

Summarizing the Data

Conventional, rather military language is used to describe the yes-no experiment. Correctly recognizing an Old item is termed a *hit*; failing to recognize it, a *miss*. Mistakenly recognizing a New item as old is a *false alarm*;
correctly responding "no" to an Old item is, abandoning the metaphor, a
correct rejection. In tabular terms:

Stimulus Class	*Response*		
	"Yes"	"No"	Total
Old (S_2)	Hits	Misses	
	(20)	(5)	(25)
New (S_1)	False alarms	Correct rejections	
	(10)	(15)	(25)

We use S_1 and S_2 as context-free names for the two stimulus classes.

Of the four numbers in the table (excluding the marginal totals), only two
provide independent information about the participant's performance.
Once we know, for example, the number of hits and false alarms, the other
two entries are determined by how many Old and New items the experimenter decided to use (25 of each, in this case). Dividing each number by

the total in its row allows us to summarize the table by two numbers: The *hit rate* (*H*) is the proportion of Old trials to which the participant responded "yes," and the *false-alarm rate* (*F*) is the proportion of New trials similarly (but incorrectly) assessed. The hit and false-alarm rates can be written as conditional probabilities[1]

$$H = P(\text{"yes"} | S_2) \tag{1.1}$$

$$F = P(\text{"yes"} | S_1), \tag{1.2}$$

where Equation 1.1 is read "The proportion of 'yes' responses when stimulus S_2 is presented."

In this example, $H = .8$ and $F = .4$. The entire matrix can be rewritten with response rates (or proportions) rather than frequencies:

Stimulus Class	Response		
	"Yes"	"No"	Total
Old (S_2)	.8	.2	1.0
New (S_1)	.4	.6	1.0

The two numbers needed to summarize an observer's performance, *F* and *H*, are denoted as an ordered (*false-alarm, hit*) pair. In our example, (*F, H*) = (.4, .8).

Measuring Sensitivity

We now seek a good way to characterize the observer's sensitivity. A function of *H* and *F* that attempts to capture this ability of the observer is called a sensitivity *measure*, *index*, or *statistic*. A perfectly sensitive participant would have a hit rate of 1 and a false-alarm rate of 0. A completely insensitive participant would be unable to distinguish the two stimuli at all and, indeed, could perform equally well without attending to them. For this observer, the probability of saying "yes" would not depend on the stimulus presented, so the hit and false-alarm rates would be the same. In interesting cases, sensitivity falls between these extremes: *H* is greater than *F*, but performance is not perfect.

---

[1] Technically, *H* and *F* are *estimates* of probabilities—a distinction that is important in statistical work (chap. 13). Probabilities characterize the observer's relation to the stimuli and are considered stable and unchanging; *H* and *F* may vary from one block of trials to the next.

The simplest possibility is to ignore one of our two response rates using, say, *H* to measure performance. For example, a lie detector might be touted as detecting 80% of liars or an X-ray reader as detecting 80% of tumors. (Alternatively, the hit rate might be ignored, and evaluation might depend totally on the false-alarm rate.) Such a measure is clearly inadequate. Compare the memory performance we have been examining with that of another group:

Stimulus Class	*Response*		
	"Yes"	"No"	Total
Old	8	17	25
New	1	24	25

Group 1 successfully recognized 80% of the Old words, Group 2 just 32%. But this comparison ignores the important fact that Group 2 participants just did not say "yes" very often. The hit rate, or any measure that depends on responses to only one of the two stimulus classes, cannot be a measure of sensitivity. To speak of sensitivity to a stimulus (as was done, for instance, in early psychophysics) is meaningless in the framework of detection theory.[2]

An important characteristic of sensitivity is that it can only be measured between two alternative stimuli and must therefore depend on both *H* and *F*. A moment's thought reveals that not all possible dependencies will do. Certainly a higher hit rate means greater, not less, sensitivity, whereas a higher false-alarm rate is an indicator of less sensitive performance. So a sensitivity measure should increase when either *H* increases or *F* decreases.

A final possible characteristic of sensitivity measures is that S_1 and S_2 trials should have equal importance: Missing an Old item is just as important an error as incorrectly recognizing a New one. In general, this is too strong a requirement, and we will encounter sensitivity measures that assign different weights to the two stimulus classes. Nevertheless, equal treatment is a good starting point, and (with one exception) the indexes described in this chapter satisfy it.

---

[2]The term *sensitivity is* used in this way, as a synonym for the hit rate, in medical diagnosis. *Specificity* is that field's term for the correct-rejection rate.

Two Simple Solutions

We are looking for a measure that goes up when H goes up, goes down when F goes up, and assigns equal importance to these statistics. How about simply subtracting F from H? The difference $H - F$ has all these characteristics. For the first group of memory participants, $H - F = .8 - .4 = .4$; for the second, $H - F = .32 - .04 = .28$, and Group 1 wins.

Another measure that combines H and F in this way is a familiar statistic, the proportion of correct responses, which we denote $p(c)$. To find proportion correct in conditions with equal numbers of S_1 and S_2 trials, we take the average of the proportion correct on S_2 trials (the hit rate, H) and the proportion correct on S_1 trials (the correct rejection rate, $1 - F$). Thus:

$$p(c) = \tfrac{1}{2}[H + (1 - F)]$$
$$= \tfrac{1}{2}(H - F) + \tfrac{1}{2} \cdot \tag{1.3}$$

If the numbers of S_1 and S_2 trials are not equal, then to find the literal proportion of trials on which a correct answer was given the actual numbers in the matrix would have to be used:

$$p(c)^* = (\text{hits} + \text{correct rejections})/\text{total trials} \; . \tag{1.4}$$

Usually it is more sensible to give H and F equal weight, as in Equation 1.3, because a sensitivity measure should not depend on the base presentation rate.

Let us look at $p(c)$ for equal presentations (Eq. 1.3). Is this a better or worse measure of sensitivity than $H - F$ itself? Neither. Because $p(c)$ depends directly on $H - F$ (and not on either H or F separately), one statistic goes up whenever the other does, and the two are monotonic functions of each other. Two measures that are monotonically related in this way are said to be *equivalent* measures of accuracy. In the running examples, $p(c)$ is .7 for Group 1 and .64 for Group 2, and $p(c)$ leads to the same conclusion as $H - F$. For both measures, Group 1 outscores Group 2.

A Detection Theory Solution

The most widely used sensitivity measure of detection theory (Green & Swets, 1966) is not quite as simple as $p(c)$, but bears an obvious family resemblance. The measure is called d' ("dee-prime") and is defined in terms of z, the inverse of the normal distribution function:

$$d' = z(H) - z(F) . \tag{1.5}$$

The z transformation converts a hit or false-alarm rate to a z score (i.e., to standard deviation units). A proportion of .5 is converted into a z score of 0, larger proportions into positive z scores, and smaller proportions into negative ones. To compute z, consult Table A5.1 in Appendix 5. The table makes use of a symmetry property of z scores: Two proportions equally far from .5 lead to the same absolute z score (positive if $p > .5$, negative if $p < .5$) so that:

$$z(1 - p) = -z(p) . \tag{1.6}$$

Thus, $z(.4) = -.253$, the negative of $z(.6)$. Use of the Gaussian z transformation is dominant in detection theory, and we often refer to normal-distribution models by the abbreviation *SDT*.

We can use Equation 1.5 to calculate d' for the data in the memory example. For Group 1, $H = .8$ and $F = .4$, so $z(H) = 0.842$, $z(F) = -0.253$, and $d' = 0.842 - (-0.253) = 1.095$. When the hit rate is greater than .5 and the false-alarm rate is less (as in this case), d' can be obtained by adding the absolute values of the corresponding z scores. For Group 2, $H = .32$ and $F = .04$, so $d' = -0.468 - (-1.751) = 1.283$. When the hit and false-alarm rates are on the same side of .5, d' is obtained by subtracting the absolute values of the z scores. Interestingly, by the d' measure, it is Group 2 (the one that was much more stingy with "yes" responses) rather than Group 1 that has the superior memory.

When observers cannot discriminate at all, $H = F$ and $d' = 0$. Inability to discriminate means having the same rate of saying "yes" when Old faces are presented as when New ones are offered. As long as $H \geq F$, d' must be greater than or equal to 0. The largest possible *finite* value of d' depends on the number of decimal places to which H and F are carried. When $H = .99$ and $F = .01$, $d' = 4.65$; many experimenters consider this an effective ceiling.

Perfect accuracy, on the other hand, implies an infinite d'. Two adjustments to avoid infinite values are in common use. One strategy is to convert proportions of 0 and 1 to $1/(2N)$ and $1 - 1/(2N)$, respectively, where N is the number of trials on which the proportion is based. Suppose a participant has 25 hits and 0 misses ($H = 1.0$) to go with 10 false alarms and 15 correct rejections ($F = .4$). The adjustment yields 24.5 hits and 0.5 misses, so $H = .98$ and $d' = 2.054 - (-0.253) = 2.307$. A second strategy (Hautus, 1995; Miller,

1996) is to add 0.5 to *all* data cells regardless of whether zeroes are present. This adjustment leads to $H = 25.5/26 = .981$ and $F = 10.5/26 = .404$. Rounding to two decimal places yields the same value as before, but d' is slightly smaller if computed exactly.

Most experiments avoid chance and perfect performance. Proportions correct between .6 and .9 correspond roughly to d' values between 0.5 and 2.5. Correct performance on 75% of both S_1 and S_2 trials yields a d' of 1.35; 69% for both stimuli gives $d' = 1.0$.

It is sometimes important to calculate d' when only $p(c)$ is known, not H and F. (Partial ignorance of this sort is common when reanalyzing published data.) Strictly speaking, the calculation cannot be done, but an approximation can be made by assuming that the hit rate equals the correct rejection rate so that $H = 1 - F$. For example, if $p(c) = .9$, we can guess at a measure for sensitivity: $d' = z(.9) - z(.1) = 1.282 - (-1.282) = 2.56$. To simplify the calculation, notice that one z score is the negative of the other (Eq. 1.6). Hence, in this special case:

$$d' = 2 \, z[p(c)] \, . \tag{1.7}$$

This calculation is *not* correct in general. For example, suppose $H = .99$ and $F = .19$, so that H and the correct rejection rate are not equal. Then $p(c)$ still equals .9, but $d' = z(.99) - z(.19) = 2.326 - (-0.878) = 3.20$ instead of 2.56, a considerable discrepancy.

Implied ROCs

ROC Space and Isosensitivity Curves

What justifies the use of d' as a summary of discrimination? Why is this measure better, according to detection theory, than the more familiar $p(c)$? A good sensitivity measure should be invariant when factors other than sensitivity change. Participants are assumed by detection theory to have a fixed sensitivity when asked to discriminate a specific pair of stimulus classes. One aspect of responding that is up to them, however, is their willingness to respond "yes" rather than "no." If d' is an invariant measure of sensitivity, then a participant whose false-alarm and hit rates are (.4, .8) can also produce the performance pairs (.2, .6) and (.07, .35); all of these pairs indicate a d' of about 1.09, and differ only in response bias.

The locus of (false-alarm, hit) pairs yielding a constant d' is called an *isosensitivity* curve because all points on the curve have the same sensitivity.

This term was proposed by Luce (1963a) as more descriptive that the original engineering nomenclature *receiver operating characteristic* (ROC). Swets (1973) reinterpreted the acronym to mean *relative operating characteristic*. We use all these terms interchangeably.

Figure 1.1 shows ROCs implied by d'. The axes of the ROC are the false-alarm rate, on the horizontal axis, and the hit rate, plotted vertically. Because both H and F range from 0 to 1, the *ROC space*, the region in which ROCs must lie, is the unit square. For every value of the false-alarm rate, the plot shows the hit rate that would be obtained to yield a particular sensitivity level. Algebraically, these curves are calculated by solving Equation 1.5 for H; different curves represent different values of d'.

When performance is at chance ($d' = 0$), the ROC is the major diagonal, where the hit and false-alarm rates are equal. For this reason, the major diagonal is sometimes called the *chance line*. As sensitivity increases, the curves shift toward the upper left corner, where accuracy is perfect ($F = 0$ and $H = 1$). These ROC curves summarize the predictions of detection theory: If an observer in a discrimination experiment produces a (F, H) pair that lies on a particular implied ROC, that observer should be able to display any other (F, H) pair on the same curve.

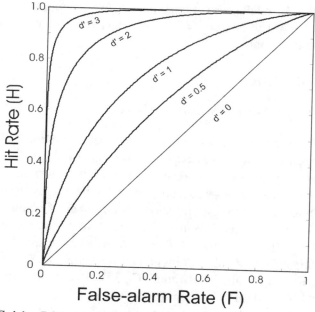

FIG. 1.1.   ROCs for SDT on linear coordinates. Curves connect locations with constant d'.

The theoretical isosensitivity curves in Fig. 1.1 have two important characteristics. First, the price of complete success in recognizing one stimulus class is complete failure in recognizing the other. For example, to be perfectly correct with Old faces and have a hit rate of 1, it is also necessary to have a false-alarm rate of 1, indicating total failure to correctly reject New faces. Similarly, a false-alarm rate of 0 can be obtained only if the hit rate is 0. Isosensitivity curves that pass through (0, 0) and (1, 1) are called *regular* (Swets & Pickett, 1982).

Second, the slope of these curves decreases as the tendency to respond "yes" increases. The slope is the change in the hit rate, relative to the change in the false-alarm rate, that results from increasing response bias toward "yes." We shall see in a later section that this systematic slope change characterizes all ROCs.

ROCs in Transformed Coordinates

The features of regularity and decreasing slope are clear in Fig. 1.1, but other aspects of ROC shape are easier to see using a different representation of the ROC, one that takes advantage of our earlier description of a sensitivity measure as the difference between the transformed hit and false-alarm rates.

Look again at Equation 1.5, which describes the isosensitivity curve for d'. To find an algebraic expression for the ROC, we would need to solve this equation for H as a function of F. A simpler task is to solve for $z(H)$ as a function of $z(F)$:

$$z(H) = z(F) + d' \ .$$

$$(1.8)$$

Equation 1.8 describes a *transformed ROC*, specifically a zROC, in which both axes are marked off in equal z scores rather than in equal proportion units. The range of values in these new units is from minus to plus infinity, although scores of more than 2.5 (i.e., 2.5 standard deviations from the mean) are rarely encountered. In these coordinates, the ROC has a particularly simple shape: It is a straight line with unit slope, as shown in Fig. 1.2.

The linearity of zROCs can be used to make a prediction about how much the false-alarm rate will go up if the hit rate increases (or vice versa). For example, suppose the false-alarm/hit pair (.2, .5) is on the ROC. Consulting Table A5.1, the z scores for F and H are –0.842 and 0. If we add the same number to each z score, the resulting scores correspond to another point on the ROC. Let us add 1.4, giving us the new z scores of 0.558 and 1.4. The table shows that the corresponding proportions are (.71, .92).

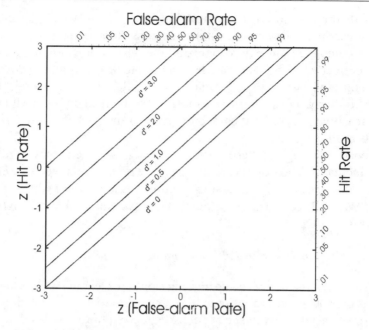

FIG. 1.2.   ROCs for SDT on *z* coordinates.

The transformed ROC of Equation 1.8 provides a simple graphical interpretation of sensitivity: d' is the intercept of the straight-line ROC in Fig. 1.2, the vertical distance in *z* units from the ROC to the chance line at the point where $z(F) = 0$. In fact, because the ROC has slope 1, the distance between these two lines is the same no matter what the false-alarm rate is, and d' equals the vertical (or horizontal) distance between them at any point.

ROCs Implied by $p(c)$

Any sensitivity index has an implied ROC, that is, a curve in ROC space that connects points of equal sensitivity as measured by that index. To extend our comparison of d' with proportion correct, we now plot the ROC implied by $p(c)$. The trick is to take the definition of $p(c)$ in Equation 1.3 and solve it for *H*:

$$H = F + [2\,p(c) - 1] \ . \tag{1.9}$$

Equation 1.9 is a straight line of unit slope. Implied ROCs for $p(c)$ are shown in Fig. 1.3 for $p(c) = .5$, .65, and .8. The intercepts equal $2p(c) - 1$, that is, 0, .3, and .6.

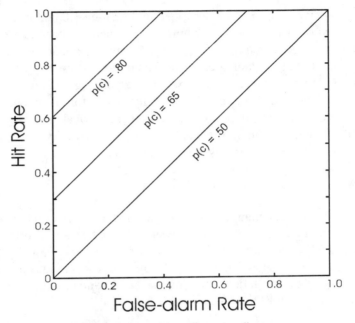

FIG. 1.3.   ROCs implied by $p(c)$ on linear coordinates.

Consider again the false-alarm/hit pair (.2, .5). If we add the same number to each of these scores (without any transformation), the resulting scores correspond to another point on the ROC. Let us add .42, giving us the new hit and false-alarm proportions of (.62, .92). Simply using $p(c)$ as a measure of performance thus makes a prediction about how much the false-alarm rate will go up if the hit rate increases, and it is different from the prediction of detection theory.

Which Implied ROCs Are Correct?

The validity of detection theory clearly depends on whether the ROCs implied by d' describe the changes that occur in H and F when response bias is manipulated. Do empirical ROCs (the topic of chap. 3) look like those implied by d', those implied by $p(c)$, or something else entirely? It turns out that the detection theory curves do a much better job than those for $p(c)$. In early psychoacoustic research (Green & Swets, 1966) and subsequent work in many content areas (Swets, 1986a), ROCs were found to be regular, to have decreasing slope on linear coordinates, and to follow straight lines on z coordinates.

One property of the zROCs described by Equation 1.8 that is *not* always observed experimentally is that of unit slope. When response bias changes, the value of d' calculated from Equation 1.5 may systematically increase or decrease instead of remaining constant. The unit-slope property reflects the equal importance of S_1 and S_2 trials to the corresponding sensitivity measure. In chapter 3, we discuss modified indexes that allow for unequal treatment.

When ROCs do have unit slope, they are symmetrical around the minor diagonal. Making explicit the dependence of sensitivity on a hit and false-alarm rate, we can express this property as

$$d'(1 - H, 1 - F) = d'(F, H) \ . \tag{1.10}$$

That is, if an observer changes response bias so that the new false-alarm rate is the old miss rate $(1 - H)$, then the new hit rate will be the old correct-rejection rate $(1 - F)$. For example, $d'(.6, .9) = d'(.1, .4)$. Mathematically, this occurs because $z(1 - p) = -z(p)$ (Eq. 1.6). Figure 1.4 provides a graphical interpretation of this relation, showing that (F, H) and $(1 - H, 1 - F)$ are on the same unit-slope ROC.

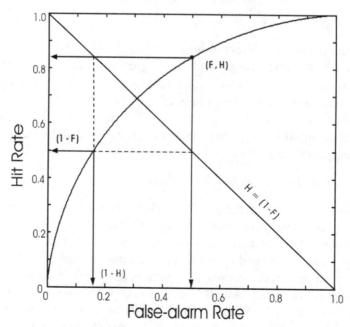

FIG. 1.4.   The points (F, H) and $(1 - H, 1 - F)$ lie on the same symmetric ROC curve.

Sensitivity as Perceptual Distance

Stimuli that are easy to discriminate can be thought of as perceptually far apart; in this metaphor, a discrimination statistic should measure perceptual distance, and d' has the mathematical properties of distance measures (Luce, 1963a): The distance between an object and itself is 0, all distances are positive (*positivity*), the distance between objects x and y is the same as between y and x (*symmetry*), and

$$d'(x, w) \leq d'(x, y) + d'(y, w) . \tag{1.11}$$

Equation 1.11 is known as the *triangle inequality*.

Because they have true zeroes and are unique except for the choice of unit, distance measures have *ratio scaling* properties. That is, when discriminability is measured by d', it makes sense to say that stimuli a and b are twice as discriminable as stimuli c and d. Suppose, for example, that two participants in our face-recognition experiment produce d' values of 1.0 and 2.0. In a second test, a day later, their sensitivities fall to 0.5 and 1.0. Although the change in d' is twice as great for Participant 2, we can say that Old and New items are half as perceptually distant, for both participants, as on the first day. No corresponding statement can be made in the language of $p(c)$.

The positivity property means that d' should not be negative in the long run. Negative values can arise by chance when calculated over a small number of trials and are not a cause for concern. The temptation to whitewash such negative values into zeroes should be resisted: When a number of measurements are averaged, this strategy inflates a true d' of 0 into a positive one.

The triangle inequality (Eq. 1.11) is sometimes replaced by a stronger assumed relation—namely,

$$d'(x, w)^n = d'(x, y)^n + d'(y, w)^n . \tag{1.12}$$

When $n = 2$, this is the Euclidean distance formula. When $n = 1$, Equation 1.12 describes the "city-block" metric; an important special case (discussed in chap. 5) arises when stimuli differ perceptually along only one dimension.

Another distance property of d' is *unboundedness*: There is no maximum value of d', and perfect performance corresponds to infinity. In practice, occasional hit rates or false-alarm rates of 1 or 0 may occur, and a correction such as one of those discussed earlier must be made to subject the data to detection theory analysis. Any such correction is predicated on the belief that the perfect performance arises from statistical ("sampling") error. If, on the contrary, stimulus differences are so great that confusions are effectively impossible then the experiment suffers from a ceiling effect, and should be redesigned.

The Signal Detection Model

The question under discussion to this point has been how best to measure accuracy. We have defended d' on pragmatic grounds. It represents the difference between the transformed hit and false-alarm rates, and it provides a good description of the relation between H and F when response bias varies. Now we ask what our measures imply about the process by which discrimination (in our example, face recognition) takes place. How are items represented internally, and how does the participant make a decision about whether a particular item is Old or New?

Underlying Distributions and the Decision Space

Detection theory assumes that a participant in our memory experiment is judging a single attribute, which we call *familiarity*. Each stimulus presentation yields a value of this decision variable. Repeated presentations do not always lead to the same result, but generate a distribution of values. The first panel of Fig. 1.5 presents the probability distribution (or likelihood distribution, or probability density) of familiarity values for New faces (stimulus class S_1). Each value on the horizontal axis has some likelihood of arising from New stimuli, indicated on the ordinate. The probability that a value above the point k will occur is the proportion of area under the curve above k (see Appendix 1 for a review of probability concepts).

On the average, Old items are more familiar than New ones—otherwise, the participant would not be able to discriminate. Thus, the whole of the distribution of familiarity due to Old (S_2) stimuli, shown in the second panel, is displaced to the right of the New distribution. There must be at least some values of the decision variable that the participant finds ambiguous, that could have arisen either from an Old or a New face; otherwise performance would be perfect. The two distributions together comprise the *decision space*—the internal or *underlying* problem facing the observer. The participant can assess the familiarity value of the stimulus, but of course does not know which distribution led to that value. What is the best strategy for deciding on a response?

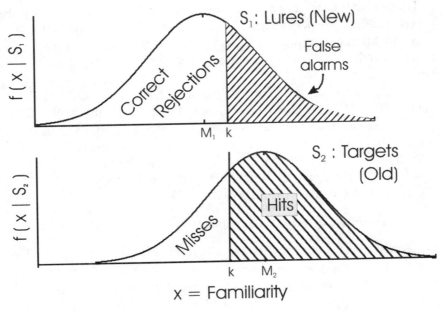

FIG. 1.5.   Underlying distributions of familiarity for Old and New items. Top curve shows distribution due to New (S_1) items; values above the criterion k lead to false alarms, those below to correct rejections. Lower curve shows distribution due to Old (S_2) items; values above the criterion k lead to hits, those below to misses. The means of the distributions are M_1 and M_2. (In this and subsequent figures, the height of the probability density curve is denoted by f.)

Response Selection in the Decision Space

The optimal rule (see Green & Swets, 1966, ch. 1) is to establish a *criterion* that divides the familiarity dimension into two parts. Above the criterion, labeled k in Fig. 1.5, the participant responds "yes" (the face is familiar enough to be Old); below the criterion, a "no" is called for. The four possible stimulus–response events are represented in the figure. If a value above the criterion arises from the Old stimulus class, the participant responds "yes" and scores a hit. The hit rate H is the proportion of area under the Old curve that is above the criterion; the area to the left of the criterion is the proportion of misses. When New stimuli are presented (upper curve), a familiarity value above the criterion leads to a false alarm. The false-alarm rate is the proportion of area under the New curve to the right of the criterion, and the area to the left of the criterion equals the correct-rejection rate.

The decision space provides an interpretation of how ROCs are produced. The participant can change the proportion of "yes" responses, and generate different points on an ROC, by moving the criterion: If the criterion is raised, both H and F will decrease, whereas lowering the criterion will increase H and F.

We saw earlier that an important feature of ROCs is regularity: If $F = 0$, then $H = 0$; if $H = 1$, then $F = 1$. Examining Fig. 1.5, this implies that if the criterion is moved so far to the right as to be beyond the entire S_1 density (so that $F = 0$), it will be beyond the entire S_2 density as well (so that $H = 0$). The other half of the regularity condition is interpreted similarly. The distributions most often used satisfy this requirement by assuming that *any* value on the decision axis can arise from either distribution.

Sensitivity in the Decision Space

We have seen that k, the criterion value of familiarity, provides a natural interpretation of response bias. What aspect of the decision space reflects sensitivity? When sensitivity is high, Old and New items differ greatly in average familiarity, so the two distributions in the decision space have very different means. When sensitivity is low, the means of the two distributions are close together. Thus, the mean difference between the S_1 and S_2 distributions—the distance $M_2 - M_1$ in Fig. 1.5—is a measure of sensitivity. We shall soon see that this distance is in fact identical to d'.

Distance along a line, as in Fig. 1.5, can be measured from any zero point; so we measure mean distances relative to the criterion k. Thus expressed, the mean difference equals $(M_2 - k) - (M_1 - k)$: Sensitivity is the difference between these two distances, the distance from the S_1 mean to the criterion and the (negative, in this case) distance from the S_2 mean to the criterion. We now show that these two mean-to-criterion distances can be estimated using the z transformation discussed earlier in the chapter.

Underlying Distributions and Transformations

Figure 1.6 shows how the distances between the means of underlying distributions and the criterion are related to the response rates in our experiment. For each value of $M - k$, the figure shows the proportion of the area of an underlying distribution that is above the criterion. When $M - k = 0$, for example, the "yes" rate is 50%; large positive differences correspond to high "yes" rates and large negative differences to low ones. The curve in Fig. 1.6 is called a *(cumulative) distribution function*; in the language of calculus, it is the integral of the probability distributions shown in Fig. 1.5.

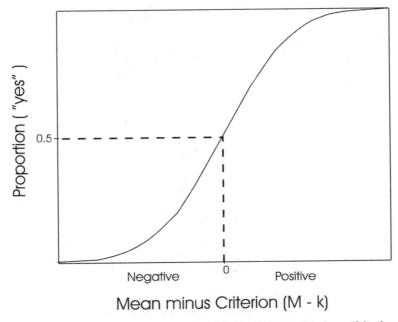

FIG. 1.6.   A cumulative distribution function (the integral of one of the densities in Fig. 1.5) giving the proportion of "yes" responses as a function of the difference between the distribution mean and the criterion.

We can use the distribution function to translate any "yes" proportion into a value of $M - k$. This is the tie between the decision space and our sensitivity measures: For any hit rate and false-alarm rate (both "yes" proportions), we can use the distribution function to find two values of $M - k$ and subtract them to find the distance between the means. The distribution function transforms a distance into a proportion; we are interested in the inverse function, from proportions to distances, denoted z. In Fig. 1.7, the hit and false-alarm proportions from our face-recognition example are ordinate values, and the corresponding values $z(H)$ and $z(F)$ are abscissa values. The distance between these abscissa points, $z(H) - z(F)$, is the distance between the S_1 and S_2 means in Fig. 1.5. It is also, by Equation 1.5, equal to d'. Because z measures distance in standard deviation units, so does d'. Thus, the sensitivity measure d' is the distance between the means of the two underlying distributions in units of their common standard deviation.

The distance between the means of distributions is a congenial interpretation of d' because it is unchanged by response bias. No matter where the

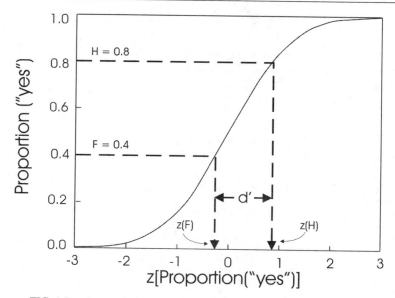

FIG. 1.7.   A cumulative normal distribution function. The inverse function can be used to transform the proportions H and F into z scores, and sensitivity is the difference between $z(H)$ and $z(F)$.

participant locates the criterion, d' equals the same number. This relation is not specific to normal-distribution SDT: Any sensitivity measure obtained by subtracting transformed hit and false-alarm rates can be represented as the distance between the means of two distributions whose shape is given by the inverse of the transformation.

We can now venture a "definition" that will at least delimit the contents of this book. By *detection theory* we mean a theory relating choice behavior to a psychological decision space. An observer's choices are determined by the distances between distributions in the space due to different stimuli (sensitivities) and by the manner in which the space is partitioned to generate the possible responses (response biases).

Calculational Methods

Calculation of d' (and other statistics yet to be introduced) can be accomplished at several levels of technical sophistication. As we have seen, a table of the normal distribution is sufficient in principle. Computer programs have been developed specifically for this job and are much more convenient when the amount of data to be analyzed is large. Appendix 6 contains one such program; it uses the most accurate algebraic approximation to z, according to Brophy (1985). A more complex program, which can also be used for the discrimination paradigms to be introduced later in the book, is *d' plus* (Macmillan & Creelman, 1997), which is available on the Internet.[3]

It is also easy to find *d'* using the "inverse normal" functions of spreadsheet programs; this is especially appealing for the many laboratories in which the data are collected or stored into spreadsheets. Basic calculations are illustrated in Table 1.1 for Excel, but are very similar in QuattroPro and other programs. The function z is written = NORMSINV. The indexes to be entered or computed are listed in Column A, and formulas are given that can be inserted in Rows 5 to 11 of Column B, then copied to subsequent columns. Sorkin (1999) explored the use of spreadsheets for SDT calculations in greater detail.

Detection theory procedures are also available as part of standard statistical packages such as Systat and SPSS. Because many users of detection theory make routine use of such packages, this is an attractive option. Data can be entered either as frequencies (number of hits, number of misses, etc.) or trial by trial, as they would be collected in an experiment. These packages can also be used when there are more than two response alternatives; we discuss them further in the context of rating designs (chap. 3).

TABLE 1.1   *Formulas for Spreadsheet (Excel) Calculation of SDT Statistics With Examples*

	A (Labels Only)	Formula (for Column B; Then Copy to C and Other Columns)	B (Set 1)	C (Set 2)
1	# hits		10	9
2	# misses		0	1
3	# false alarms		2	0
4	# correct rejections		8	10
5	H (hit rate)	= IF(B2>0, B1/(B1 + B2), (B1 – 0.5)/(B1 + B2))	.950	.900
6	F (false-alarm rate)	= IF(B3>0, B3/(B3+B4), 0.5/(B3+B4))	.200	.050
7	$z(H)$	= NORMSINV(B5)	1.645	1.282
8	$z(F)$	= NORMSINV(B6)	–0.842	–1.645
9	d'	= B7 – B8	2.486	2.926
10	c	= –0.5*(B7 + B8)	–0.402	0.182
11	β	=EXP(B9*B10)	0.368	1.702

[3]The site is http://psych.utoronto.ca/~creelman/.

Essay: The Provenance of Detection Theory

Psychophysics, the oldest psychology, has continually adapted itself to the substantive concerns of experimentalists. In particular, detection theory is well suited to cognitive psychology and might indeed be considered one of its sources. No grounding in history is needed to use this book, but some appreciation of the intellectual strains that meet here will help place these tools in context.

The term *psychophysics* was invented by Gustav Fechner (1860), the 19th-century physicist, philosopher, and mystic. He was the first to take a mathematical approach to relating the internal and external worlds on the basis of experimental data. Some present-day psychophysicists directly pursue Fechner's interest in relating mental experience to the physical world, usually in simple perceptual experiments. Measuring the way in which the reported experience of loudness grows with physical intensity is a psychophysical problem of this sort; we consider a detection theory approach to this problem in chapter 5.

This book is part of a second Fechnerian legacy, also methodological, but more general than the first. Fechner developed, tested, and described experimental methods for estimating the *difference threshold*, or *just noticeable difference (jnd)*, the minimal difference between two stimuli that leads to a change in experience. Fechner's assumption that the jnd could be the unit of measurement, the fundamental building block or atom of experience, was central to Wundt's and Titchener's structuralism, the first experimentally based theory of perception. The analogy to 19th-century chemistry was close: Theory and experiment should focus on uncovering the basic units and the laws of combination of those units.

Fechner's methods were adopted and became topics of investigation in their own right; they still form the backbone of experimental psychology. Attempts to measure jnds led to two complications: (a) The threshold appeared not to be a fixed quantity because, as the difference between two stimuli increases, correct discrimination becomes only gradually more likely (Urban, 1908); and (b) different methods produced different values for the jnd.

The concept of the jnd survived the first problem by redefinition: The jnd is now considered to be the stimulus difference that can be discriminated on some fixed percentage of trials (see chaps. 5 and 11). Two early reactions to the problem of continuity in psychophysical data are recognizable in modern research (see Jones, 1974).

One line of thought retained the literal notion of a sensory threshold, building mechanical and mathematical models to explain the gradual nature of observed functions (see chap. 4 for the current status of such models). The threshold idea was congenial with early 20th-century behaviorist and operationist attitudes: Sensory function could be studied and measured without invoking unpopular notions of mental content (Garner, Hake, & Eriksen, 1956). The threshold, in this view, was a construct derived from data and did not have to relate to any internal and unobservable mental process. The solution to method dependence was merely to subscript thresholds to indicate the method by which they were obtained (Graham, 1950; Osgood, 1958).

The second response to the variability problem, instigated, according to Jones (1974), by Delboef (1883), substituted a continuum of experience for the discrete processes of the threshold; it is this view that informs most contemporary psychophysics. One approach to measuring such continuous experience was Stevens' (1975) magnitude estimation, which used direct verbal estimates. Detection measurement, in contrast, relies on underlying random variation or noise. Psychologists' realization of the importance of random variation dates at least to Fullerton and Cattell (1892), who invoked it in a rigorous quantitative way to account for inconsistency in response with repetitions of identical stimuli. Variability later served as the key building block for the pioneering work of Thurstone (1927a, 1927b) in measuring distances along sensory continua indirectly.

The idea of variability or noise as an explanatory concept also arose in engineering, with the development and evaluation of radar detection apparatus. Radar and sonar are limited in performance by intrinsic noise in the input signal. Any input from an antenna or sensor can be due to noise alone or to a signal of interest embedded in the background noise. Groups at the University of Michigan (Peterson, Birdsall, & Fox, 1954), MIT (van Meter & Middleton, 1954), and in the Soviet Union (Kotel'nikov, 1960) recognized that the physical noise that was mixed with all signals, and that could mimic signal presence, was a major limitation to detection performance.

Knowing that stimulus environments are noisy does not, in itself, tell an observer how best to cope with them. An approach to this problem was contributed by another applied science: statistical decision theory. Decision theorists pointed out that information derived from noisy signals could lead to action only when evaluated against well-defined goals. Decisions (and thus action) should depend not only on the stimulus, but on the expected outcomes of actions. The viewer of a radar display that might or might not contain a blip, for example, should consider the relative effects of failing to detect a real bomber and of detecting a phantom before deciding on a response to that display.

W. P. Tanner, Jr., working with J. A. Swets at the University of Michigan, realized that these engineering notions could be applied to psychology and appropriated them directly into the psychophysical experiment (Tanner & Swets, 1954). By separating the world of stimuli and their perturbations from that of the decision process, detection theory was able to offer measures of performance that were not specific to procedure and that were independent of motivation. Procedure and motivation could influence data, but affected only the decision process, leaving measurable aspects of the internal stimulus world unchanged and capable of being evaluated separately.

According to detection theory, the observer's access to the stimuli being discriminated is indirect: An intelligent, not entirely reliable process makes inferences about them and acts according to the demands of the experimental situation. One might say that detection theory "deals with the processes by which [a decision about] a perceived, remembered, and thought-about world is brought into being from [an] unpromising beginning" (Neisser, 1967, p. 4). Neisser's landmark book linked perception and cognition into a unified framework after a hiatus of many decades. The constructionist (although not complicated) decision processes of detection theory mark it as an early example of cognitive psychology. The ideas behind detection theory are the everyday assumptions of behavioral experimenters in the cognitive era, and the theory itself is central to a wide range of research areas in cognitive science. Perhaps Estes' (2002) assessment is not an overstatement: "... [SDT is] the most towering achievement of basic psychological research of the last half century" (p. 15).

Summary

The results of a one-interval discrimination experiment can be described by a hit and a false-alarm rate, which in turn can be reduced to a single measure of sensitivity. Good indexes can be written as the difference between the hit and false-alarm rates when both are appropriately transformed. The sensitivity measure proposed by detection theory, d', uses the normal-distribution z transformation. The primary rationale for d' as a measure accuracy is that it is roughly invariant when response bias is manipulated; simpler indexes such as proportion correct do not have this property. The use of d' implies a model in which the two possible stimulus classes lead to normal distributions differing in mean, and the observer decides which class occurred by comparing an observation with an adjustable criterion.

Conditions under which the methods described in this chapter are appropriate are spelled out in Chart 2 of Appendix 3.

Problems

1.1.   Suppose you are measuring the sensitivity of a polygraph ("lie detector"). What are "hits," "misses," "false alarms," and "correct rejections"?

1.2.   The following tables give the number of trials in three conditions of a detection experiment on which participants responded "yes" or "no" to S_1 or S_2. (a) Calculate H and F. (b) Find $H - F$, $p(c)$, and $p(c)^*$. For these data sets, can $H - F$ be greater than $p(c)$ in one case and the reverse ordering occur in another, or is one index *always* greater than the other?

(a)	"yes"	"no"
S_2	9	6
S_1	7	8

(b)	"yes"	"no"
S_2	55	45
S_1	5	25

(c)	"yes"	"no"
S_2	45	55
S_1	25	5

1.3   (a). In Problem 1.2(a), the numbers of S_1 and S_2 trials are equal, but in (b) and (c) they are not. Does this matter computationally? experimentally?

(b). Is it possible to calculate $p(c)$ for S_2 trials only? What would this statistic measure?

1.4.   Compute d' for the following (F, H) pairs:

(a) (.16, .84), (.01, .99), (.75, .75).

(b) (.6, .9), (.5, .9), (.05, .9).

1.5   (a). If $p(c) = .8$ and H and F are unknown, estimate d'.

(b). If $p(c) = .8$, the numbers of S_1 and S_2 trials are equal, and $F = .05$, find H and d'.

1.6   (a). Suppose $d' = 1$. What is H if $F = .01, .1, .5$?

(b) Plot the ROC from these points on linear and z coordinates, and use the zROC to confirm the value of d'.

1.7.   For the data matrixes of Problem 1.2, find d' from H and F and also from $p(c)$. Is there a pattern to the results?

1.8.   Are the points $(.3, .9)$ and $(.1, .7)$ on the same ROC according to detection theory (i.e., do they imply the same value of d')? Do they imply the same value of $p(c)$?

1.9.   Suppose $(F, H) = (.2, .6)$. If F is unchanged, what would H have to be to double the participant's sensitivity, according to detection theory? If H is unchanged, what would F have to be?

1.10.   Plot the ROCs implied by the following measures, on both linear and z coordinates: $H^2 - F^2$, $H^{1/2} - F^{1/2}$, H/F^2, H^2/F. Which measures are best? worst?

1.11.   Suppose a face-recognition experiment yields 20 hits and 10 false alarms in 45 trials. Can you compute d'? If not, is it possible to narrow down the possibilities? *Hint*: The stimulus–response matrix looks like this:

20		
10		
		45

What happens if there are 0 misses, or 0 correct rejections?

2

The Yes-No Experiment:
Response Bias

In dealing with other people, "bias" is the tendency to respond on some basis other than merit, showing a degree of favoritism. In a correspondence experiment, *response bias* measures the participant's tilt toward one response or the other.

The sensitivity measure d' depends on stimulus parameters, but is untainted by response bias: To a good approximation, it remains constant in the face of changes in response popularity. We now adopt the complementary perspective, seeking an index of response bias that is uncolored by sensitivity. Conceptually, d' corresponds to a fixed aspect of the observer's decision space, the difference between the means of underlying distributions; a measure of bias should also reflect an appropriate characteristic of the perceptual representation. How can we assign a value to the participant's preference for one of the two responses?

Two Examples

Example 2a: Face Recognition, Continued

Consider again the face-recognition experiment of chapter 1, in which viewers discriminated Old from New faces. Suppose the investigator now repeats the experiment, this time hypnotizing the participants in an effort to improve their memory, and obtains the following results from a representative observer:

	Normal		Hypnotized	
	"Yes"	"No"	"Yes"	"No"
Old	69	31	89	11
New	31	69	59	41

Applying the analyses of chapter 1 reveals that hypnosis has not affected sensitivity: d' is approximately 1.0 in both the normal and hypnotized conditions.

Hypnosis *does* appear to affect willingness to say "yes"; there are many more positive responses in the hypnotized condition than in the control data. (For a discussion of whether hypnotism actually has this effect, see Klatzky & Erdelyi, 1985.) In this example, therefore, an experimental manipulation affects bias, but not sensitivity. In the next example, a single variable affects both.

Example 2b: X-ray Reader Training

Apprentice radiologists must be trained to distinguish normal from abnormal X-rays (see Getty, Pickett, D'Orsi, & Swets, 1988, for a description of one training program). In this field, a hit is conventionally defined to be the correct diagnosis of a tumor from an X-ray, and a false alarm is the incorrect labeling of normal tissue as tumorous. Consider three readers who before training are equally able to distinguish X-rays displaying real tumors from X-rays of normal tissue, attaining exactly the same performance, but emerge from training with different scores on a posttest:

	Before Training	*After Training*
Trainee 1	$H = .89$	$H = .96$
	$F = .59$	$F = .39$
Trainee 2	$H = .89$	$H = .993$
	$F = .59$	$F = .68$
Trainee 3	$H = .89$	$H = .915$
	$F = .59$	$F = .265$

The trained readers are more sensitive—two of them show both a higher proportion of hits and a lower proportion of false alarms than before training. But has there also been a change in willingness to say "yes"? In the hypnotic recognition experiment, a response bias change merely masked the constancy of sensitivity; in this second example, there is clear evidence for a sensitivity change, but an interesting response-bias question remains.

Measuring Response Bias

Characteristics of a Good Response-Bias Measure

Because a response-bias index is intended to measure the participant's willingness to say "yes," we expect it to depend systematically on both the hit and false-alarm rates and in the same direction—either increasing or decreasing in both. Sensitivity measures, remember, increase with H and decrease with F, an analogous property. A response-bias index should depend on the *sum* of terms involving H and F, whereas the sensitivity statistic d' depends on the *difference* of H and F terms.

Response-bias statistics can reflect either the degree to which "yes" responses dominate or the degree to which "no" responses are preferred. All the measures in this book index a leaning in the same direction: A positive bias is a tendency to say "no," whereas a negative bias is a tendency to say "yes." The rationale for these apparently illogical pairings will become clear when we discuss the representation.

Criterion Location (c)

The basic bias measure for detection theory, called c (for *criterion*), is defined as:

$$c = -\tfrac{1}{2}[z(H) + z(F)] \ . \tag{2.1}$$

When the false-alarm and miss rates are equal, $z(F) = z(1 - H) = -z(H)$ and c equals 0. Negative values arise when the false-alarm rate exceeds the miss rate, and positive values arise when it is lower. Extreme values of c occur when H and F are both large or both small: If both equal .99, for example, $c = -2.33$, whereas if both equal .01, $c = +2.33$. The range of c is therefore the same as that of d', although 0 is at the center rather than an endpoint. Figure 2.1 shows the locus of positive, negative, and 0 values of response bias in the part of ROC space where sensitivity is above chance.

Table A5.1, which was introduced in chapter 1 as a tool for calculating d', can also be used to compute the bias measure c. Spreadsheets accomplish the table-lookup task automatically (see Table 1.1, which includes some bias measures). Analyzing the face-recognition results, we find that c shifts from 0 to -0.73 under hypnosis, reflecting an increase in "yes" responses.

To interpret these numbers according to our model, consider the decision space in Fig. 2.2. The familiarity decision axis is labeled in standard deviation units, 0 being the point midway between the two distributions. Because $d' = 1.0$, the mean of the Old distribution is at 0.5, the mean of the New at -0.5. The participant's decision rule is to divide the familiarity axis into "yes" and "no" regions at a *criterion*.

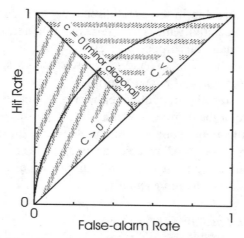

FIG. 2.1.   The representation of criterion location in ROC space. Points in the shaded regions arise from criteria that are positive (below the minor diagonal) and negative (above the minor diagonal). Points in the unshaded region below the major diagonal result from negative sensitivity.

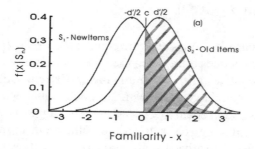

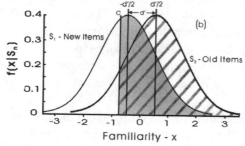

FIG. 2.2.   Decision spaces for the Normal and Hypnotized conditions of Example 2a, according to SDT. Shaded area corresponds to F, diagonally striped area to H. (a) Normal controls have a symmetric criterion, $d' = 1.0$. (b) Hypnotized participants display identical sensitivity but a lower criterion, and thus have higher hit and false-alarm rates.

A simple calculation shows that the value of this criterion, in standard deviation units from the midpoint, is the bias parameter c. In chapter 1, we saw that the z score of the "yes" rate corresponds to the mean-minus-criterion distance. For the S_1 distribution, this implies

$$-d'/2 - c = z(F) \,, \tag{2.2a}$$

and for the S_2 distribution

$$d'/2 - c = z(H) \,. \tag{2.2b}$$

Adding these two equations produces Equation 2.1.

The different values of response bias in the normal and hypnotized conditions of our face-recognition experiment, therefore, correspond to different criterion locations. In the control condition (Fig. 2.2a), the criterion is located at 0, exactly halfway between the two distributions, and the participant is said to be "unbiased." Under hypnosis (Fig. 2.2b), the participant's criterion is much lower, below the mean of the New distribution. Because it is 0.73 standard deviations below the zero-bias point, $c = -0.73$.

Analysis of the radiology training data from Example 2b is equally straightforward. All trainees improve in sensitivity: d' about doubles. Values of c can be calculated from Equation 2.1. Trainee 1 maintains the same criterion location after training as before ($c = -0.74$). Trainee 2 has a more extreme bias (-1.46), and Trainee 3 has a less extreme one (-0.37). The degree to which the criteria differ among trainees is easily seen in Fig. 2.3, which shows the decision space and criterion settings for each reader: The first row represents the pretraining decision space of all trainees, and the other rows represent the posttraining spaces of each one individually.

Alternative Measures of Bias

Detection theory offers one measure of sensitivity (for two-response experiments), but is more generous with bias parameters. Besides criterion location, just described, bias can be specified by *relative criterion* location and *likelihood ratio*.

Relative Criterion Location (c′)

In this measure of bias, we scale the criterion location relative to performance. A rationale for such scaling is that with easier discrimination tasks a

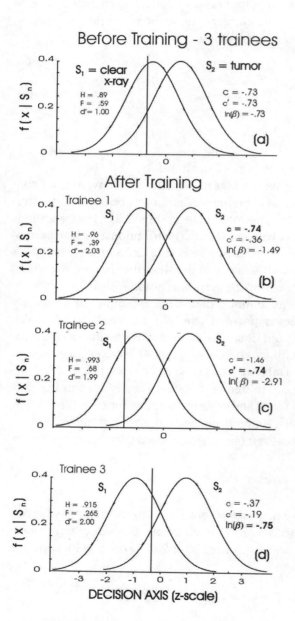

FIG. 2.3. Decision spaces for the three radiology trainees of Example 2b. In each case the hit rate, false-alarm rate, sensitivity, and three alternative criterion measures are shown. (a) Before training, $d' = 1.0$. The criterion c, the relative criterion c', and log likelihood ratio equal -0.73 for all trainees. (b) Trainee 1, after training; increased sensitivity and approximately the same criterion location c as before training. (c) Trainee 2, after training; increased sensitivity and approximately the same relative criterion location c' as before training. (d) Trainee 3, after training; increased sensitivity and approximately the same value of log likelihood ratio [$\ln(\beta)$] as before training.

more extreme criterion (as measured by c) would be needed to yield the same amount of bias.

Look again at the radiography training data of Example 2b. The first radiologist's criterion location is indeed the same distance from 0 (the equal-bias point) before and after training, but whether this is to be called "no change" can be argued. The criterion was initially below the mean of the S_1 distribution, but is above it afterward. If distance from the criterion to a distribution mean is the key to bias, this observer's bias has become less extreme. Would it not be sensible to calculate the criterion distance as a proportion of sensitivity distance? The alternative bias measure suggested by this reasoning is:

$$c' = \frac{c}{d'} = -\frac{1}{2} \frac{[z(H) + z(F)]}{[z(H) - z(F)]} \ . \tag{2.3}$$

Calculated values for c' are given in Fig. 2.3. It happens in this example that before training, $c = c'$, but only because $d' = 1.0$. After training, c' is half the magnitude of c because $d' = 2$. When d' varies, one must decide whether in discussing "bias" one wishes to take account of sensitivity. Of the three radiologists, it is Trainee 2 who maintains the same bias in the sense of c' and Trainee 1 whose bias is unchanged in the sense of c.

Likelihood Ratio (β)

The third measure of bias is found by an apparently different strategy. In the decision space, each value x on the decision axis has two associated "likelihoods," one for each distribution. Each likelihood is the height of one of the distributions; we denote this height at the location x by $f(x)$, and to distinguish the two distributions we refer to the heights of S_1 and S_2 as $f(x|S_1)$ and $f(x|S_2)$. The relative likelihood of S_2 versus S_1, obtained by dividing these, is called the *likelihood ratio*:

$$\text{LR}(x) = f(x|S_2)/f(x|S_1) \ . \tag{2.4}$$

Each point x has an associated value of likelihood ratio: It is 1.0 at the center (where the two distributions cross), greater than 1.0 to the right, and between 0 and 1.0 to the left. One measure of response bias, therefore, is the value of likelihood ratio at the criterion.

Equation 2.4 suggests an interesting interpretation of likelihood ratio in terms of the ROC. Consider two points very close together on the decision axis—imagine they are a small value ε units apart, as shown in Fig. 2.4a. The change in the hit rate between the two points is approximately $f(x|S_2)\varepsilon$, the height of the S_2 distribution multiplied by the width of a tiny rectangle. The change in the false-alarm rate, by the same token, equals $f(x|S_1)\varepsilon$. The ratio of these changes, which is the slope of the ROC, is $f(x|S_2)/f(x|S_1)$. Notice that this slope exactly equals the likelihood ratio. The assertion in chapter 1 that the slope of the ROC continuously decreases follows from the equivalence of likelihood ratio and ROC slope. As the criterion goes from large to small values of c, the likelihood ratio must decrease, and so therefore must the slope.

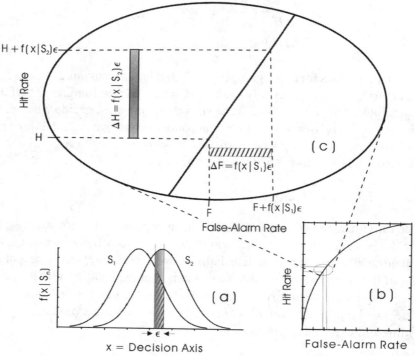

FIG. 2.4.   Geometric demonstration that the slope of the ROC at any point is the likelihood ratio at the criterion value that yields that point. (a) In the decision space, two criteria are shown that differ by a small amount ε. For the lower criterion, the hit rate is greater by an amount equal to the area of the filled rectangle, and the false-alarm rate is greater by an amount equal to the area of the diagonally shaded rectangle. (b) The two criteria correspond to two points on an ROC curve. (c) An expanded view of the relevant section of the ROC. The lower point (higher criterion) is (F, H). At the higher point, the hit and false-alarm rates increase by the areas of the rectangles in (a). The slope of the ROC, the ratio of these two increments, is $f(x|S_2)/f(x|S_1)$, which is the likelihood ratio.

This conclusion does not depend on any assumptions about the shape of the underlying distributions, but actual calculation of likelihood ratio does require such a commitment. In the normal-distribution model we have been exploring, the height of the likelihood function, denoted ϕ, depends on x and on the distribution's mean μ and standard deviation σ according to the equation

$$\phi(x) = \frac{1}{\sqrt{2\pi}\,\sigma} e^{-\frac{1}{2}\left(\frac{x-\mu}{\sigma}\right)^2} . \tag{2.5}$$

Values of ϕ are given in Table A5.1.

The general strategy for finding the likelihood ratio can now be applied to the normal model. The likelihood function f in Equation 2.4 equals ϕ, and the likelihood ratio is the ratio of two values of $\phi(x)$—one for the S_2 distribution and one for S_1. A little calculation (to be found in the Computational Appendix) shows that the likelihood ratio, usually called β in the normal model, depends on sensitivity and the criterion location in a simple way:

$$\beta = e^{cd'}.$$

An equivalent form can be found by taking logarithms:

$$\ln(\beta) = cd' = -\tfrac{1}{2}[z(H)^2 - z(F)^2] . \tag{2.6}$$

Likelihood ratio can be calculated either directly from likelihoods (given by Eq. 2.5 or Table A5.1) or from its relation to c and d' (Eq. 2.6). For Trainee 3, the likelihood ratio equals $\phi(.915)/\phi(.265) = .1556/.3276 = 0.475$, and $\ln(\beta) = -0.75$. Alternatively, because $d' = 2.00$ and $c = -0.373$ for this observer, $\ln(\beta) = cd' = -0.75$. By this measure, Trainee 3 maintained the same response bias before and after training, whereas the other trainees adopted more extreme criteria (−1.49 and −2.91) after training. Summarizing bias using β [or $\ln(\beta)$] leads us to a different conclusion about our radiologists than did c or c'.

Isobias Curves

The isosensitivity curve, which describes the relation between the hit and false-alarm rates when bias (but not sensitivity) changes, is useful in evaluating measures of accuracy. A function relating H and F for changing sensitivity (but not bias) is equally important in understanding bias statistics. The locus of points in ROC space that reflect equal bias is called an *isobias curve*.

For a particular bias parameter, the isobias curve is a prediction about how performance changes when bias is held fixed while sensitivity varies. Consider Fig. 2.5, which locates the performance of all the radiologists of Example 2b in ROC space. Trainee 1, remember, displayed the same value of c before and after training, so the points B and T_1 lie on the isobias contour for c defined by $c = -0.73$. Other points for which c takes on this value are connected by a continuous curve. Similarly, Trainees 2 and 3 generate two points on the isobias curves for c' and β, respectively. Clearly the three measures predict very different patterns of performance when sensitivity changes and bias remains the same.

To derive the form of an index's isobias curve, it is necessary to solve the equation defining the measure for H as a function of F. For example, the isobias function for c is found from Equation 2.1 to be

$$z(H) = -2c - z(F) . \tag{2.7}$$

As can be seen in Fig. 2.6 (upper right panel), this relation is a straight line in z coordinates. Families of curves for all three measures are shown in Fig. 2.6, in both linear and z coordinates.

Comparing the Bias Measures

How can a choice be made among the bias statistics available? The three bias measures, all quite plausible, are simply related. The criterion location relative to the zero-bias point, c, is *divided* by d' to obtain the relative bias c', and *multiplied* by d' to obtain the likelihood ratio measure $\ln(\beta)$. Because the logarithm is a monotonic function, $\ln(\beta)$ is equivalent to likelihood ratio itself.

Likelihood ratio is the most general of these three concepts: Unlike absolute or relative criterion location, it is meaningful for representations of any complexity. Early writers on detection theory (Licklider, 1959; Peterson, Birdsall, & Fox, 1954), therefore, placed great stress on the likelihood ratio as the basis for decision. When sensitivity is constant, d' serves as an arbitrary scale factor on the interval-scaled decision axis, and one may fairly say that log likelihood ratio *is* the decision variable.

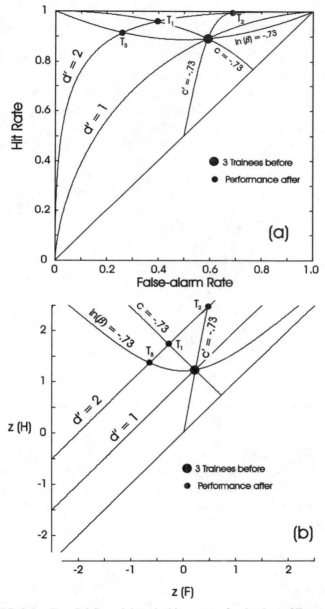

FIG. 2.5.   Two ROCs and three isobias curves for the data of Example 2b. One ROC describes the sensitivity for all three trainees before training ($d' = 1$), the other sensitivity after ($d' = 2$). The isobias curves are for constant c (Trainee 1), constant c' (Trainee 2), and constant β (Trainee 3). Linear coordinates are used in (a), z coordinates in (b).

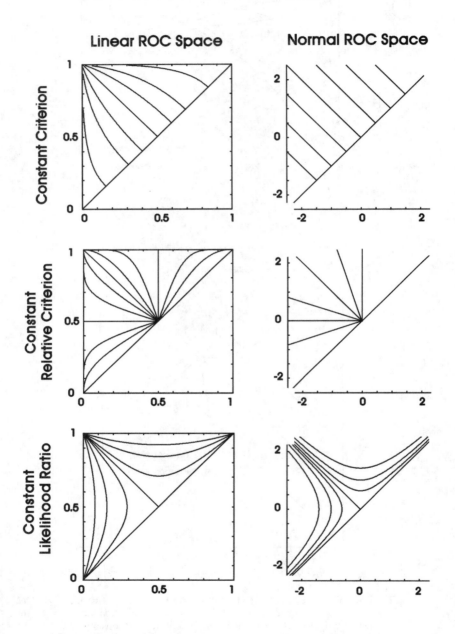

FIG. 2.6.   Families of isobias curves (hit rate vs. false-alarm rate, d' varying) for constant criterion c, relative criterion c', and likelihood ratio β, on linear and z axes.

But this does not mean that $\ln(\beta)$ is necessarily to be preferred as a response bias measure for comparing experimental conditions—when detectability is constant, any measure will do. In Example 2a, d' equals 1.0 in both face-recognition conditions; thus, the bias parameters c, c', and $\ln(\beta)$ give the same values. If d' were constant but not equal to 1.0, and the three statistics therefore differed numerically, they would still lead to the same conclusion in any comparison between conditions. But when d' varies, as in Example 2b, the three measures of bias support different interpretations.

We consider three standards to which a candidate bias measure might be held: (a) Its isobias curve should be supported empirically, (b) it should depend monotonically on H and F in the same direction, and (c) it should be independent of the sensitivity index. The second and third standards favor the criterion location c; the first has not provided clear-cut support for any one measure.

Form of Empirical Isobias Curves

The curves in Fig. 2.6 are similar, but they differ substantially in certain parts of the space (at the corners and near the major diagonal), and the reader may eagerly expect that, as with sensitivity, we shall be able to decide among the implied measures on the basis of data. Dusoir (1975, 1983) compared isobias curves for c, c', β, and several other potential bias statistics with data from an auditory detection experiment. Sensitivity was varied by changing tone intensity and bias via instructions, allowing isobias curves to be constructed.

Dusoir found great individual differences in shapes of isobias curves and could not support any one measure as superior to any other. This lack of unanimity reflects an important asymmetry between sensitivity and bias: To derive measures that yield constant sensitivity requires, at most, a theory of the stimuli; to do the same for bias indexes requires, at least, a theory of the instructions. Theories of tone detection, and even theories of memory, are a good deal more advanced than the kind of theory of language understanding needed to predict isobias performance. Dusoir (1983) concluded that participants may vary in their understanding of bias-inducing instructions so that different observers may all be holding some—but not the same—bias parameter fixed.

Two more recent applications produced more internally consistent results, although they are not in agreement with each other. See, Warm, Dember, and Howe (1997) examined a "vigilance" situation in which signals occur infrequently and trials are not defined by the experimenter. Such tasks are of interest as models of, for example, radar monitoring, and observers typically show a decrement in performance after less than 1 hour (Davies & Parasuraman, 1982). See et al. asked observers to detect small increases in the height of lines on a computer screen, and they manipulated response bias by varying the probability of signal occurrence and monetary payoffs. In a critical test, they chose two levels of "salience" (detectability), and were thus able to construct two-point isobias curves; these curves had the form predicted for c.

Recognition memory data relevant to the isobias question have been reported by Stretch and Wixted (1998), who reanalyzed the data of Ratcliff, McKoon, and Tindall (1994) and also conducted new experiments. Sensitivity was manipulated by varying the time or rate of exposure of words, or the number of times items were presented. Response bias was evaluated through a rating design; we discuss the details of this procedure in the next chapter. Although they did not plot isobias curves per se, their data come closest to the form predicted for β.

It is tempting to conclude that the bias manipulations used by See et al. (1997) and Stretch and Wixted (1998) are superior to the instructional method used by Dusoir (1983), at least in the sense that individual differences may be smaller. But changes in presentation probability have their own problems, as we shall see in chapter 3. In any case, these two studies reach different conclusions. Perhaps this is not something to worry about: Although one would like theories of response bias to be oblivious to the stimulus domain being studied, such a goal may be too optimistic.

Monotonicity of Theoretical Isobias Curves

Our general condition on bias measures, according to which an increase in either the hit rate or false-alarm rate should mean a decrease in bias, imposes a restriction on isobias curves: As F increases, H must decrease. All measures satisfy this condition in the upper left quadrant of ROC space, where $H \geq .5$ and $F \leq .5$, but both relative criterion and likelihood ratio violate it elsewhere.

Two other regions of ROC space, in which the curves of Fig. 2.6 show different behaviors, are the chance line $H = F$ and the area below it. When sensitivity is 0, it is still meaningful to talk about bias: An observer for whom $H = F = .1$ clearly has a different bias from one operating at $H = F = .9$. The criterion location c does take on different values along the diagonal, but c' and β do not. In fact, when underlying distributions of likelihood are identical, both of these statistics are undefined.[1]

Below-chance behavior may seem uninteresting or even illogical, but statistical fluctuations can easily lead to such performance. Two points in ROC space that are symmetrically located across the chance line—(F, H) and (H, F)—should intuitively show the same or similar biases. By this test, criterion location is again the best measure, giving the same value for the two points. Both c' and $\ln(\beta)$ show discontinuities, changing sign as they cross the diagonal (see Macmillan & Creelman, 1990, for more detail). These measures can be salvaged by multiplying their values by -1 below the chance line [as has been suggested for $\ln(\beta)$ by Waldmann & Göttert, 1989, and for other, "nonparametric" bias measures by Aaronson & Watts, 1987, and Snodgrass & Corwin, 1988].

Independence of Bias From Sensitivity

That response bias be independent of accuracy is clearly a desirable outcome, but we must be careful what we wish for, because *independence* has multiple meanings. First, consider statistical independence, the condition that in repeated tests neither of two measures affects the other. Only c is independent of d' in this sense (see the Computational Appendix for proof that it is).

Second, we can examine the dependence of bias measures on stimulus strength—there should be none. An analogous strategy, finding noneffects of bias on sensitivity, provided some of the earliest support for the use of d'. Dusoir (1983) applied this test, but the results were inconclusive: Of 21 comparisons (each representing a single observer in a single experimental condition), c, c', and β were each significantly correlated with sensitivity 12 times. Other statistics considered by Dusoir (some of which we encounter in chap. 4) were at least equally unsuccessful. In the See et al. (1997) experiment, c showed a slight dependence on d' in one of three experiments, but the correlations between β and d' were both more widespread and stronger.

A final type of independence can be seen intuitively in the ranges of these variables: The range of c does not depend on d', whereas the range of the other measures does (Banks, 1970; Ingham, 1970). When d' is large, c' has a small range and β a large one; when d' is small, the reverse is true. The criterion location c is the only index whose magnitude can be interpreted with-

---

[1]In one interpretation, likelihood ratio equals 1 for the zero-sensitivity case. As noted earlier, however, the decision axis itself may be considered to be likelihood ratio, so the decision space collapses to a single point.

out knowledge of d' (Macmillan & Creelman, 1990; Snodgrass & Corwin, 1988). A caveat remains, however: Perhaps the range of biases is truly *not* the same at different levels. Thus, Stretch and Wixted (1998) concluded from the nature of the relation between criterion and stimulus strength in their memory experiment that the range of biases was narrower at high levels of accuracy.

How Does the Participant Choose a Decision Rule?

Whatever the best response bias measure is, the decision process leading to it is not in dispute.[2] The participant establishes a criterion at some point on a relevant internal dimension and uses it to partition the dimension into regions of "yes" and "no" responses. Two questions remain: (a) Is this always the best thing to do? If so, (b) where should the criterion be located?

As long as the stimuli (and thus sensitivities) are fixed, using a criterion to determine responses is, indeed, always the right strategy, and for an interesting reason. As we have seen, the decision axis is a monotonic function of likelihood ratio in the fixed-sensitivity case, so the question becomes whether it is optimal to use likelihood ratio to make decisions.[3] To answer this question requires consensus about what the "best" decision rule should accomplish, something about which reasonable people can agree. Green and Swets (1966: ch. 1) nominated four decision goals; for each a likelihood ratio decision rule is indicated and the optimal likelihood ratio at the criterion can be calculated:

1. *Maximize proportion correct.* When presentation probabilities are equal, a participant who wishes to maximize proportion correct must treat the two stimulus classes symmetrically, preferring to make neither false alarms nor misses more often. This is accomplished by setting c to equal 0, the zero-bias point. Likelihood ratio at this point is 1. If S_2 is presented more often than S_1, however, it will pay the participant to be more willing to respond "yes," and a lower criterion should be set. If $p(S_1)$ and $p(S_2)$ are the a priori probabilities of presenting the two stimuli, then the optimal value of likelihood ratio is $p(S_1)/p(S_2)$.

---

[2]Well, not in *much* dispute. One alternative interpretation of detection data rejects the whole idea of criterion shifts (Balakrishnan, 1999) in favor of changes in the distributions themselves.

[3]In chapter 3, we encounter a case in which the likelihood ratio is not monotonic with the decision axis; even then the likelihood ratio rule is best.

2. *Maximize a weighted combination of hits and correct rejections.* An observer may be more interested in hits than in false alarms, or vice versa, for reasons other than presentation probability. For example, the X-ray readers of Example 2b should be much more willing to make a false alarm (detecting a tumor when none is there) than a miss (failing to detect). To maximize a weighted average—say, three times the hit rate, plus the correct rejection rate—the observer should set a criterion at the "importance ratio," in this example, three. That is, only if the X-ray under examination is at least three times as likely to be normal as pathological should the observer say, "no, there is no tumor." If the importance ratio equals the ratio of presentation probabilities, this objective is the same as maximizing proportion correct.

3. *Maximize expected value.* The decision rule suggested for the X-ray reader, just above, was based on the relative value of the two kinds of correct decisions. This can be made explicit, at least in experimental situations: Participants can be rewarded for hits and correct rejections, or they can be penalized for false alarms and misses. In the laboratory, the rewards are sometimes small financial ones, sometimes merely "points" (see chap. 3).

The ideal value of likelihood ratio in such a situation depends on the reward function R that specifies the payoff for each experimental outcome:

$$LR(x) = \beta = \frac{[R(correct\ rejection) - R(false\ alarm)]}{[R(hit) - R(miss)]} \cdot \frac{p(S_1)}{p(S_2)} \quad (2.8)$$

Normally, the "rewards" for false alarms and misses are negative. If an observer is paid 10 cents for each correct response and is penalized 1 cent for misses and a dollar for false alarms, the optimal value for the criterion (assuming equal presentation probabilities) is $[0.10 - (-1.00)]/[(0.10 - (-0.01)] = 10$; that is, the observer should insist that the odds favoring S_2 given the data be 10 to 1 or larger before hazarding a "yes" response.

People rarely adopt such extreme criteria; when payoffs are changed to favor "no," criteria generally shift, but not to the degree prescribed by Equation 2.8. The theoretical analysis is sometimes salvaged by reference to "subjective" rewards, which are presumed to lag behind real ones. We are aware of no attempt to verify that the subjective criteria actually used by participants in payoff-driven experiments are, in any sense, optimal. Many practical matters, such as participants' (usually negative) attitudes toward piece work, competition among participants, and a tendency to see performance (in what is frequently, after all, a professor's laboratory) as a measure of intelligence, all suggest that Equation 2.8 captures only some of the real basis for human decision making.

4. *Test a statistical hypothesis.* A decision maker is often instructed, explicitly or implicitly, to obtain as high a hit rate as possible while holding the false-alarm rate to some predetermined level, a goal called the *Neyman–Pearson objective.* Thus, our X-ray reader might be advised to keep the false-alarm rate below .5; for an air traffic controller, an acceptable value of F might also be quite high, because it is the misses—failures to notice impending collisions—that have to be minimized. Clients undergoing audiological testing often adopt a much more severe criterion, being unwilling to make more than a few false alarms, which they view as lies.

Satisfaction of the Neyman–Pearson objective also requires a likelihood ratio criterion decision rule, with the value of likelihood ratio set to produce the desired false-alarm rate. Jerzy Neyman and Karl Pearson were among the founding fathers of modern statistics, and their objective is exactly that met by conventional statistical hypothesis testing. False alarms in that context are called Type I errors, and the false-alarm rate is arbitrarily set to a small value, typically .05 or .01. Observations (sample means, sample mean differences, etc.) above the criterion lead to rejection of the null hypothesis, either correctly (hits) or, with fixed low probability, incorrectly (false alarms).

Coda: Calculating Hit and False-Alarm Rates From Parameters

The outcome of a yes-no discrimination experiment, we have seen, can be characterized by either of two pairs of parameters: the hit and false-alarm rates, or sensitivity and bias. Detection theory asserts that the latter pair is more illuminating. These first two chapters have therefore focused on expressions for sensitivity and bias in terms of H and F. When solved for H, these expressions describe isosensitivity and isobias curves.

It is sometimes useful, however, to reverse this process and calculate H and F from detection theory parameters. We do this here according to the following plan. In the decision space, the hit and false-alarm rates—both proportions of "yes" responses—correspond to the area under a probability function above the criterion. The (cumulative) distribution function at the criterion gives the complementary probability of an observation falling below criterion. This distribution function can be easily calculated, because it is the inverse of the z transformation that converts proportions to distances. The calculation is illustrated in Fig. 2.7.

The z transformation converts a proportion to a standardized distance from the mean. The inverse of z, which gives the "no" rate when the criterion is at z, is the normal distribution function, denoted $\Phi(z)$. The value of $\Phi(z)$ can be found from a normal table, but Table A5.1 is not ideally arranged for this purpose. In that table, p values are given in units of .01, which is helpful when p is known, as in data analysis. Table A5.2 gives the same information, but for z scores in units of .01, which is more convenient when z is known. The probability p corresponding to a z score is $\Phi(z)$.

The "yes" rate is $1 - \Phi(z)$; because the normal distribution is symmetric, this equals $\Phi(-z)$. Expressed as a z score, the criterion equals $c - d'/2$ for the S_2 distribution and $c - (-d'/2)$ for S_1; so

$$H = \Phi(d'/2 - c) \qquad\qquad (2.9)$$

$$F = \Phi(-d'/2 - c) \,.$$

For an unbiased observer, $c = 0$, $H = \Phi(d'/2)$, and $F = \Phi(-d'/2)$. In this case, the hit and correct rejection rates both equal proportion correct, so

$$p(c) = \Phi(d'/2) \,. \qquad\qquad (2.10)$$

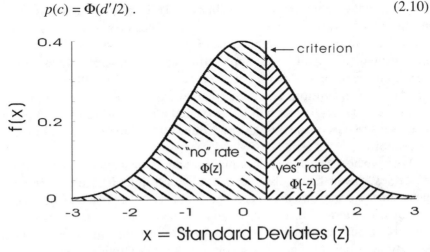

FIG. 2.7. Relation between underlying distributions and "yes" rates (hit and false-alarm rates). When the criterion is at z, the yes rate is $\Phi(-z)$.

Essay: On Human Decision Making

Much of the large literature on decision making by human beings (see, e.g., Kahnemann, Slovic, & Tversky, 1982) asks how closely our behavior corresponds to what we "should" do. The decision problem described in this chapter is in many ways rather simple: Only one dimension is relevant, the stimuli are presented at predictable times (in most applications), and repeated trials allow the observer to focus on relevant aspects of the stimulus display. Does the observer indeed deal with this problem in the "right" way—by establishing a criterion and using it?

At least two nonoptimal strategies have occurred to most psychophysicists who have studied (and, frequently, served as participants in) correspondence experiments: *inattention* and *inconsistency.* An inattentive observer dozes off, or at least drifts into reverie, on some proportion of trials; because failing to respond is usually discouraged, this leads to an unknown number of $d' = 0$ trials, ones on which the observer responds despite not having paid attention, mixed in with the others. An inconsistent participant uses a criterion, but changes the location of the cutoff from trial to trial; because the criterion must be compared to a sensory event, the movement adds an unknown amount of variance to the underlying distributions (Wickelgren, 1968). Both strategies, if they may be called that, serve to reduce observed sensitivity.

Do these effects occur? Almost certainly, but little is known about how badly they contaminate experiments. Training provided before experimental data are collected may serve to reduce these errors; observers who fail to improve during practice may be suspected of persisting in a nonoptimal strategy. In most applications, small amounts of inattention or inconsistency matter little. Stimulus pairs that yield high performance levels are an exception: The experimenter who wishes to make a precise estimate of a d' of 4 or so will be frustrated by even an occasional lapse. If lapses are part of the human condition, such estimates are doomed to unreliability.

We have been speaking of optimal strategies; what about optimal *use* of strategies? Given that an observer is using a criterion in the manner we suppose, are there ways we can encourage "unbiased" decision making, that is, symmetric criterion placement? Arguments are sometimes put forward that one or another experimental technique will accomplish this goal, which is sometimes a valuable one (see especially chap. 11). Often, however, there is no reason to aim for a symmetric criterion. After all, the sensitivity measure with which detection theory provides us is unaffected by bias, so why worry? Perhaps only because in common parlance (but not in psychophysics) *bias* is a pejorative term, something worth avoiding.

Another appeal of unbiased responding is that it makes almost any measure of sensitivity satisfactory, eliminating the need for complex psychophysics. The search for unbiased responding may thus be a vestige of the belief that, really, simple, untransformed measures are to be trusted more than theoretical ones. We shall critically evaluate this possibility in chapter 4.

Finally, the concept of bias in detection theory has sometimes been misunderstood in a way that makes neutral bias qualitatively different from other values. The location of the criterion can, we have seen, be manipulated by instructions: Apparently, then, observers can consciously choose to change it. If no instructions are given, however, observers are not aware of the possibility of varying a criterion. Thus, the argument goes, instructions to change bias provide conscious interference with a normally unconscious process. In our view, the distinction between consciousness and its lack has nothing to do with either the existence or location of a criterion. Detection theory takes no stand on the conscious status of a criterion, and in any case observers do *not* naturally choose a neutral value. We shall encounter this issue again in chapter 10 when we briefly discuss the alleged phenomenon of subliminal perception. An observer who responds "no" when a stimulus is presented because of a high criterion is not necessarily aware of the possibility that a "yes" response would have been possible had the criterion been set lower.

Summary

Whereas a good sensitivity statistic is the difference between the transformed hit and false-alarm rates (chap. 1), a good measure of response bias is the sum of the same two quantities. In the decision space, this index describes the location of a criterion that divides the decision axis between values that lead to "yes" and "no" responses. Other measures—relative criterion and likelihood ratio—are equivalent when sensitivity is unvarying, but not when accuracy changes across conditions. Criterion location has advantages, both logically and, in some cases, empirically. Using a criterion to partition the decision axis is an optimal response strategy. The optimal location of the criterion can be calculated if the performance goal is specified.

Conditions under which the methods described in this chapter are appropriate are spelled out in Chart 3 of Appendix 3.

Computational Appendix

Derivation of Equation 2.6

The likelihood ratio is the ratio of the values of the S_2 and S_1 normal likelihood functions at the location $x = c$. The function $\phi(x)$ is defined by Equation 2.5. For both distributions, the standard deviation $\sigma = 1$; the mean μ equals d' in the numerator and 0 in the denominator. The resulting likelihood ratio, called β in SDT, is

$$\beta = e^{cd'}, \text{ or}$$

$$\ln(\beta) = cd' = -\tfrac{1}{2}[z(H) + z(F)][z(H) - z(F)] \tag{2.11}$$

$$= \tfrac{1}{2}[z(H)^2 - z(F)^2] \, .$$

Statistical Independence of d' and c

That c is statistically independent of d' can be seen as follows: Hit and false-alarm rates are independently computed (from data on S_2 and S_1 trials, respectively); so $z(H)$ and $z(F)$ are independent, and thus uncorrelated across repeated estimates. The sum (c) and difference (d') of uncorrelated variables are also uncorrelated and, because $z(H)$ and $z(F)$ are normally distributed, independent. Neither c' nor β is independent of d' in this sense.

Problems

2.1.   For the data of Example 2b, calculate $p(c)$ for each trainee. Do all readers improve, according to this measure?

2.2.   For the data of Example 2b, suppose all trainees adopted symmetric criteria, both before and after training. (a) What values of $p(c)$ would they obtain? (b) How do c, c', and $\ln(\beta)$ compare?

2.3.   For the matrixes of Problem 1.2, find c, c', and β.

2.4.   (a) Suppose an investigator decides to use F itself to measure response bias. What is the isobias curve for this statistic? (b) Another simple statistic is the yes rate, $(H + F)/2$. Find the isobias curve.

2.5.   Suppose $(F, H) = (.2, .6)$. If d' doubles and the observer maintains the "same bias," what will the new (F, H) point be? (Interpret "same bias" to mean same criterion, same relative criterion, and same likelihood ratio; you will have three answers.)

2.6. Ekman, O'Sullivan, and Frank (1999) videotaped men either lying or telling the truth about social issues on which they held strong beliefs, and played the tapes to four groups of observers: trained interrogators, Los Angeles County sheriffs, clinical psychologists interested in deception, and academic psychologists. The proportions of correct responses to lies and truths were:

Experimental group	P("lie"|lie)	P("truth"|truth)
Interrogators	.80	.66
Sheriffs	.77	.56
Clinical psychologists	.71	.64
Academic psychologists	.57	.58

The authors concluded that "... the most accurate groups did especially well in judging the lies compared with the truths ...," and that this could not be attributed to response bias. What would a detection theory analysis have to say about bias in these four groups? About sensitivity?

2.7. Suppose an observer is paid 10 cents for all correct responses. (a) What does the payoff matrix look like? (b) What is the optimal value of likelihood ratio if the proportion of S_2 trials is .5? .25? .1? (c) If $d' = 1$, find the criterion location c for each case. (d) Again assume $d' = 1$. In each case, what would the hit and false-alarm rates be? (*Hint*: Use the "coda" section.)

2.8. In a grating detection experiment, observers try to distinguish the presence of a pattern of alternating light and dark stripes from a uniform grey patch. There are two experimental conditions, with the stimuli differing in *contrast*, such that the stripes are better defined in the high-contrast than the low-contrast condition. Two groups of participants each view both conditions, but differ in that one group is given feedback (told after their response whether the grating was present) and the other is not. For each set of observers, how would you expect criterion placement to differ with contrast? Would it make a difference if the high- and low-contrast gratings were presented in different blocks of trials, or mixed together in a single block?

2.9. In this chapter, an analogy between detection theory and conventional statistical hypothesis testing is presented. According to this analogy, statistical results usually are summarized by a "false-alarm rate." Why are they not instead summarized by a sensitivity measure such as d'?

2.10.   Suppose $d' = 2$. (a) What are H and F if $c = 0.5$? -0.5? What are H and F if $c' = 0.5$? -0.5?

3

The Rating Experiment
and Empirical ROCs

In the last two chapters, we described correspondence experiments in which people report which of two events (such as seeing a New or Old face) had occurred. According to detection theory, they do this by comparing the strength of evidence, which we called *familiarity*, with a criterion. Observations of more than criterial familiarity are called "old," and those below criterion are called "new." The criterion is placed at a location of the observer's choice: Strict criteria serve to minimize false alarms, lax criteria to minimize misses.

If observers can set *different* criteria in different experimental conditions, they must know more about events in their experience than is needed to make a simple yes-no judgment. In this chapter, we see how observers can make graded reports about the degree of their experience by setting multiple criteria simultaneously. Our two primary examples are both tests of recognition memory, but for rather different materials: odors and words.

Design of Rating Experiments

Example 3a: Recognition Memory for Odors

How is memory for odors affected by the passage of time? Rabin and Cain (1984) presented participants with 20 odors to remember, then tested them at a delay of 10 minutes, 1 day, and 7 days. At each test, a different set of 20 New odors was intermixed with the Old stimuli.

Observers labeled each smell as "old" or "new" and also rated their confidence in these answers on a 5-point scale, which we have reduced to a 3-point scale for illustrative purposes. Thus, there are two kinds of stimuli and six possible responses (three ratings for each binary judgment). In our simplification, there are 375 trials of each sort (Old and New). Table 3.1 shows the data matrixes for two conditions: Each entry is the number of trials on which an observer used a specific response (old/new plus rating) for one stimulus class.

TABLE 3.1   *Frequency of Each Response for Each Stimulus (Odor Recognition)*

	"Old"			"New"			Total
	"3"	"2"	"1"	"1"	"2"	"3"	
10-Minute Delay							
Old	112	112	72	53	22	4	375
New	7	38	50	117	101	62	375
7-Day Delay							
Old	49	94	75	60	75	22	375
New	8	37	45	60	113	113	375

Response Alternatives in Rating Experiments

Like the examples in chapters 1 and 2, this experiment employs a one-interval design with two possible stimuli (Old and New odors). Rating experiments differ from yes-no experiments only in the response set available to the observer.

The rating task offers the observer a set of responses varying from great confidence in one alternative, through relative indifference between the two choices, to great confidence in the second alternative. Notice that this is the ordering in Table 3.1. Three different, equivalent response sets meet this requirement. The simplest is a set of numerals, "1" to "10," "1" to "5," or some other range. The lowest numbered response indicates high certainty that, for example, the test odor was in the study set, whereas the highest number indicates high certainty that it was not. As a second possibility, the response set may consist of verbal categories ranging from "certain it is not old" through "fairly sure it is not old," through some intermediate categories, to "very certain it is an old word." A third way to organize the response set for participants, the method used in the Rabin and Cain (1984) experiment, is to require two subresponses. Observers first judge the stimulus as old or new and then grade the certainty of the response with a number or verbal category.

ROC Analysis

As in chapter 1, we wish to know how sensitive the participants were to the experimental distinction, but we have more information than in two-response examples. If no confidence judgments had been used, the two conditions could be easily summarized. By ignoring the distinctions among confidence levels, Table 3.1 could be reduced to the following:

	10-Minute Delay		7-Day Delay	
	"Yes"	"No"	"Yes"	"No"
Old (S_2)	296	79	218	157
New (S_1)	95	280	90	285

Then d' could be calculated for each condition as follows:

10-Minute Delay	7-Day Delay
$H = 296/375 = .79$	$H = 218/375 = .58$
$F = 95/375 = .25$	$F = 90/375 = .24$
$d' = 1.48$	$d' = 0.91$

How can we use the supplementary ratings to improve our assessment of the participants' performance? We first discuss the calculation of hit and false-alarm rates from the data, and then we provide a graphical presentation of the results.

Calculating Hit and False-Alarm Rates

The first steps in analyzing rating data parallel the treatment of yes-no data in chapter 1. The number of responses of each type for each stimulus is tabulated (Table 3.1), and each cell frequency is divided by the total number of stimulus presentations in its row to estimate a conditional probability. Table 3.2 shows the results of this transformation for both matrixes of our example. Each entry in the new table shows the proportion of trials on which the stimulus yielded that response, so each row totals 1.0 (except for rounding error).

The third step is special to rating designs. For each response, we find the proportion of trials leading to that response, or any response to the left of it, by summing the conditional probability table, left to right. The cumulative probabilities are shown, for our data, in Table 3.3. For both conditions, each column of the table now contains a hit rate and false-alarm rate. We have already pointed out that if ratings had been omitted, the participants in the 7-day delay condition could have produced a false-alarm/hit pair of (.24, .58). But observers also could have reached the pair (.4, .74) by using "yes"

TABLE 3.2   *Proportion of Each Response for Each Stimulus (Odor Recognition)*

	"Old"			"New"			Total
	"3"	"2"	"1"	"1"	"2"	"3"	
			10-Minute Delay				
Old	.299	.299	.192	.141	.059	.011	1.00
New	.019	.101	.133	.312	.269	.165	1.00
			7-Day Delay				
Old	.131	.251	.200	.160	.200	.059	1.00
New	.021	.099	.120	.160	.301	.301	1.00

TABLE 3.3   *Hit and False-Alarm Rates: Cumulative Proportions for Each Stimulus (Odor Recognition)*

	"Old"			"New"		
	"3"	"2"	"1"	"1"	"2"	"3"
			10-Minute Delay			
Old	.299	.598	.790	.931	.990	1.00
New	.019	.120	.253	.565	.834	1.00
			7-Day Delay			
Old	.131	.382	.582	.742	.942	1.00
New	.021	.120	.240	.400	.701	1.00

to include both "old" (all levels of confidence) and "new but with a confidence rating of 1." Each of the other pairs also corresponds to a partition of the response categories into two subsets and reflects a possible yes-no decision rule.[1]

As in the two-response experiment, we now use Table A5.1 to find z scores for each cumulative conditional probability; these are shown for our illustration in Table 3.4. The result is a whole set of d' measures, one for each possible partition of the response set. A 2×2 table produces one estimate of sensitivity; a 2×6 table, five. The multiple estimates of d' obtained from the rating data are not independent of each other. In moving from one partition to the adjacent one to the right, all the same data are used, plus one new column. In this example, the estimates of d' obtained at different criteria agree rather well.

[1]Notice that if we had summed right to left, the entries would have become miss and correct rejection rates.

TABLE 3.4 *Transformed Hit and False-Alarm Rates, z Scores, and d' Estimate for Each Response (Odor Recognition)*

	"Old"			"New"	
	"3"	"2"	"1"	"1"	"2"
10-Minute Delay					
Old	−0.527	0.232	0.807	1.484	2.327
New	−2.081	−1.175	−0.665	0.164	0.970
d'	1.554	1.407	1.472	1.320	1.357
7-Day Delay					
Old	−1.121	−0.301	0.207	0.649	1.573
New	−2.037	−1.175	−0.706	−0.253	0.527
d'	0.916	0.874	0.913	0.902	1.046

Graphing Discrimination—The Empirical ROC

To make sense of these data, it is useful to represent them graphically. We can do so using the ROC curve, the function that relates hit and false-alarm rates. In chapter 1, we described the *implied ROC*, the set of (hit, false-alarm) pairs consistent with a specific value of a sensitivity measure. The implied ROC makes a theoretical statement.

The data in Table 3.3 can be used to plot *empirical ROCs*, as illustrated in Fig. 3.1a. Each possible (F, H) pair in Table 3.3 is a point in ROC space: The leftmost pair falls near the lower left corner of the graph in the first panel of the figure, and each succeeding pair is farther to the right and higher. The points *must* increase in this way because each point is obtained by adding the data from one new column to the previous one. The second panel shows the same data as the first, but uses the values in Table 3.4 to locate points on z coordinates rather than linear probability axes.

The shape of an empirical ROC is predicted by detection theory. For example, the ROC implied by the use of d' as a sensitivity index is $z(H) = d' + z(F)$ (Eq. 1.8). This is the equation of a straight line with a slope of 1.0, displaced upward from the origin by the constant sensitivity. Such a line has been drawn through each set of data in Fig. 3.1b; the curves in Fig. 3.1a were produced by transforming back from z values to probabilities. The five points on each ROC are separate (although not independent) estimates of the discriminability of the two stimulus sets. A single underlying d' can yield any data point that gives the same difference of the transformed values on the two axes of the ROC plot.

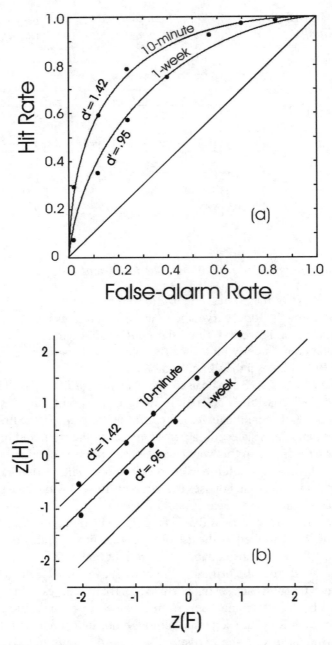

FIG. 3.1.   ROCs for odor recognition (Example 3a) on (a) linear coordinates, and (b) *z* coordinates. Upper curves are for a 10-minute delay, lower curves for a 1-week delay.

The data show that odor memory is initially good, but declines over a 7-day period (for Rabin and Cain's observers, most of the decline occurred in just 1 day). Recognition memory for smell is, of course, a different quality than the ability of smells to evoke other memories. Proust's classic *Remembrance of Things Past* celebrates this long-term power of olfactory memory.

ROC Analysis With Slopes Other Than 1

All points on each unit-slope line in Fig. 3.1b represent the same detectability. What would we make of a line that did not have a slope of 1 so that sensitivity, as estimated by $z(H) - z(F)$, was different for each point on the line? How can we best summarize such data?

Example 3b: Recognition Memory for Words of Varying Frequency

Models of memory for words are often tested by examining ROC data (Ratcliff, McKoon, & Tindall, 1994). The next example compares experimental conditions in which the studied items have either high frequency of occurrence (i.e., are common words) or low frequency. Participants choose from a set of five responses, which might be numerical or verbal judgments, as shown in Table 3.5. The data, also given in the table, are typical of such experiments in that recognition is better for low- than for high-frequency words.

TABLE 3.5   *Frequency of Each Response for Each Stimulus (Word Recognition)*

	"1" = "Sure Old"	"2" = "Maybe Old"	"3" = "Uncertain"	"4" = "Maybe New"	"5" = "Sure New"
Low-Frequency Words					
Old	61	15	15	5	4
New	2	8	37	23	30
High-Frequency Words					
Old	37	25	18	11	9
New	4	18	28	21	29

Plotting the ROC

The cumulated proportions based on these data are plotted in Fig. 3.2 on both linear probability and normal (*z* score) axes. (The lines and summary

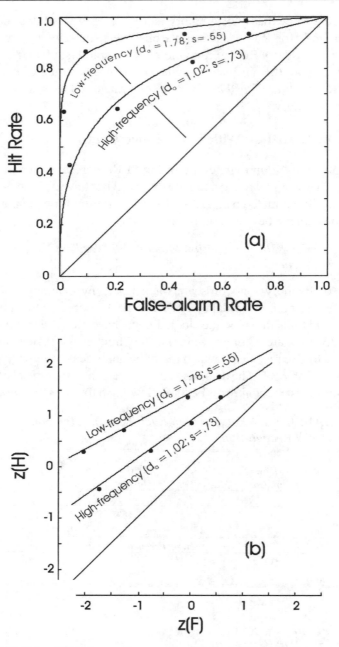

FIG. 3.2.   ROCs for word recognition (Example 3b) on (a) linear coordinates, and (b) *z* coordinates. Upper curves are for low-frequency words, lower curves for high-frequency words.

statistics in the figure will be explained presently.) The z-transformed hit and false-alarm rates are presented in Table 3.6. The last line in the table gives the difference of the normal deviates, as in Table 3.4. Here the values of $z(H) - z(F)$ vary considerably and systematically; for low-frequency words, for instance, they range from 2.33 to 1.23. If we had conducted a yes-no experiment, our sensitivity estimate could have been anything between these two values depending on the response criterion chosen by the observer. We need a way to summarize this set of data by a single value, and it is clear that the z score difference does not provide one.

TABLE 3.6 *Cumulative Proportions in z Score Units (Word Recognition)*

	"1"	"2"	"3"	"4"
	Low-Frequency Words			
Old	0.279	0.706	1.341	1.751
New	−2.054	−1.282	−0.075	0.524
$z(H) - z(F)$	2.333	1.988	1.416	1.227
	High-Frequency Words			
Old	−0.332	0.306	0.842	1.341
New	−1.751	−0.772	0.0	0.553
$z(H) - z(F)$	1.419	1.078	0.842	0.788

Estimating Sensitivity: d_a, *ROC Slope, and Related Indexes*

When the slope of the ROC equals 1, d' is both the horizontal distance and the vertical distance between the zROC and the chance line. When the slope is not equal to 1, however, these two distances differ, as shown in Fig. 3.3. A large value of d' (call it d'_1) can be measured by taking the horizontal distance from the ROC to the major diagonal at the point where $z(H) = 0$. A small value (d'_2) is the vertical distance between them where $z(F) = 0$. Examination of the figure reveals that s, the slope of the zROC, is equal to

$$s = d'_2/d'_1 . \tag{3.1}$$

The equation of the ROC can be written

$$d'_1 = (1/s)z(H) - z(F), \quad \text{or} \tag{3.2}$$

$$d'_2 = z(H) - sz(F) . \tag{3.3}$$

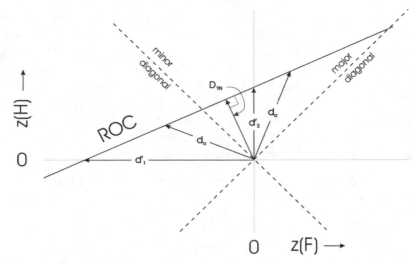

FIG. 3.3.   Nonunit-slope ROC, showing alternative indexes of sensitivity: d'_1 (unit is the standard deviation of S_1), d'_2 (unit is the standard deviation of S_2), and d_a (unit is the rms average of the two standard deviations). The distance d_a is $\sqrt{2}$ times as long as D_{YN}, the perpendicular from the origin to the ROC.

Some insight into these measures can be gained by considering the underlying distributions implied by this ROC. What does it mean that the ROC slope is less than 1? It means that moving one z unit, or standard deviation, on the F-axis produces a change of less than one standard deviation (s units to be exact) on the H axis. That is, the standard deviations of the S_1 and S_2 distributions are in the ratio $s:1$. Pairs of distributions having this characteristic (with $s = 0.5$) are shown in Fig. 3.4.

Our alternate measures of sensitivity (Eqs. 3.2 and 3.3) each rely on one of these standard deviations, and each corresponds to a different distance in the decision space. The index d'_1 is measured at the point $z(H) = 0$, that is, for a criterion set at the mean of the S_2 distribution, as in Fig. 3.4a. At that point, d'_1 equals $-z(F)$ (Eq. 3.2) and thus only depends on the standard deviation of the S_1 distribution. On the other hand, d'_2 is measured where $z(F) = 0$, so that the criterion is at the mean of the S_2 distribution (Fig. 3.4b). Here, d'_2 equals $z(H)$ and depends only on the standard deviation of the S_2 distribution.

The best single measure of sensitivity in this situation is neither d'_1 nor d'_2, but a compromise. In ROC space (Fig. 3.3), it should be a distance between the ROC and the chance line that is shorter than d'_1 but longer than

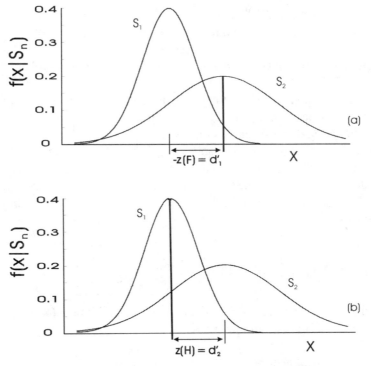

FIG. 3.4. Decision space for nonunit-slope ROC, standard deviation of S_2 double that of S_1: (a) criterion at mean of S_2, difference between the means measured in units of the S_1 standard deviation; (b) criterion at mean of S_1, difference between the means measured in units of the S_2 standard deviation.

d'_2. In the decision space, it should measure the mean distance between distributions in units of some kind of average of the two standard deviations.

Figure 3.3 illustrates such a measure. Instead of selecting the horizontal or vertical distance from the origin to the ROC, consider the shortest distance between them. This statistic, termed D_{YN} by its inventors (Schulman & Mitchell, 1966), is a principled choice, but gives values of the wrong magnitude, always smaller than either d'_1 or d'_2. The correct adjustment is to multiply D_{YN} by $\sqrt{2}$. The new distance, called d_a (Simpson & Fitter, 1973), is shown in Fig. 3.3 as the length of the hypotenuse of the equilateral triangle whose legs have length D_{YN}.

The index d_a has the properties we want. First, it is intermediate in size between d'_1 and d'_2. Second, it is equivalent to d' when the ROC slope is 1, for then the perpendicular line of length D_{YN} coincides with the minor diagonal, and the two lines of length d_a coincide with the d'_1 and d'_2 segments. Third, d_a turns out to be equivalent to the difference between the means in units of the *root-mean-square (rms)* standard deviation, a kind of average equal to the square root of the mean of the squares of the standard deviations of S_1 and S_2.

To find d_a from an ROC that is linear in z coordinates, it is easiest to first estimate d'_2 (the vertical intercept), d'_1 (the horizontal intercept), and s (from Eq. 3.1). Because the standard deviation of the S_1 distribution is s times as large as that of S_2, we can set the standard deviation of S_1 to s and that of S_2 to 1. Then the standard deviation used for measuring d_a is $[\frac{1}{2}(1 + s^2)]^{\frac{1}{2}}$, the rms average of the two. When sensitivity is measured in units of the S_2 distribution, it equals d'_2; to convert to the right units for d_a, divide d'_2 by the rms standard deviation:

$$ d_a = \frac{d'_2}{[\frac{1}{2}(1+s^2)]^{\frac{1}{2}}} = \left(\frac{2}{1+s^2}\right)^{\frac{1}{2}} d'_2 . \tag{3.4}$$

To find d_a directly from a point on the ROC (once the slope is known), Equations 3.3 and 3.4 are combined to yield

$$ d_a = \left(\frac{2}{1+s^2}\right)^{\frac{1}{2}} [z(H) - sz(F)] . \tag{3.5}$$

When a single number is desired to characterize an ROC, d_a is a good one; by also reporting the slope s, the investigator can completely describe the curve. For the two conditions of the word-recognition experiment (Fig. 3.2 and Tables 3.5 and 3.6), $s = 0.55$ for low-frequency words and 0.73 for high-frequency words. Values of d_a for the two conditions are approximately 1.78 and 1.02.

Another commonly used distance measure, d'_e, is named for James Egan, who conducted several important early experiments establishing the usefulness of detection theory. This index is based on the *arithmetic average* of the standard deviations and is defined by

$$ d'_e = \frac{d'_2}{\frac{1}{2}(1+s)} = \frac{2}{(1+s)} d'_2 \tag{3.6}$$

and can be found from a hit/false-alarm pair by

$$d'_e = \left(\frac{2}{1+s} [z(H) - sz(F)] \right). \tag{3.7}$$

Like d_a, d'_e is a distance between the ROC and the chance line that is intermediate in length between d'_1 and d'_2. In fact, as Fig. 3.5 shows, it is the distance between the two lines measured at the minor diagonal. Any measure based on an average of the two underlying standard deviations is an improvement over d'_1 or d'_2 (the geometric mean has also been proposed; see Grey & Morgan, 1972).

Area Under the ROC: A_z and A_g

Sometimes a measure of performance expressed as a proportion is preferred to one expressed as a distance. The index A_z, which is simply D_{YN} transformed by the normal distribution function Φ, is appropriate in such circumstances (Swets & Pickett, 1982):

$$A_z = \Phi(D_{YN}) = \Phi(d_a / \sqrt{2}). \tag{3.8}$$

In the current examples, $A_z = .9$ and .76. This statistic equals the area under the normal-model ROC curve, which increases from .5 at zero sensitivity to 1.0 for perfect performance.

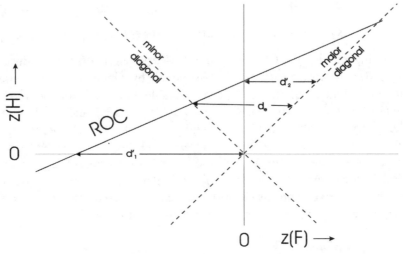

FIG. 3.5.   Nonunit-slope ROC, showing alternative indexes of sensitivity: d'_1 (unit is the standard deviation of S_1), d'_2 (unit is the standard deviation of S_2), and d'_e (unit is the arithmetic average of the two standard deviations).

Area under the ROC is a good index of sensitivity and can be measured without any model assumptions—the first truly nonparametric measure we have encountered. Pollack and Hsieh (1969) suggested estimating this area in a straightforward way. Using the linear-coordinate ROC, connect the successive (F, H) points and draw vertical lines from each point to the F-axis, creating a series of trapezoids (and one triangle). Each of these figures has an area equal to the difference in the F values times the average H value, and the total area (which Pollack and Hsieh called A_g) is found by summing these areas:

$$A_g = \tfrac{1}{2}\Sigma(F_{i+1}-F_i)(H_{i+1} + H_i) . \tag{3.9}$$

The index i tracks the ROC points so that (F_1, H_1) equals $(0, 0)$, (F_2, H_2) is the first point to the right, and the last point is $(1,1)$.

This measure is best with a large number of responses: The polygon form of the ROC is systematically lower than the "true" ROC, and this difference is greatest for curves with few points. Donaldson and Good (1996) proposed a measure, $A'r$ (r for rating), that increases A_g to approximately compensate for this discrepancy. Of course if the ROC is consistent with the normal-distribution model, A_z *exactly* compensates, so the nonparametric A_g and $A'r$ measures are most useful when this model does not hold.

Estimating Bias

Decision Space for the Rating Experiment

In deriving empirical ROCs, the stimulus–response matrix is partitioned many different ways, once for each possible rule by which the observer could have reduced the matrix to a simple yes-no table. Each partition yields a different (false-alarm, hit) pair, as shown in Figs. 3.1 and 3.2, and thus implies a different criterion. One criterion is enough to generate one ROC point; to produce an ROC curve with n points, the observer must maintain n criteria simultaneously. No paraphernalia beyond those introduced in chapter 2 are needed to find the locations of these criteria. The rating matrix is reanalyzed as separate yes-no matrixes, and any desired bias measure can then be computed. We illustrate the calculations involved for our two examples.

Unit-Slope ROCs

First, consider the 7-day delay condition in the Rabin and Cain (1984) odor-recognition experiment. Table 3.7 extends that part of Table 3.4 in which d'

was computed for this condition to the bias parameter c. For each column, $c = -\frac{1}{2}[z(H) + z(F)]$, just as in two-response experiments. The highest values of the criterion c indicate reluctance to say "old," and therefore correspond to the left-most responses in the table. Figure 3.6 shows the locations of the criteria vis-à-vis the underlying distributions.

The likelihood ratio, another measure of bias included in Table 3.7, can be computed in two ways: either directly from the heights of the underlying densities or from the product of the criterion location and d' (Eq. 2.6). We illustrate both methods for the criterion dividing "old" from "new" responses. At this point, $z(H) = 0.207$ and $z(F) = -0.706$, that is, $H = .58$ and $F = .24$. Using Table A5.1, we find that the likelihood ratio β is $0.391/0.311 = 1.26$ and $\ln(\beta) = 0.23$. And, according to Table 3.2, $d' = 0.913$ and $c = 0.250$, so $\ln(\beta) = d'c = 0.23$.

TABLE 3.7   *Transformed Hit and False-Alarm Rates, d', and Bias for Each Response in the Odor Recognition Experiment (7-Day Delay Condition)*

	"Old"			"New"	
Old	−1.121	−0.301	0.207	0.649	1.573
New	−2.037	−1.175	−0.706	−0.253	0.527
d'	0.916	0.874	0.913	0.902	1.046
c	1.579	0.738	0.250	−0.198	−1.050
$\ln(\beta)$	1.446	0.645	0.230	−0.179	−1.098

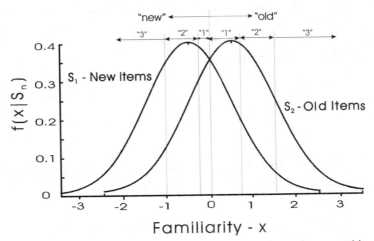

FIG. 3.6.   Decision space and response criteria for the odor-recognition rating experiment of Example 3a (1-week delay condition).

Notice that the likelihood ratio β decreases as the hit and false-alarm rates increase (moving through the table from left to right). As we saw in chapter 2, the likelihood ratio is the slope of the ROC curve (on linear, not z score coordinates), so the slope of the ROC must also continually decrease. Because of variability, however, empirical ROCs do not always have monotonically decreasing slope. One particularly glaring violation of monotonicity occurs when a cell not in an end column of the data matrix contains a 0. A 0 in the upper (S_2) row implies that two adjacent points on the ROC fall on the same horizontal line, and a 0 in the lower row implies two points on the same vertical line. The first yields a likelihood ratio of zero, the second of infinity. Because such values are inconsistent with most models, some experimenters "smooth" them in plotting their data. The simplest method for doing this is to merge any column with a 0 in the S_2 row with the column to its left, and any column with a 0 in the S_1 row with the column to its right. This procedure selects the more sensitive of two horizontally or vertically paired points, and it eliminates the one displaying less sensitivity.

Nonunit-Slope ROCs

If the slope of the ROC on z coordinates does not equal 1, a decision must be made about the unit in which c is to be measured. Let us first consider using the standard deviation of the S_2 distribution in this role, calling the resulting criterion location index c_2. Figure 3.7 shows an unequal-variance decision space; the S_1 distribution has standard deviation s and the S_2 distribution standard deviation 1. We wish to calculate the criterion location c_2 relative to the zero-bias point. As we saw in chapter 1, the z coordinate of each "yes" rate is a difference between a distribution mean and the criterion location expressed in standard deviation units:

$$z(H) = M_2 - c_2$$
$$z(F) = \frac{M_1 - c_2}{s} \tag{3.10}$$

The analysis differs from that of chapter 1 only because the standard deviation of the S_1 distribution is not 1. Combining Equations 3.10 leads to[2]

$$c_2 = \frac{-s}{(1+s)}[z(H) + z(F)] . \tag{3.11}$$

[2]See Computational Appendix for derivation.

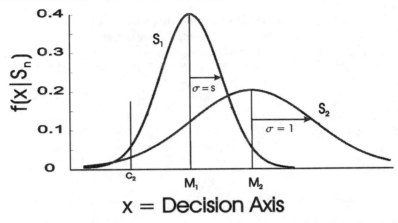

FIG. 3.7. Unequal-variance decision space portraying criterion location c_2 (measured in terms of the S_2 standard deviation).

Three other possible bias statistics employ the standard deviation of the S_1 distribution, the rms standard deviation, and the average standard deviation, and can be calculated from

$$c_1 = \frac{-1}{(1+s)}[z(H)+z(F)] \tag{3.12}$$

$$c_a = \frac{-\sqrt{2}\,s}{(1+s^2)^{\frac{1}{2}}(1+s)}[z(H)+z(F)] \tag{3.13}$$

$$c_e = \frac{-2s}{(1+s)^2}[z(H)+z(F)] . \tag{3.14}$$

Values for all four measures in the low-frequency condition of the word-recognition experiment (Example 3b) are given in Table 3.8. Because the measures differ only in unit, they are related to each other by multiplicative constants. The isobias curves of all measures are the same, and the same as that for c (Fig. 2.6). When $s = 1$, all the indexes are equal to each other and to c.

Two other classes of bias measures were considered in chapter 2: relative distances and likelihood ratio. In the unequal-variance case, relative distances (c') can be computed by combining the criterion values of Equations

TABLE 3.8   *Transformed Hit and False-Alarm Rates and Bias Measures for the Word-Recognition Experiment (Low-Frequency Condition)*

	"1"	"2"	"3"	"4"
Old	0.279	0.706	1.341	1.751
New	−2.054	−1.282	−0.075	0.524
$z(H) + z(F)$	−1.775	−0.576	1.266	2.275
c_1	1.145	0.372	−0.817	−1.468
c_2	0.630	0.204	−0.449	−0.807
c_a	0.780	0.253	−0.557	−1.000
c_e	0.813	0.264	−0.580	−1.042

3.12 to 3.14 with the corresponding sensitivity values. Likelihood ratio can be calculated, as in the previous example, by finding the heights of the S_1 and S_2 densities that correspond to H and F in Table A5.1. However, the simple relation between likelihood ratio and c (Eq. 2.6) no longer applies. In fact, the use of likelihood ratio in describing nonunit-slope ROCs presents a difficulty. Two normal densities that differ in both mean and variance intersect each other at two points (Luce, 1963a), as shown in the first panel of Fig. 3.8. Because each intersection point reflects a likelihood ratio of 1, the decision axis can no longer be monotonic with likelihood ratio. Indeed, if the observer uses a cutpoint decision rule on this axis, H will be *less* than F near one corner of the ROC (Fig. 3.8b).

Although ROCs with nonunit slope are common, we are not aware of any data for which H is systematically less than F. There are two possible reasons for this nonphenomenon. First, the reversal occurs at extreme points in ROC space. If $d'_1 = 2$ and $s = 0.5$ (as in the figure), the reversal occurs at $H = F = .995$, an ROC point rarely encountered in application. Even if $d'_1 = 0.5$, the critical point is $H = F = .98$. The small magnitude of the potential reversals makes them hard to distinguish from chance ($H = F$) performance.

Second, no observer is forced to use the cutpoint response rule, which is not ideal in this situation. The optimal decision maker establishes *two* cutpoints, one at each location having the critical value of likelihood ratio, and responds "yes" for ratios greater than that value. Such observations are either above the upper cutpoint or below the lower one. The third panel of Fig. 3.8 illustrates this rule for a likelihood ratio of 1.0. The corresponding ROC, portrayed in the second panel, differs most from the cutpoint rule in the upper corner, where the likelihood ratio rule does *not* produce below-chance performance.

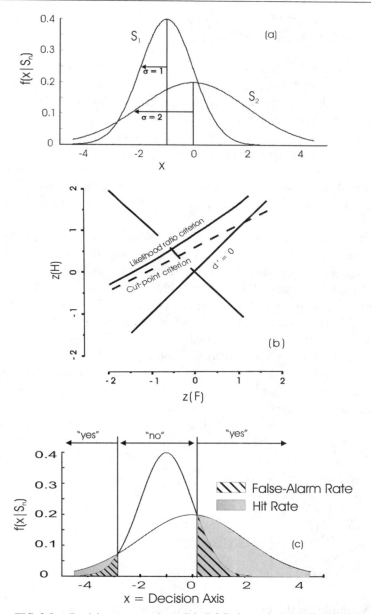

FIG. 3.8.   Decision space and possible ROCs for underlying distributions of un-equal variance. (a) Normal distributions with standard deviations 1 and 2. The distributions intersect at two points. (b) Two possible ROCs. Cutpoint rule yields a linear ROC ($s = 0.5$) and leads to below-chance performance at low criteria; likelihood ratio rule has nonlinear shape. (c) Decision space of panel (a) showing the two cutpoints required by a likelihood ratio response rule, here set at $\beta = 1$.

Systematic Parameter Estimation
and Calculational Methods

Statistical methods are used to find the ROC curve of a given shape that best fits the data. Fitting curves to points is a common procedure in behavioral research, but the ROC presents a peculiar problem. Whereas most experimental data plots have a dependent variable (something measured) on the ordinate and an independent variable (something varied by the experimenter) on the abscissa, in an ROC *both* axes are dependent variables. For a linear fit when only one dimension is free to vary, one aims to minimize the total discrepancy between the data points and the line on that dimension. This tactic is inappropriate for ROC data (see Appendix 1 for more discussion).

A solution to this difficulty is provided by the statistical curve-fitting procedure of *maximum-likelihood estimation* (discussed in chap. 13 and in Appendix 1). A program that uses this method to calculate ROCs was developed by Dorfman and Alf (1969); a modified version, called ROCKIT, and several extensions have been made available by Metz and his colleagues (e.g., Metz & Kronman, 1980). These programs are available online.[3] The output gives the value of d'_2 (called A), s (called B), and A_z.

Statistical packages with detection theory modules can also be used for both the normal model and a variety of other distributional assumptions (some of which we discuss in chap. 4). To use the signal detection module in Systat, first enter the data into three columns, one for the stimulus (0 or 1), the second for the response (any integers in the range –6 to +6), and the third for the frequency of that stimulus–response combination, which we call FREQ. Give this data set a name, say PROBLEM1, and then issue the following instructions:

> USE PROBLEM1
> SIGNAL
> MODEL RATING=SIGNAL
> FREQ=COUNT
>   ESTIMATE

The output gives the value of d'_1 (confusingly, this is called d'), d_a, $1/s$, A_z, and the bias measures β and $\ln(\beta)$. The ROC is plotted, and its goodness of fit is measured by chi-square. SPSS provides a similar module.

---

[3]The Web site is http://www.radiology.uchicago.edu .

Alternative Ways to Generate ROCs

There is more than one way to gather ROC data. Although the rating method is the most efficient, it does not even have historical priority as a procedure (Tanner & Swets, 1954). All other methods use the same stimulus alternatives under several different experimental conditions at different times. Under the different conditions, observers are encouraged in one way or another to change their willingness to say "yes"; we expect any change in such willingness to change both the hit and false-alarm rates, but not sensitivity.

Monetary Rewards (Payoffs)

Experimenters may reward observers, trial by trial, for their performance. Use of rewards mimics some real-life discrimination situations. An automotive quality-control inspector should perceive the cost of failing to detect faulty work to be large relative to the cost of a false alarm. On the other hand we may hope that those who can start a war in response to intelligence information are made cautious by the very high cost of a false alarm.

In a simple yes-no experiment with two possible stimuli and two responses, there are four values to manipulate: the amounts paid (or debited) by the experimenter for hits, misses, false alarms, and correct rejections. Different runs use the same stimuli, but a different set of financial rewards or *payoff matrix*. Each set of payoffs produces a separate 2×2 data matrix, and the set of (F, H) pairs defines an ROC. The optimal value of the criterion under each payoff can be calculated from Equation 2.8.

Verbal Instructions

Explicit financial incentives can often be effectively replaced by verbal instructions. Participants are urged during some experimental runs to be lax in reporting, for instance, that a stimulus is Old, whereas during other runs they are urged to be strict. Well-trained participants seem able to understand these instructions—perhaps a bit of support for the notions of SDT in itself—and can also use "neutral" as a criterion, as well as degrees of strictness or laxness. This procedure is just as time-consuming as paying money because each verbal criterion must be set in a separate session to establish an ROC.

Exactly what terms can be used, either in separate sessions, or in rating experiments, is not always obvious. In recognition memory research, for example, participants are sometimes asked to distinguish items they can *remember* encountering in the experiment from those they *know* were presented even though the specific episode is not available. The original motivation for such experiments (Tulving, 1985) was to tap distinct explicit versus implicit memorial processes, but participants may also treat "remember" and "know" as different levels of confidence along the same dimension. Donaldson (1996) proposed the latter interpretation, supporting it with an analysis in which sensitivity is calculated separately for the two putative levels of confidence represented by these responses. More recently, Rotello, Macmillan, and Reeder (2004) argued that "remember" and "know" responses reflect multiple sensitivities as well as different response rules. The key point is that not every manipulation that affects hit and false-alarm rates generates a true ROC; whether a particular set of points in ROC space reveals isosensitivity is a substantive theoretical question.

Manipulating Presentation Probability

Another way to alter the willingness of people to say "yes" as opposed to "no" is to change the relative likelihood of presenting the two stimuli. Viewers who are aware of the presentation probabilities are more willing to report the more likely stimulus. Experimenters can change the *presentation probability* of one stimulus from session to session and keep separate records for each probability condition. This strategy is even more tedious than the last two, especially when one of the stimuli is very unlikely, because estimating a hit or false-alarm rate requires many runs simply to get a sufficiently large sample of trials. However, some intrepid souls have collected ROC data this way (Creelman, 1965).

Use of presentation probability to trace ROCs encounters two other problems. First, if feedback is not used, so that the observers are unaware of the a priori probabilities, decreasing the probability of presenting S_2 may actually *increase* the number of "yes" responses (T. Tanner, Haller, & Atkinson, 1967; T. Tanner, Rauk, & Atkinson, 1970). One interpretation of this result is that participants tend to believe the presentation probabilities to be equal (similar effects in identification experiments are discussed in chap. 5). The last difficulty with changing presentation probabilities is that doing so may influence sensitivity as well as bias (Markowitz & Swets, 1967; see also Dusoir, 1983). In a detection task, the higher the proportion of S_2 (Signal) trials, the better the observer will be able to remember the Signal. Balakrishnan (1999) also found that changes in presentation probability—and even changes in payoffs—can affect sensitivity. These difficulties, and the tedium of collecting enough data at low presentation probabilities, generally make this strategy for collecting ROCs unattractive.

Another Kind of ROC: Type 2

An empirical ROC curve plots ratings conditional on one *stimulus class* against ratings conditional on another, but an analogous curve can be constructed from ratings conditional on the *correctness of responses*. In the odor-recognition task (Example 3a), observers first labeled stimuli as old or not and then expressed their confidence in their answers. A *Type-2 ROC curve* relates their confidence judgments on correct trials to confidence judgments on incorrect trials. Type-2 ROCs, first analyzed by Clarke, Birdsall, and Tanner (1959), provide a perspective on the decision space different from that of their Type-1 siblings, as shown in Fig. 3.9.

Clarke et al. (1959) supposed that the observer's initial response was based on an unbiased criterion placement ($c = 0$) and that the later confidence judgments would be high for extreme observations in either direction. Only two levels of confidence are shown in the figure: The observer reports "sure" for observations that are either above a positive cutoff k or below the negative cutoff $-k$, and "unsure" otherwise. Thus, the initial response depends on a binary partition of the decision axis, and the later confidence rating relies on a multiple partition applied to the absolute value.

The hit and false-alarm rates for a Type-2 ROC are the proportions of rating responses up to a particular level of confidence given truly correct and

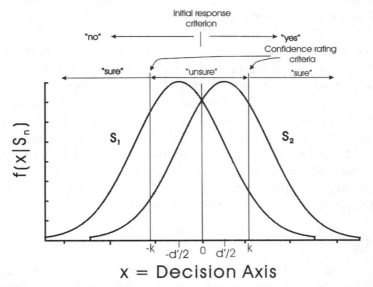

FIG. 3.9.   Analysis of the Type-2 ROC curve. Subjects first respond "yes" or "no," then provide a confidence rating.

incorrect initial responses. Because the initial response is assumed to be symmetrically determined, these response rates can be calculated by considering only one of the two stimuli, say S_2. A conditional probability equals the probability of a joint event (say rating a correct response as correct) divided by the probability of a marginal one (an initial correct response). Thus,

P(rating correct | initial correct)

$$= P(|x| > k \,|\, x > 0) = \frac{P(x > k)}{P(x > 0)} = \frac{\Phi\left(\dfrac{d'}{2} - k\right)}{\Phi\left(\dfrac{d'}{2}\right)}$$

P(rating correct | initial incorrect)

$$= P(|x| > k \,|\, x < 0) = \frac{P(x < -k)}{P(x < 0)} = \frac{\Phi\left(-\dfrac{d'}{2} - k\right)}{\Phi\left(-\dfrac{d'}{2}\right)} \;. \tag{3.15}$$

Equations 3.15 contain only two unknowns (d' and k) and can be solved by iteration. The apparent sensitivity index of the Type-2 ROC (i.e., the intercept of the ROC on z axes) is less than d'. The slope is less than the assumed Type-1 slope of 1.0.

Although the analysis of Clarke et al. (1959) is almost as old as the standard Type-1 method, it is little used. This is unfortunate, because many experiments that are unsuitable for Type 1 are susceptible in principle to Type-2 description. Consider, for example, the traditional recall task: A participant hears a series of words and later lists as many as possible. With each item recalled, the participant provides a rating of confidence that it was in fact on the original list. A Type-2 ROC clearly can be constructed from data of this sort, but because there is only one *stimulus* class (the words on the original list), no Type-1 curve is possible. To our knowledge, no one has tested the usefulness of this approach to recall.

Essay: Are ROCs Necessary?

When an empirical ROC has unit slope, any point on the curve provides the same estimate of sensitivity. Rating responses are in this case unnecessary if we are only interested in sensitivity: The entire ROC can be inferred from one point, and a yes-no experiment suffices to find that point. Indeed, not every researcher employing detection theory collects ROCs, a procedure that may appear to introduce unnecessary tedium and complexity into experiments. Are ROCs worth the effort?

It is even possible that the rating procedure distorts "true" yes-no behavior. Perhaps maintaining several criteria is a more taxing cognitive chore than setting only one (Wickelgren, 1968). Most existing data are reassuring: Early in the history of detection theory, Egan and his colleagues (Egan & Clarke, 1956; Egan, Schulman, & Greenberg, 1959) obtained equivalent measures of detectability with and without ratings in an auditory task.

The difficulty in *not* collecting ROCs is, of course, that if ROC slopes are not equal to 1, then comparisons of observed sensitivity estimates may be misleading. Consider the two points shown in Fig. 3.10, $(F = .31, H = .83)$ and $(F = .62, H = .96)$. For both, $d' = z(H) - z(F) = 1.45$. Yet if the true ROCs have slopes of, say, 0.5, the second point reflects much greater sensitivity than the first.

Collection of full ROCs could be avoided, even if slopes did not equal 1, if slopes were known a priori. Several theorists have offered models in which slope is systematically related to sensitivity. Green and Swets (1966, ch. 4) proposed that the slope $s = 4/(4 + d'_1)$, so that ROCs reflecting low sensitivity have slopes near 1 and those measuring good performance have

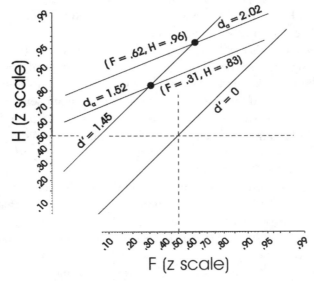

FIG. 3.10. Two points [(.62, .96) and (.31, .83)] that lie on different nonunit-slope ROCs but the same unit-slope ROC. (The axes are scaled in z score units.)

increasingly shallow slopes. Other families of ROCs that show a negative correlation between sensitivity and slope are those that take the underlying distributions to be (a) chi-square, for which the mean equals half the variance; (b) Poisson, for which the mean equals the variance; or (c) exponential, for which the mean equals the standard deviation. This last possibility generates ROCs of a simple type, power functions of the form $H = F^n$, where n is a number between 0 and 1. Egan (1975) provided a thorough description of each of these ROC families, and Laming (1986) provided a theoretical rationale for various ROC slopes and shapes.

The key question is, of course, whether the slope and sensitivity of ROCs (equivalently, the mean and variance of the underlying distributions) are actually related in a predictable pattern. Early enthusiasm for this idea was based on psychoacoustic models of the ideal observer in which sensitivity was limited by statistical characteristics of the stimuli (see chap. 12). More recently, Ratcliff, Sheu, and Gronlund (1992) pointed out that different theories of recognition memory make distinct predictions about ROC shape, and the ROC is a popular tool for testing models in that field.

There are several psychophysical models, not tied to specific stimulus sets, that attempt to account for ROC slopes. Graham, Kramer, and Yager (1987) have shown that if detection of a known signal leads to a unit-slope ROC, then detection of a signal whose characteristics are unknown (see chap. 8) leads to a curve with a shallower slope. The stimulus-based model of Laming (1986) predicts that discrimination ROCs should have unit slopes and detection ROCs shallow ones.

But in many fields in which ROCs have been collected, no theories exist for predicting their shape. In a survey of a wide range of content areas, Swets (1986b) concluded that the slopes of empirical ROCs vary from about 0.5 to 2.0, and that they are not predictable from sensitivity or any stimulus characteristic. A similar conclusion is reached in Swets and Pickett's (1982) survey of detection theory applications to diagnostic systems in medicine and elsewhere. This finding leads directly to their recommendation that ROCs should always be collected.

Meanwhile, the user of detection theory who does not collect ratings is at risk. For purposes of comparing two points in ROC space, the risk is least in some important special cases: (a) If two points have the same value of F but different values of H (or vice versa), there is no question which represents the greater sensitivity, and (b) two points with the same bias can always be compared. We found in chapter 2 that "the same bias" is an ambiguous phrase, but there is less doubt about what "neutral bias" means: $H = 1 - F$. Thus, if bias is minimal, ROCs are minimally necessary. Conditions under which bias is neutral are not easy to specify either, but experience may be an adequate guide.

Summary

In a single session of a one-interval experiment we can collect data that can be interpreted as multiple (false-alarm, hit) pairs. This is accomplished by asking observers to provide a graded rather than a binary response, rating their experience on an ordered scale. The result is an empirical ROC curve.

The data are interpreted as if the observer maintained several response criteria simultaneously. Sensitivity can be estimated separately for each criterion. If the empirical ROC has unit slope on z coordinates (so that the variances of the underlying distributions are equal), the sensitivity measure will be the same at all criteria. If the slope of the ROC does not equal 1, apparent sensitivity changes along the decision axis; the slope can be interpreted as the ratio of the standard deviations of the underlying distributions. Sensitivity can be measured in units of either standard deviation or, most commonly, some sort of average.

Response criteria can be estimated as in the yes-no design except that multiple criteria are now found. When variances are unequal, the criterion location c can be measured in any of the units used for sensitivity.

Alternative ways to get multiple points on an ROC are to conduct separate sessions with different a priori probabilities or apply different payoffs and penalties for the various outcomes.

Conditions under which the methods described in this chapter are appropriate are spelled out in Chart 4 of Appendix 3.

Computational Appendix

Derivation of Equation 3.11

Combining Equations 3.10 yields

$$-c_2\left(1+\frac{1}{s}\right) = z(H) + z(F) - \left(\frac{M_1}{s} + M_2\right). \qquad (3.16)$$

The point of equal bias, where the criterion must equal zero, occurs where $z(H) = -z(F)$. In this case, the left side of Equation 3.16 equals zero, and the right side equals $-[(1/s)M_1 + M_2]$. Thus, M_1 has the opposite sign from M_2 and is s times as far from zero, and $-[(1/s)M_1 + M_2]$ always equals zero. The last term in Equation 3.16 can therefore be dropped, leading to Equation 3.11.

Calculation of the Point Where H = F in the
Unequal-Variance ROC

In the example, the ROC curve has a slope of 0.5 so that the S_2 distribution has a standard deviation twice that of S_1. Using Equation 2.5, normal densities with means of 0 and d'_1 and standard deviations of 1 and 2 can be written as

$$\phi_1(x) = be^{-\frac{1}{2}x^2}$$

$$\phi_2(x) = \left(\frac{b}{2}\right)e^{-\frac{1}{2}\left(\frac{x-d'_1}{2}\right)^2} . \tag{3.17}$$

Setting $\phi_1 = \phi_2$ yields a quadratic equation, which we can solve for the (two) values at which the S_1 and S_2 curves cross. In units of the S_1 distribution, the intersections are at

$$\frac{1}{3}(-d'_1 \pm \sqrt{4d'^2 + 24\ln(2)}) ,$$

the negative solution being the point below which $H < F$.

Problems

3.1. In music perception experiments, listeners are sometimes asked to discriminate between chords (combinations of notes played together) that are in tune versus out of tune. Consider a three-response experiment in which 25 trials of each type are presented, and the response set is "sure it was in tune," "sure it was out of tune," and "not sure." For in-tune stimuli, the number of responses in these categories is 13, 5, and 7; for out-of-tune stimuli, they are 3, 16, and 6. (a) Find the two ROC points implied by these data. (b) Calculate d' for both points. (c) Plot the points on a graph (linear coordinates).

3.2. In the previous problem, suppose the listeners refuse to use the "uncertain" category, but distribute those responses evenly between the "sure" categories. How would this affect the analysis?

3.3. Here are some data from a one-interval auditory detection experiment. The participants made a binary ("yes" or "no") detection response followed by a binary ("definitely" or "probably") confidence judgment.

	"Definitely Yes"	"Probably Yes"	"Probably No"	"Definitely No"
		Condition 1		
Signal	162	22	2	14
No signal	6	22	12	160
		Condition 2		
Signal	76	70	14	40
No signal	8	52	22	118

(a) How do Conditions 1 and 2 probably differ, experimentally?

(b) Use statistical software (Systat or SPSS) or ROCKIT to fit ROCs to these data.

Find d_a, s, A_z, and the bias measures β and $\ln(\beta)$.

3.4. Suppose an ROC has an intercept $d'_e = 1.0$ and a slope $s = 0.7$. Calculate $z(H) - z(F)$ for F proportions of .1, .5, and .99.

3.5. A computer program informs you that $d'_1 = 2.5$ and $s = 0.8$. Calculate d_a, d'_e, and A_z.

3.6. For the data of Example 3b (word recognition), calculate A_g. How do the values compare with those for A_z?

3.7. Two experimenters conduct studies of auditory pitch discrimination with the same two stimuli, a 200-Hz and a 202-Hz tone, using a rating design. One defines a hit to be a correct response to the 200-Hz tone, the other defines it as a correct response to the 202-Hz tone. If the first investigator finds $d_a = 1.5$ and $s = 0.5$, what results will the second investigator obtain?

3.8. In a subliminal perception experiment, a geometric figure is presented very briefly on S_2 trials but not on S_1 trials. Both underlying distributions have the same mean, but the standard deviation is twice as great for S_2 trials. (a) Sketch the ROC on z coordinates that would be obtained in a rating task. (b) If the task is yes-no (rather than rating), how will estimated d' differ if the participant adopts a strict versus a lenient criterion?

3.9. List possible experimental situations or areas of interest that might most reasonably be studied using rating methods with the following response sets: (a) a set of 5 numbers, (b) binary response plus three categories of certainty, and (c) verbal categories instead of numerical ones to signify degrees of certainty.

4

Alternative Approaches: Threshold Models and Choice Theory

Detection theory models have faced three classes of competitors. First, before the advent of detection theory, much of psychophysics was concerned with measuring "thresholds," below which stimuli were thought not to be perceived. Second, in the 1950s and early 1960s, as Tanner, Green, and Swets were developing signal detection theory, Luce (1959, 1963a) proposed Choice Theory, a conceptually different analysis of a similar range of experiments. Third, one reaction to detection theory has been an attempt to avoid the "parametric" assumption that the underlying distributions are Gaussian.

These three lines of work occupy distinct psychophysical niches in the current research environment. Choice Theory differs only slightly from detection theory in the simplest cases, but its quite different framework allows for a wide range of application. Threshold concepts lead to models that describe most data less well than detection theory; there are exceptions, however, and threshold ideas have been extended usefully to "multinomial" models of complex tasks. "Nonparametric" measures have turned out, on examination, to be related to threshold theory, Choice Theory, or both, and they are just as theory-bound as other statistics. We discuss explicit threshold models first, then Choice Theory, and then "nonparametric" analysis.

Threshold theory (Krantz, 1969; Luce, 1963a) assumes that the decision space is characterized by a few discrete states, rather than the continuous dimensions of detection theory. Different threshold models propose different connections between stimulus classes and discrete internal states, and between internal states and responses. For most models, we develop a *state diagram* that spells out these connections and defines the model. From the state diagram, the form of the implied ROC can be deduced; for those models that have a single sensitivity parameter, this ROC is the locus of points with the same value of sensitivity.

A single isosensitivity curve is, however, consistent with more than one set of underlying distributions, and we use the threshold ROC also to describe the decision space in the alternative terms of continuous rectangular distributions. These continuous distributions are not mathematically equivalent to the corresponding state diagrams, but because both are consistent with the ROC, both are legitimate representations of the decision space. The continuous version has the important advantage of enabling comparison with the detection theory models of chapters 1 and 2.

In this chapter, we deviate from the example-based structure used so far because few data support the use of threshold measures to summarize discrimination data. We do not organize the presentation around data sets for which these approaches are appropriate because, in most cases, there are none.

The thresholds discussed here are theoretical, referring to internal states. A second use of the term is empirical, denoting stimuli: A threshold *stimulus* is one that can barely be discriminated from the background or another stimulus. We discuss this *empirical threshold* in chapters 5 and 11; its use as a dependent measure is not challenged by the arguments in this chapter.

Single High-Threshold Theory

Sensitivity Measure q

The first model we consider proposes that sensitivity be measured by the adjusted hit rate:

$$q = (H - F)/(1 - F) . \tag{4.1}$$

In chapter 1, we discussed $H - F$ as a possible sensitivity measure. The statistic q, sometimes said to "correct" the hit rate for "guessing," adjusts this index so that it ranges from 0 to 1, rather than from 0 to $1 - F$. In effect, Equation 4.1 deflates H to take account of the tendency to make false alarms. The correction for guessing lowers the hit rate more when the false-alarm rate is higher. Thus, if $H = .75$ and $F = .1$, the adjusted hit rate q is .72; but for the same hit rate and $F = .5$, q is only .5.

Underlying Representation

The essential tenet of high-threshold theory is that "yes" responses to S_1 must be guesses based on no information. A state diagram incorporating this idea is shown in Fig. 4.1a. The diagram contains only two internal states, which we call D_1 and D_2, and it specifies the possible paths from stimulus to internal state and from state to response, together with the probabilities of each path. The adjusted hit rate q is the probability that S_2 leads to the D_2 state; if observers could be relied on to report their internal states accurately, q would equal the hit rate H, and the false-alarm rate F would equal

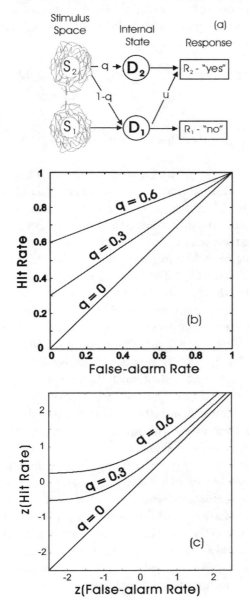

FIG. 4.1. (a) State diagram for single high-threshold theory. Stimuli in class S_2 lead to state D_2 with probability q; "yes" responses (guesses) are made from state D_1 with probability u. (b) ROCs implied by high-threshold theory, on linear coordinates, for three values of q. Changing u maps out the ROC. (c) ROCs in z coordinates. Panels (a) and (b) adapted from Macmillan and Creelman (1990) by permission of the publisher. Copyright 1990 by the American Psychological Association.

zero. Instead observers respond "yes" on some occasions even when in state D_1; these contaminating guesses occur with probability u and make the correction recommended by Equation 4.1 necessary.

The dependence of H and F on the adjusted hit rate and the guessing rate can be calculated directly from the state diagram. The probability of each path through the diagram is the product of the probabilities of the segments, and the total probability of a response given a stimulus is the sum of the probabilities of the possible paths. Thus:

$$H = P(\text{"yes"}|S_2) = q + u(1 - q) \tag{4.2}$$

$$F = P(\text{"yes"}|S_1) = u \, .$$

Eliminating the guessing parameter u from these equations leads back to Equation 4.1.

The ROC implied by q is obtained by solving Equation 4.1 for H in terms of F. As shown in Fig. 4.1b, it is a straight line from $(0, q)$ to $(1, 1)$. Unlike the isosensitivity curves of SDT and Choice Theory, it is nonregular: A false-alarm rate of zero can be obtained with a nonzero hit rate. On z coordinates, the ROC is not straight, but strongly concave upward.

How can we construct continuous underlying distributions that are consistent with the single high-threshold ROC? To allow for the point $(0, q)$, there must be a region on the decision axis where events only occur due to S_2—otherwise the false-alarm rate would not be 0. Such a region is drawn on the right side of Fig. 4.2a. In the rest of the decision space, corresponding to the ROC segment $(0, q)$ to $(1, 1)$, the S_1 and S_2 distributions could have any shape. They must be proportional to each other, however, because the ratio of their heights—the likelihood ratio—is constant when the ROC has constant slope. Thus, the decision space is divided into two regions, one with a likelihood ratio that is some constant less than 1, the other with a likelihood ratio of infinity. The boundary between these two areas is the "threshold," the decision-axis value above which only S_2 events occur.

Figure 4.2b eliminates some unnecessary complexity by representing the underlying distributions in a simple rectangular form. A changing value of the parameter u (the proportion of D_1 trials on which the observer responds "yes") is modeled in this decision space by a shift in the criterion, but not by a change in the value of likelihood ratio. The criterion can be sensibly located only on the below-threshold segment of the decision axis (a higher location reduces the hit rate without any compensating reduction in F, and it corresponds to a point along the vertical ROC axis below the intercept).

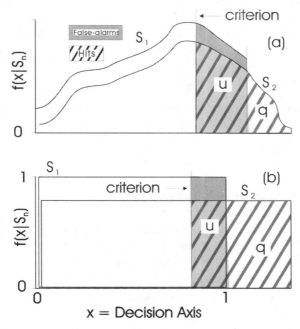

FIG. 4.2. Two representations of a decision space for single high-threshold theory consistent with the ROCs of Fig. 4.1: (a) arbitrary distributions, and (b) rectangular distributions. Panel (b) adapted from Macmillan and Creelman (1990) by permission of the publisher. Copyright 1990 by the American Psychological Association.

This model is traditionally termed *high threshold* because of the asymmetry between hits and false alarms. The threshold—the dividing line between the internal states—is "high" because S_1 stimuli cannot hurdle it, although S_2 stimuli can. In the original application of this model to detection experiments, the model captured the (now discredited) intuition that background noise could never lead to a "true detection," so that errors on noise trials arose only from guessing.

Bias Measures F *and* u

Because the observer can control response bias only by changing the guessing rate, u is the natural bias index for single high-threshold theory. Because u equals F (Eq. 4.2), the false-alarm rate itself is the model's bias statistic. In terms of underlying rectangular distributions (Fig. 4.2b), u (and F) measures the location of the response criterion relative to the upper end of the S_1 distribution.

Its association with the single high-threshold model is one count against *F* as a bias index. A more serious charge is its failure to depend at all (much less monotonically, as we have been requiring) on the hit rate *H*.

Low-Threshold Theory

In low-threshold theory (Luce, 1963b), asymmetric treatment of hits and false alarms is abandoned. To compare the two theories, consider the low-threshold state diagram in Fig. 4.3. As before, there are two internal states, but now S_2 as well as S_1 can lead to either state. There are two "sensitivity" parameters: q_2, the probability that S_2 leads to state D_2, the "true" hit rate; and q_1, the probability that S_1 leads to state D_2, the "true" false-alarm rate.

The transition paths from internal state to response take one of two forms, depending on which of two response strategies the observer uses. A response of "yes" may be given to all D_2 states plus a proportion *u* of D_1 states (panel a), or to only some proportion *t* of D_2 states and no D_1 states (panel b). These strategies are called, for reasons that will soon be evident, "upper limb" and "lower limb" responding.

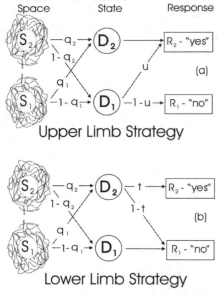

FIG. 4.3.   State diagrams from low-threshold theory. Stimuli in classes S_1 and S_2 lead to state D_2 with probabilities q_1 and q_2, respectively. (a) In the upper limb strategy, "yes" responses are always made from state D_2 and with probability *u* from state D_1. (b) In the lower limb strategy, "yes" responses are never made from state D_1 and with probability *t* from state D_2.

The state diagram leads directly to expressions for the hit and false-alarm rates:

$$H = q_2 + u(1 - q_2) \tag{4.3}$$

$$F = q_1 + u(1 - q_1) \qquad \text{(upper limb)}$$

$$H = tq_2$$

$$F = tq_1 . \qquad \text{(lower limb)}$$

The bias parameters t and u vary from 0 to 1.

The ROC for low-threshold theory is shown in Fig. 4.4. On linear coordinates, it consists of two straight lines, or "limbs," of different slopes, meeting at the point (q_1, q_2). The lower limb arises from the conservative lower limb strategy, the upper limb from the more lax upper limb response rule. The theory predicts regular ROCs that are only moderately nonlinear in z coordinates. Despite the tell-tale "corner" predicted by low-threshold the-

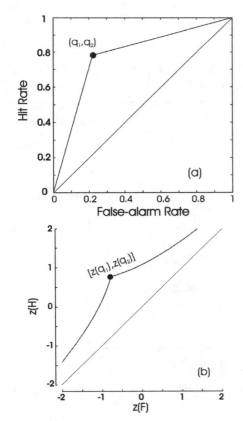

FIG. 4.4. (a) ROC implied by low-threshold theory, in linear coordinates. Changing u maps out the upper limb and changing t the lower limb. (b) Same ROC in z coordinates.

ory, it has been experimentally difficult to distinguish this theory from normal-distribution detection theory.

To find continuous underlying distributions corresponding to the two-limbed ROC, we follow the same logic as for high-threshold theory. Because the ROC has only two slopes, there are two possible values of likelihood ratio. In each state, however, the likelihood ratio is finite, so each of the two distributions takes on two different heights, as shown in Fig. 4.5. The criterion can be located in either state, depending on whether the observer uses an upper or a lower limb response strategy.

Low-threshold theory retains the appealing intuitions of high-threshold theory, but avoids the unpalatable nonregularity prediction. Its primary disadvantage is its lack of a single sensitivity measure that can be calculated from one (F, H) pair. Despite this drawback, the theory has been of substantive interest as a model of auditory detection and, before Luce described it, of categorical perception in speech (Liberman, Harris, Hoffman, & Griffith, 1957).

Double High-Threshold Theory

Double high-threshold theory is most often encountered not as a proposal about a discrete underlying process, but indirectly via its sensitivity parameter: This theory justifies the use of proportion correct to measure performance. It was first explicitly proposed by Egan (1958; summarized in Green & Swets, 1966, pp. 337–341).

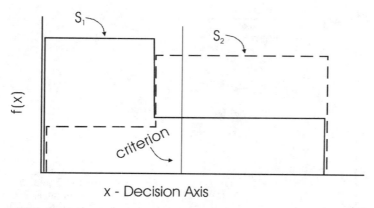

FIG. 4.5.   Decision space consistent with the low-threshold ROC of Fig. 4.4, using rectilinear distributions.

The Sensitivity Measure p(c)

In chapter 1, we contrasted $p(c)$ with d' as a measure of performance. In general, $p(c)$ is found by averaging H and $1 - F$ using presentation probabilities as weights:

$$p(c) = p(S_2)H + p(S_1)(1 - F) \tag{4.4}$$

$$= p(S_1) + p(S_2)H - p(S_1)F ,$$

where $p(S_i)$ is the probability that S_i is presented. Proportion correct equals a constant plus the difference between weighted hit and false-alarm rates, with different weights (multiplicative constants) applied to each. When the number of trials for each type of stimulus is equal, the weights are the same and proportion correct only depends on the difference between H and F:

$$p(c) = \tfrac{1}{2}(1 + H - F) . \tag{4.5}$$

Early on, Woodworth (1938) suggested $H - F$ as a performance measure for recognition memory experiments.

Underlying Rectangular Distributions

Like all sensitivity measures, $p(c)$ implies a decision theory: To use $p(c)$ to summarize performance is to say that when bias is manipulated, $p(c)$ should remain constant. The state diagram of the underlying model is shown in Fig. 4.6a. There are three discrete states: D_1 arises only when S_1 occurs, D_2 can be triggered only by S_2, and an intermediate state $D_?$ can occur for either stimulus. The model specifies two "high" thresholds, each of which can be crossed by only one of the two stimuli. The special case in which the D_1 state is omitted is equivalent to single high-threshold theory.

As with both high- and low-threshold theories, the sensitivity parameter in this model is a "true" detection rate. The proportion of S_2 presentations leading to the D_2 state equals the proportion of S_1 presentations leading to the D_1 state; both equal $2p(c) - 1$. If $p(c)$ equals .8, for example, the proportion of trials falling in the "sure" D_1 and D_2 states is .6. Other trials lead to the uncertain state $D_?$, where they are assigned "yes" and "no" responses according to the observer's response bias v.

The ROCs for $p(c)$ were shown in chapter 1 to be straight lines with unit slope when plotted on probability coordinates. Like the ROCs for single high-threshold theory, they are curved on z coordinates. Underlying distri-

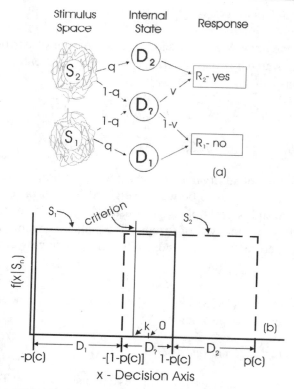

FIG. 4.6.   (a) State diagram implied by double high-threshold theory. Stimuli in class S_i lead to state D_i with probability q and to state $D_?$ with probability $1-q$. The uncertain state leads to a "yes" response with probability v. (b) Underlying rectangular distributions consistent with double high-threshold theory. The criterion can be located anywhere in the $D_?$ region. Adapted from Macmillan and Creelman (1990) by permission of the publisher. Copyright 1990 by the American Psychological Association.

butions consistent with double high-threshold theory are shown in Fig. 4.6b; as the state diagram (Fig. 4.6a) shows, S_1 presentations can lead either to D_1 or $D_?$, S_2 presentations to either D_2 or $D_?$. There are three values of likelihood ratio—zero, infinity, and one value between. The use of proportion correct makes very strong assumptions about the internal representation of stimuli.

For sensory detection experiments, these assumptions are not very plausible, but some memory studies have produced linear ROCs. Yonelinas (1997) conducted an associative recognition experiment: Participants were presented with pairs of words in both the study and test phases; the question was whether the test pairs had occurred *together* in the study phase. The ROC data, presented in Fig. 4.7, are clearly linear and consistent with double-high threshold theory. Notice that the ROCs do not have slope 1; instead they are consistent with a representation in which S_2 presentations are detected as Old at a different rate than S_1 presentations are detected as New. In the state diagram of Fig. 4.6a, the parameter q is replaced by separate parameters q_1 and q_2; in Fig. 4.6b, there are still three values of likelihood ratio, but the intermediate value is not 1.

What accounts for ROC data of this sort? Yonelinas argued that decisions in associative recognition cannot be based on familiarity because familiar words may not have occurred together in the study phase. Instead participants must "recollect" the specific episode in which they last encountered the pair, and recollection is a threshold process. The two limbs of the state diagram reflect different types of recollection: A pair may be recollected as Old, or the participant may recollect that one of the two words had a differ-

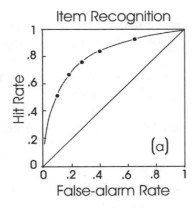

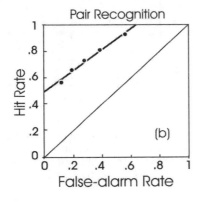

FIG. 4.7.   ROCs for recognition memory from Yonelinas (1997). (a) Item (single-word) recognition, and (b) associative (word-pair) recognition. Adapted with permission.

ent partner in the study phase. The accuracy of these two strategies may differ, accounting for the different intercepts of the ROC.

Bias Measures

Two bias measures that appear to make no distributional assumptions are actually consistent with the double high-threshold model. These are the *yes rate*, $\frac{1}{2}(H + F)$, and the *error ratio* $(1 - H)/F$.

Yes Rate.   To see the connection between the yes rate and the double high-threshold model, consider again the model's decision space, shown in Fig. 4.8. The center of the region of overlap is set to zero, and the criterion k is measured with respect to this origin. Then $H = p(c) - k$ and $F = 1 - p(c) - k$; solving these equations yields

$$k = \frac{1}{2}[1 - (H + F)] \ . \tag{4.6}$$

The criterion is thus a simple linear transformation of the yes rate; like c in detection theory, the yes rate reflects the location of the criterion relative to the halfway point between the S_1 and S_2 distributions.

The relation between k and $p(c)$ is suggested by the similarity between Equation 4.6 and the corresponding expression for sensitivity when an unweighted average of H and F is used (Eq. 4.5). The false-alarm and hit rates are added in Equation 4.6 and subtracted in Equation 4.5, and the same

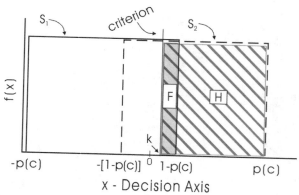

FIG. 4.8.   Decision space for the double high-threshold model as in Fig. 4.6b. Shaded area is the false-alarm rate, diagonal area is the hit rate. The criterion k is monotonic with the overall yes rate. Adapted from Macmillan and Creelman (1990) by permission of the publisher. Copyright 1990 by the American Psychological Association.

transformation is applied to the result. We encountered a similar relation for detection theory models (e.g., compare Eqs. 1.5 and 2.1).

Error Ratio.   Like c, the yes rate measures the same distance along the decision axis whether the sensitivity measure is large or small. If we linearly transform k into a new variable k' that varies from 0 to 1, no matter what $p(c)$ is, we obtain

$$k' = \left(1 + \frac{F}{1-H}\right)^{-1} , \tag{4.7}$$

that is, something that only depends on the error ratio. The parameter k' is, in fact, equal to $1 - v$ (see Fig. 4.6) and is therefore equivalent as a bias measure to v, which was proposed as a bias index by Snodgrass and Corwin (1988).

Comparison of Indexes.   Isobias curves for the yes rate and error ratio are shown in Fig. 4.9. As might be expected from their decision-space interpretation, the two indexes share attributes with analogous detection theory statistics. Curves for the yes rate, like curves for c, are parallel, but on linear rather than z coordinates. Curves for the error ratio, like detection theory curves for relative criterion location, converge at a point ($H = F = 1$ and $H = F = .5$, respectively).

What about the likelihood ratio? As noted earlier, there are only three different values of likelihood ratio in the proportion correct model. Variation of criterion within the overlap region does not change likelihood ratio, which is therefore of little use as a bias statistic for this (or any other) threshold theory.

Evaluating the yes rate and the error ratio as measures of bias is more difficult than passing judgment on threshold sensitivity indexes. A long history of collecting empirical ROCs (Green & Swets, 1966; Swets, 1986b) has suggested limits on the shape of implied ROCs, whereas the much shorter history of collecting empirical isobias curves has been inconclusive.

By some theoretical standards, however (Macmillan & Creelman, 1990), the two measures fare well. Both change in the same direction with increases in H and F, behave well when sensitivity is at chance, and are undistorted if computed by averaging across participants or conditions. Because its isobias curves are parallel rather than divergent, the yes rate is independent of $p(c)$ and acts sensibly when sensitivity is below chance; the error ratio only approximately meets these desiderata.

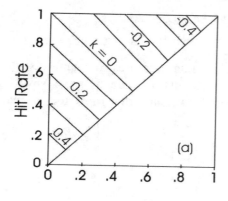

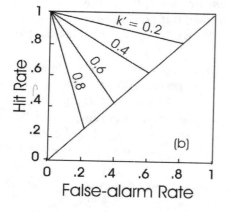

FIG. 4.9.   Isobias curves for (a) the yes rate, and (b) the error ratio. Adapted from Macmillan and Creelman (1990) by permission of the publisher. Copyright 1990 by the American Psychological Association.

There is an argument for preferring c over the yes rate: When sensitivity and bias indexes are both reported, they should derive from the same model. Although there is little to choose between detection theory and double high-threshold bias measures, the sensitivity statistic of detection theory is superior.

Choice Theory

Luce (1959) conjectured that the odds of choosing one stimulus over a second are unaffected by other possible stimuli, and this *choice axiom* is the basis for the structure of Choice Theory. Although this starting point does not sound related to the principles of detection in noise that led to the models of chapters 1 to 3, we shall see that the two theories are formally very similar. The idea of a decision continuum, and the form of underlying distributions, can be derived from the choice axiom. Choice Theory predictions look much like those from the normal-distribution model in simple detection tasks and are sometimes easier to generate for more complex experiments. Because Choice Theory is a close cousin of signal detection theory in many applications, from now on we include it under the phrase "detection theory." We continue to use the abbreviation "SDT" to refer to normal distribution models.

Sensitivity Measures

In Choice Theory (Luce, 1959), the sensitivity measure α is found by

$$\alpha = \left[\frac{H(1-F)}{(1-H)F} \right]^{\frac{1}{2}} . \tag{4.8}$$

In chapter 1, we noted that the sensitivity measures d' and $p(c)$ amounted to differences between transformed values of H and F. Choice Theory also has such an index, obtained by taking the logarithm of α (and thus equivalent to it):[1]

$$\ln(\alpha) = \frac{1}{2} \ln\left(\frac{H}{1-H} \right) - \frac{1}{2} \ln\left(\frac{F}{1-F} \right) . \tag{4.9}$$

In Choice Theory, the transformation applied to H and F is the *log-odds* transform, which converts a proportion p to $p/(1-p)$ (the odds in favor) and then takes logarithms.

To give an idea of the magnitude of α: If $F = .4$ and $H = .8$, then $\alpha = [(.8 \times .6)/(.2 \times .4)]^{\frac{1}{2}} = 2.45$ and $\ln(\alpha) = 0.90$. The (F, H) pair $(.1, .4)$ leads to the same values; these points give similar (although not identical) values of d'. Total inability to discriminate ($H = F$) leads to $\alpha = 1$, $\ln(\alpha) = 0$. When $H = .99$ and $F = .01$, $\alpha = 99$ and $\ln(\alpha) = 4.60$. A proportion correct of .75 on both types of trials yields $\alpha = 3$, $\ln(\alpha) = 1.10$; $p(c) = .73$ corresponds to $\ln(\alpha) = 1.0$.

These examples suggest that d' and $\ln(\alpha)$ are similar as measures of sensitivity, and Fig. 4.10 shows that they are very nearly proportional to each other for low to moderate values. The relation between them can be approximated by $\ln(\alpha) = 0.81\, d'$, with deviations from this equation being greatest for hit rates near 1 or false-alarm rates near 0. Figure 4.10 encourages us to choose between the two accuracy indexes on the basis of convenience; the two analyses are not likely to support discrepant conclusions.

---

[1]Luce (1959, 1963a) assigned sensitivity the symbol α, with slightly different meanings in two versions of Choice Theory. In memory research, one of the areas in which Choice Theory is most widely used, $\ln(\alpha)$ is sometimes called d_L (Hintzman & Curran, 1994) to highlight its similarity to d'.

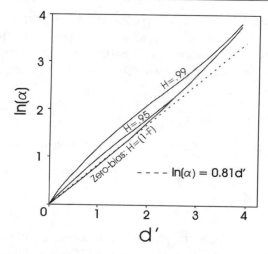

FIG. 4.10.   Relations between $\ln(\alpha)$ and d' for the zero-bias case, and for two cases of bias to respond "yes."

Implied ROC Curves

What is the form of the ROC implied by the Choice Theory measure α? To answer this question with a question, what transformation would render these curves straight lines? As Equation 4.9 makes clear, the required function is log odds because

$$\ln\left(\frac{H}{1-H}\right) = \ln\left(\frac{F}{1-F}\right) + 2\ln(\alpha) . \tag{4.10}$$

If we were to plot the ROC in log-odds coordinates, $(\ln[H/(1 - H)]$ vs $\ln[F/(1 - F)])$, then the (constant) distance between the ROC and the chance line would be $2 \ln(\alpha)$. The analogy to SDT correctly suggests that $2 \ln(\alpha)$ plays the role of mean difference in the decision space.

To get a feel for the relation between the log-odds and z transformations, consider Fig. 4.11, in which ROCs for constant α and constant d' are plotted. It is hard to distinguish the two sets of curves, which differ systematically only for very small or large proportions. An important difference between Choice Theory and SDT, however, is that the ROCs implied by α are always symmetric (like those implied by d'), but there is no measure analogous to d_a that allows for ROC curves that are not of unit slope.

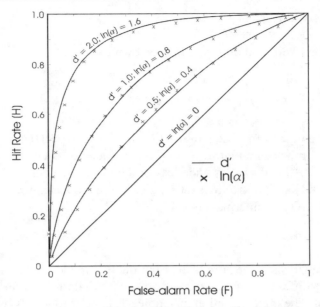

FIG. 4.11.   ROCs for SDT and Choice Theory on linear coordinates. Curves connect locations with constant d', and ×s are points of constant α.

Bias Measures

Choice Theory's bias measure b (for "bias") can be computed from

$$b = \left(\frac{(1-H)(1-F)}{HF} \right)^{\frac{1}{2}}. \tag{4.11}$$

Taking logarithms reveals that $\ln(b)$, like c in SDT, is the sum of the transformed hit and false-alarm rates, the transformation in this case being log odds:

$$\ln(b) = -\frac{1}{2}\left[\ln\left(\frac{H}{1-H}\right) + \ln\left(\frac{F}{1-F}\right) \right]. \tag{4.12}$$

As we shall see shortly, $\ln(b)$ is a measure of criterion location. Division by the sensitivity parameter $2\ln(\alpha)$ yields a measure of relative criterion analogous to c' :

$$b' = \ln(b)/[2 \ln(\alpha)] \ .$$ (4.13)

Finally, the likelihood ratio β_L can be shown to equal[2]

$$\beta_L = \frac{H(1-H)}{F(1-F)} \ .$$ (4.14)

The algebraic form of Equation 4.12 leads one to expect the isobias curve for b to be much like that for c, and this conjecture is correct. Although Equations 4.13 and 4.14 provide less of a hint, isobias curves for relative criterion and likelihood ratio in the Choice Theory model are also very similar to their SDT counterparts (see Fig. 2.7).

Decision Space

From our analysis of the normal distribution SDT model, we know that sensitivity is a difference of transformed hit and false-alarm rates and response bias a sum. The transformation, in Choice Theory, is log-odds, which converts a proportion p to $\ln[p/(1-p)]$. Figure 4.12 shows how this operation is used to convert the false-alarm/hit pair (.4, .8) to the sensitivity statistic $2\ln(\alpha)$ and the bias statistic $\ln(b)$.

The decision space implied by these Choice Theory measures contains two underlying distributions whose form is *logistic*, rather than normal. The logistic distribution is symmetric and only subtly different in shape from the normal when plotted on a log-odds axis (see Fig. 4.13). As in the SDT model, the distance between the means of the S_1 and S_2 distributions is a sensitivity measure; its value is $2\ln(\alpha)$. If we define 0 as the point at which the two distributions cross, then the distribution means are at $\pm\ln(\alpha)$ and the criterion is located at $\ln(b)$.

In the normal model, the transformation from p to location on the decision axis is $z(p)$; the reverse operation, to find hit and false-alarm rates from a z-score axis location, is Φ. Both are found using the normal table. In the logistic model, the log-odds transformation is used to find log-odds axis locations, called *logits*; to find hit and false-alarm rates from a log-odds value requires solving the equation

$$x = \ln[p/(1-p)]$$ (4.15)

---

[2]See Computational Appendix.

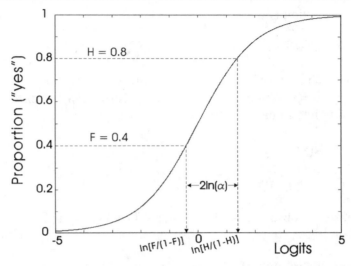

FIG. 4.12. A logistic distribution function. The inverse function can be used to transform proportions into logits. Sensitivity $[2\ln(\alpha)]$ is the difference between $\ln[H/(1 - H)]$ and $\ln[F/(1 - F)]$.

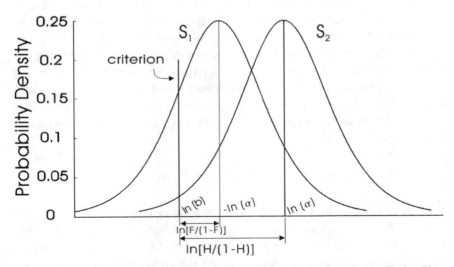

FIG. 4.13. Decision space for the yes-no experiment according to Choice Theory (logistic distributions).

for p. The solution is

$$p = 1/(1 + e^x) .\qquad(4.16)$$

To find H and F, x must be expressed as a distance from the mean. For the S_2 distribution, $x = \ln(b) - \ln(\alpha)$, and for the S_1 distribution, $x = \ln(b) + \ln(\alpha)$. Substituting into Equation 4.16 yields

$$H = \alpha/(\alpha + b)\qquad(4.17)$$

$$F = 1/(1 + \alpha b) .\qquad(4.18)$$

For an unbiased observer, $b = 1$, $H = \alpha/(\alpha + 1)$, and $F = 1/(\alpha + 1)$. Again, $H = 1 - F$, and

$$p(c) = \alpha/(\alpha + 1) .\qquad(4.19)$$

Measures Based on Areas in ROC Space: Unintentional Applications of Choice Theory

An appealing measure of sensitivity is the area under the ROC, which increases from .5 for chance performance to 1.0 for perfect responding. We saw in chapter 3 that if the underlying distributions are normal, the estimated area A_z is simply related to the mean difference index d_a; in addition, the area can be estimated nonparametrically from ROC data. If only a single (F, H) point is available, however, we are forced to assume that the underlying distributions are normal, logistic, rectangular, or something specific. In this section, we consider measures of sensitivity and bias for single ROC points that were developed without recourse to detection theory. We shall find, however, that most of them are equivalent to parameters of the logistic model.

Sensitivity: Area Under the One-Point ROC

If only one point in ROC space is obtained in an experiment, there are many possible ROCs on which it could lie, and some assumptions must be made to estimate the area under the ROC. One possibility is to find the smallest possible area consistent with that point. As shown in Fig. 4.14, this is equivalent to finding the area under the low-threshold ROC for which the ob-

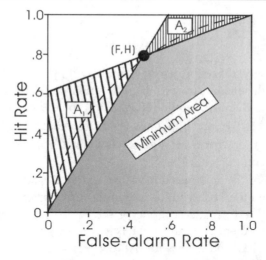

FIG. 4.14. Calculation of the area under the one-point ROC. The minimum area is shaded; the statistic A' is the minimum area plus one half the sum of regions A_1 and A_2. The dashed line is an example of an ROC that bounds an area greater than the minimum but less than the maximum (minimum plus A_1 and A_2).

tained point forms the corner. This area turns out to equal proportion correct, a measure with which we have already dealt harshly.

A better estimate, proposed by Pollack and Norman (1964), is also diagramed in Fig. 4.14. Their measure A' is a kind of average between minimum and maximum performance and can be calculated (Grier, 1971) by[3]

$$A' = \frac{1}{2} + \frac{(H - F)(1 + H - F)}{4H(1 - F)} \quad \text{if } H \ge F . \tag{4.20}$$

If performance is below chance, so that $H < F$, the equation must be modified (Aaronson & Watts, 1987):

$$A' = \frac{1}{2} - \frac{(F - H)(1 + F - H)}{4F(1 - H)} \quad \text{if } H \le F . \tag{4.21}$$

Macmillan and Creelman (1996) have shown that A' (for above-chance performance) can be written as a function of sensitivity measures we have already encountered:

[3]Smith (1995) pointed out that the maximal area under the ROC is less than that assumed by Pollack and Norman (1964), and defined a corrected measure A''. Zhang and Mueller (in press) improved on this measure. Implied ROCs for these indexes are similar to those for A'.

$$A' = \frac{1}{2} + \frac{1}{2} p(c) \left(1 - \frac{1}{[2\ln(\alpha)]^2} \right) . \tag{4.22}$$

At low sensitivity, this expression is dominated by the logistic term $\ln(\alpha)$, whereas at high sensitivities $p(c)$ is more important. The shift is shown in Fig. 4.15, which displays implied ROCs for A' on the same plot as those for α (panel a) and $p(c)$ (panel b). At low levels a constant-A' ROC is similar to a constant-α curve, whereas at high levels a constant-A' ROC is similar to a constant-$p(c)$ curve.

One appeal of the area measure is that, unlike d' and α, it can be calculated even when the observed hit or correct-rejection rate is 1.0. Unfortunately, perfect performance on one of the two stimulus classes tends to mean high performance overall, and it is for high values that A' has undesirable, threshold-like characteristics. At low performance levels, A' is much

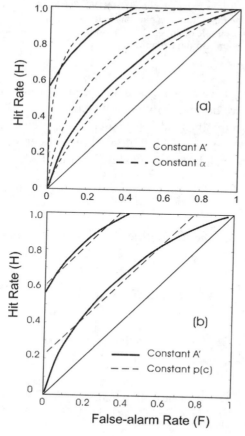

FIG. 4.15. ROCs implied by A', which are (a) similar to those implied by α at low sensitivity and (b) similar to those implied by $p(c)$ at high sensitivity.

like α (and thus much like d'). In neither case is it assumption-free, nor does it have any other perceptible advantages over detection theory measures.

Another constraint imposed by A' is that it is symmetric, in the sense that the pair $(1 - H, 1 - F)$ leads to the same value and is thus on the same ROC as (F, H). Use of this statistic thus implies an equal-variance representation, and it is inappropriate for content areas (e.g., item-recognition memory) in which the slope of the zROC is known not to be 1.

Two other "nonparametric" strategies for estimating sensitivity from H and F, both derived from statistics, share the threshold characteristics of $p(c)$ and A'. Proportions are sometimes transformed by taking the arcsine of the square root to equalize the variance for proportions of different magnitude. Mosteller (cited in Torgerson, 1958) suggested that this transformation be used in analyzing discrimination data as well, but Macmillan and Kaplan (1985) showed that the ROCs implied by this transformation have the same shape as those of $p(c)$ and A'. The contingency coefficient ϕ equals $(\chi^2/N)^{\frac{1}{2}}$, where χ^2 is the chi-square statistic (see Appendix 1), and N is the total number of trials. Swets (1986a) derived ROCs for this index, which are also similar to those of A'. The family resemblance to A' is an argument against taking either of these routes to sensitivity measurement.

Bias Measures Based on ROC Geometry

Hodos (1970) proposed a bias measure, B'_{H}, that is also based on the geometry of ROC space. This particular index is little used because Grier (1971) suggested an equivalent statistic (one with the same isobias curve) that is easier to calculate. Like A', the measure B'' must be modified if performance is below chance:

$$B'' = \frac{H(1-H) - F(1-F)}{H(1-H) + F(1-F)} \text{ if } H \geq F ,$$

$$(4.23)$$

$$B'' = \frac{F(1-F) - H(1-H)}{H(1-H) + F(1-F)} \text{ if } H \leq F .$$

The Hodos and Grier statistics are often paired with A', just as β and d' are paired by users of SDT or b and α by users of Choice Theory. But there is no model that unifies A' and B'', as there is for the others. In fact B'' is only superficially related to A', but it is strongly connected to an entirely different bias index, logistic likelihood ratio. Equation 4.23 is a monotonic function of, and therefore equivalent to, β_L (Eq. 4.14). A different measure based on ROC geometry, proposed by Donaldson (1992), turns out to be equivalent to the logistic criterion b. As Macmillan and Creelman (1996) pointed out, all these measures are based on the sums, differences, products, and ratios of areas of triangles, and the sides of the triangles equal H, F, $1 - H$, or $1 - F$. That the statistics end up being dependent on odds ratios, the stuff of Choice Theory, is not surprising.

Nonparametric Analysis of Rating Data

The admirable goal of measuring sensitivity nonparametrically is quite possible, but only if the yes-no design is abandoned in favor of ratings. We have seen that the true area under the ROC is a nonparametric index of accuracy, and if there are enough data points this can be estimated without fitting a theoretical model. A related measure has been developed by Balakrishnan (1998) for the dual-response version of the rating paradigm. The separate distributions of confidence ratings for Signal and Noise take over the role of the hypothetical distributions in SDT. The difference between the cumulative distributions of these ratings measures the discrepancy between the hit and false-alarm rates at each level of confidence. The sum of these differences is S', an estimate of the difference between the two confidence distributions under the assumption that the criteria used by the observer are equally spaced. In simulations, Balakrishnan showed that S' did a better job than d' of rank ordering conditions that differed slightly in sensitivity. A similar strategy, applied to the two-response rating design ("yes" or "no" followed by a confidence judgment), leads to a nonparametric measure of response bias.

Essay: The Appeal of Discrete Models

Mark Twain once remarked that there are two kinds of people: those who believe there are two kinds of people, and those who do not. Similarly, there are two kinds of performance measures and two kinds of psychophysical models: those that imply two (or at most a few) internal states, and those that envision a continuum. We have argued that discrete models, especially single and double high-threshold theory, are less successful than continuous models of detection theory.

The continued appeal of discrete models (which is broader than psychophysics) is worth consideration, and in this essay we raise one possible reason for this popularity. Discrete models offer *identifiability*: The response of a participant on a single trial may either directly indicate the current internal state or reduce the possibilities to a very small number. In single high-threshold theory, for example, a "no" response implies state D_1. In double high-threshold theory, a "no" response eliminates the possibility of D_2, and a "yes" response eliminates the possibility of D_1 on that trial.

According to detection theory, on the other hand, any point in the decision space can arise from either stimulus alternative. Further, an observer responding appropriately to instructions can assign any point in the space to either response. Thus, the observer's response on a single trial never reveals the internal state evoked, only whether the observation is above or below an adjustable criterion.

Discrete models, then, promise a direct access into the mind of the observer that detection theory denies. Because much behavioral research aims to either understand internal states or use them to explain actions, the attractiveness of these models is great. The threshold differs importantly from the criterion of detection theory models: If responses are determined by comparing events with fixed thresholds, they inform us about sensations; if they are determined by comparing events with adjustable criteria, they inform us only about the confluence of sensation and the decision process, leaving much work to be done.

We now consider briefly two domains in which discrete models have played an important role: subliminal perception and the classification of speech sounds. A detection theory treatment of both areas is offered later in the book (in chaps. 10 and 5, respectively). A final topic is the relation between discrete thinking and statistical hypothesis testing.

Subliminal Perception

A widespread use of the threshold concept is in the popular distinction between normal and subliminal (below threshold) perception (see Holender, 1986, for a review). Putative demonstrations of subliminal perception typically present a participant with a stimulus that is "below threshold," but find that the ability to identify the stimulus in some way remains.

Such a result seems surprising because stimuli that are not perceived are nonetheless effective. Indeed the finding is contrary to high-threshold theory: The "no" response implies that the observer is in state D_1, where no information about the stimulus exists. There is, however, a natural detection theory interpretation. The "no" response only means that the stimulus led to an event below the detection criterion. On a later identification test, an event above the (possibly different) criterion occurs. Because criteria are adjustable, this is not remarkable.

One reason for the appeal of threshold approaches to this topic may be the interpretation of "threshold" as the dividing line between consciousness and its absence. Indeed the threshold idea has been extended to include both "objective" and "subjective" variants (Reingold & Merikle, 1988). Detection theory, however, has no construct corresponding to consciousness.

Classification of Speech Sounds

In a popular approach to the perception of speech sounds, stimuli from a synthetic continuum between two syllables are offered one at a time to be classified as one of the two endpoints. Responses are taken to reveal sensory states directly, and the pattern of data is used to find the "boundary" between the two speech categories (e.g., Diehl, 1981; Liberman et al., 1957).

The experiment differs from those we have described in having more than two stimuli, but two internal states are still postulated. When classification changes (e.g., because of context), the result is interpreted as a change in "perception." Applications of detection theory to such data, however, have shown that context effects can affect criterion location rather than sensitivity (Elman, 1979; Macmillan, Goldberg, & Braida, 1988). In chapter 8, we consider experiments of this type more fully as tests of perceptual interaction. Speech classification experiments are often combined with discrimination experiments to study "categorical perception." We discuss both threshold and detection theory approaches to this problem in chapter 5.

Statistical Hypothesis Testing

There is an analogy between the observer in a detection experiment and an experimenter deciding about the source of data. The observer must determine whether the pattern of stimulation arose from S_2 (a Signal) or S_1 (Noise); the experimenter must decide whether the data are best explained with an "alternative" hypothesis (a real difference in the state of the world) or a null hypothesis (an apparent difference arising from sampling variability). Standard hypothesis-testing methods are used to decide which interpretation is more appropriate.

What information does an investigator take away from such data analysis? A binary decision ("significant" or "not significant") is usually required, just as a "yes" or "no" response is required by a detection subject. In addition, however, there is evidence that some experimenters try to make a direct inference to the state of the world, in the manner that discrete models permit and continuous ones do not.

Discrete thinking has been demonstrated, in this context, by Tversky and Kahneman (1971). They asked two samples of psychologists the following question: "Suppose you have run an experiment on 20 subjects, and have obtained a significant result which confirms your theory ($z = 2.23$, $p < .05$, two-tailed). You now have cause to run an additional group of 10 subjects. What do you think the probability is that the results will be significant, by a one-tailed test, separately for this group?" (p. 105). The median answer from Tversky and Kahneman's respondents was .85; a more reasonable estimate is .48. One interpretation of this result (similar but not identical to that of Tversky & Kahneman) is that psychologists view an effect as present or absent. The first experiment demonstrated an effect, which should therefore reveal itself, independent of factors like the power of the test.

Discrete thinking appears to be a decision-making heuristic of some generality. It offers the advantage of reducing cognitive complexity: The stimulus was perceived or not, the experimental hypothesis was true or false. Like high-contrast film, however, discrete models convert shades of gray into black and white at the expense of fidelity.

Summary

Threshold theories of discrimination postulate a small number of internal states, rather than a continuum. In such models, sensitivity is related to the probability that stimuli lead to the appropriate state(s). We have considered three such models:

1. Single high-threshold theory assumes that one stimulus (S_1) always leads to the correct state and defines sensitivity as the adjusted hit rate, the probability that the other stimulus (S_2) also leads to the correct state. This model predicts a nonregular ROC and almost always has been rejected when tested.

2. Low-threshold theory assumes that either stimulus can lead to either internal state. It predicts a regular ROC, and it is often as consistent with data as are detection theory models. But because it has no single sensitivity measure, this model is not widely applied.

3. Double high-threshold theory assumes three internal states, so that neither stimulus ever leads to the extreme state corresponding to the other stimulus. This model is implied when proportion correct is used

as an accuracy measure. The nonregular ROC shape that it predicts is unusual, but some data have been reported that are consistent with it.

Choice Theory is, in the applications discussed so far, very similar to normal distribution detection theory, and thus consistent with a wide range of data. A limitation of the theory is that it predicts unit-slope ROCs.

Some measures described as nonparametric, such as proportion correct and area under the one-point ROC, make threshold assumptions and predict nonregular ROCs. As with explicit threshold theories, these measures limit the underlying distributions and are not truly nonparametric. "Nonparametric" measures of bias are, in fact, the theoretical indexes of threshold or detection theory models: The false-alarm rate derives from single high-threshold theory, the yes rate and error ratio from double high-threshold theory, and area-based indexes from Choice Theory.

The implications of using the measures described in this chapter are summarized in Charts 2 and 3 of Appendix 3.

Computational Appendix

Logistic distributions have the form

$$\lambda(x) = \frac{e^{x-\mu}}{\left[1 + e^{x-\mu}\right]^2} .$$

(4.24)

It is convenient to work with the logarithm of this quantity. The criterion location $x = \ln(b)$, so

$$\ln(\lambda) = \ln(b) - \mu - 2\ln(1 + e^{x-\mu})$$

$$= \ln(b) - \mu - 2\ln(1 + be^{-\mu}) .$$

(4.25)

The likelihood ratio β_L is the ratio of two values of λ, one for $\mu = \ln(\alpha)$, the other for $\mu = -\ln(\alpha)$. The logarithm of the likelihood ratio is

$$\ln\left(\frac{\lambda|_{\mu=\ln(\alpha)}}{\lambda|_{\mu=-\ln(\alpha)}}\right) = \ln(\lambda|_{\mu=\ln(\alpha)}) - \ln(\lambda|_{\mu=-\ln(\alpha)}) = 2\ln\left(\frac{1+\alpha b}{\alpha + b}\right). \quad (4.26)$$

The likelihood ratio equals e to this power:

$$\beta_L = e^{2\ln\left(\frac{1+\alpha b}{\alpha+b}\right)} = \left(\frac{1+\alpha b}{\alpha+b}\right)^2 \qquad (4.27)$$

Equation 4.27 expresses the likelihood ratio in terms of logistic model parameters; to write it as a function of data, we substitute from Equations 4.17, 4.18, and 4.8 to find

$$\beta_L = \frac{H(1-H)}{F(1-F)} \qquad (4.28)$$

Logistic likelihood ratios are more extreme (farther from unity) than Gaussian ratios for the same values of H and F. For moderate values, $\ln(\beta_L)$ and $\ln(\beta)$ are roughly proportional.

Problems

4.1. Suppose single high-threshold theory is correct. (a) If $H = .8$, what is the "true" hit rate q if $F = .2$? $.5$? $.8$? (b) If $F = .2$, what is the value of the bias parameter u if $H = .8$? $.5$? $.2$?

4.2. (a) If you observe $(F = .4, H = .7)$ and wish to assume low-threshold theory, how can you tell whether the point is on the upper or lower limb? (b) Suppose you assume that the "corner" in low-threshold theory is on the minor diagonal (where $H = 1 - F$). If you now observe the point $(.4, .7)$, which limb is it on?

4.3. Suppose that high-threshold theory is true for a certain observer and $q = .4$ for some stimulus pair. What are the largest and smallest values of $p(c)$ this observer can obtain, assuming equal presentation probabilities? For H and F between $.01$ and $.99$, what is the largest and smallest possible value of d' ?

4.4. (a) For two experimental participants, $(F, H) = (.4, .9)$ and $(.2, .9)$. For each, compute $p(c)$, the yes rate, and the error ratio. (b) If double high-threshold theory is correct, what is the lowest hit rate and highest false-alarm rate that could be obtained by these participants?

4.5. Suppose $(.25, .75)$ is a point on an ROC. Find the area under the ROC (using geometry) assuming (a) high-threshold theory; (b) low-threshold theory, letting the observed point be the corner; and (c) double high-threshold theory.

4.6. If $p(S_2) = .80$, what is the highest and lowest value of $p(c)$ a participant can get, in a visual experiment, without looking?

4.7. What is the decision space (in either underlying distribution or state diagram form) implied by the ROC in Fig. 4.16?

4.8. Suppose $H = .8$ and $F = .3$ in a yes-no experiment. What is the area under the two-limbed ROC curve determined by this point? What is A'? Find the area under the complete ROC (assume equal-variance SDT).

4.9. Suppose you observe an ROC point (F, H) and decide to measure A'. Is the area you get most similar to the area under the ROC of high-threshold, low-threshold, or double high-threshold theory?

4.10. Suppose a yes-no experiment yields the following detection data:

condition 1: $H = .8$, $F = .4$
condition 2: $H = .6$, $F = .2$.

(a) For each condition, compute d', α, $p(c)$, and A'. According to each measure of sensitivity, in which of these two conditions is performance better?

(b) Assuming high-threshold theory to be correct, find the "true" hit rate q. Do both points have the same value?

(c) There is a unique low-threshold ROC curve that is consistent with both of these points. Find it, that is, find the location of the "corner" in ROC space.

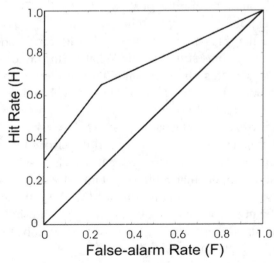

FIG. 4.16.   ROC for the model proposed by Krantz (1969), on linear coordinates.

4.11. The high-threshold model says that false alarms arise only from guesses, but misses can be due to imperfect encoding. An alternative threshold model could be constructed by making an opposite assumption, namely that misses arise only from guesses but false alarms can be due to imperfect encoding. What would the ROC look like? What would be appropriate measures of sensitivity and bias?

4.12. Show that the bias measures k'(Eq. 4.7) and v [P("yes"$|D_?$) in the state diagram of Fig. 4.6] add up to 1. (*Hint*: First use the state diagram to write H and F as functions of v and the sensitivity parameter q, then solve for v as a function of H and F.)

4.13. For the ROC points (.7, .9) and (.3, .9), show that β_L and B'' are the same for both points, but that $\ln(b)$ is different. Generalize this result.

5

Classification Experiments For One-Dimensional Stimulus Sets

Successful participants in discrimination experiments can distinguish two stimulus classes, but in most paradigms they need not be able to name them. In this chapter, we extend detection theory to encompass experiments in which stimuli drawn from large sets are named or classified by the observer. These sets are "one-dimensional," that is, they contain stimuli that differ for the participant in just one characteristic. As in earlier chapters, we are interested in sensitivity and bias, but multiple parameters must be estimated, and their interpretation is somewhat different.

One-dimensional classification experiments are among the oldest psychophysical tasks, and they take on many aliases. Accordingly, this chapter has an unusually large number of examples, but one basic strategy for data analysis fits all.

Design of Classification Experiments

In *classification* experiments, observers use M responses to sort N stimuli into categories. If there are two stimuli and two responses ($N = M = 2$), the task is the familiar one-interval yes-no discrimination. If there are more possible stimuli than responses ($N > M$), the design is traditionally called *category scaling*, but is now often called *categorization*. We consider the important special case in which $M = 2$ in detail first. When N equals M but both are greater than two, the experiment is *absolute judgment*, *absolute identification*, or simply *identification*; the second part of the chapter concerns this task.

Classification experiments can be modified by the addition of a *standard* stimulus. The stimuli being judged are called *comparisons*, and a (standard, comparison) pair is offered on each trial. The presence of standards makes

113

no difference to our analysis of classification because the standard gives no information regarding which response is appropriate. As examples, we use both tasks with standards and tasks without.

Perceptual One-Dimensionality

What is a "one-dimensional" stimulus set? In the examples used so far in this book, some stimulus sets are *physically* one-dimensional (or, to borrow Klein's [1985] phrase, can be produced with a "single knob"). Examples in sensory work include intensity and frequency. The stimuli in face recognition and X-ray reading, on the other hand, clearly vary in many physical dimensions.

The question of the *perceptual* dimensionality of a stimulus set is distinct from that of physical structure. Stimuli differing in one dimension can produce multi-dimensional perceptual changes. A dimension that seems to behave in this way is the phase relation between components in a visual grating. Data suggest that changing the relative phase of components of a stimulus from negative to zero to positive may yield two-dimensional ("monopolar" and "bipolar") effects (Klein, 1985). Conversely, stimuli differing in complex ways can produce internal representations differing along a single continuum. Cases in which two variables appear to contribute to a common dimension of judgment, called *trading relations*, occur in such disparate fields as lateralization of binaural stimuli (Moore, 2003) and speech recognition (Repp, 1982). We consider a speech example later in the chapter.

A detection-theory characterization of perceptual one-dimensionality is shown in Fig. 5.1. The sensitivity statistic d' is a distance measure, as we saw in chapter 1, and distances along a single dimension add up. Thus, if stimuli S_1, S_2, and S_3 give rise to distributions along a continuum, with their means in the order $\mu_1 < \mu_2 < \mu_3$, then

$$d'(1,3) = d'(1,2) + d'(2,3) \ . \tag{5.1}$$

Equation 5.1 can be viewed as a prediction about the result of three different two-stimulus experiments or one experiment in which all three possible stimulus pairs occur. The sensitivity distance between any stimulus and the endpoint stimulus is a useful measure, *cumulative d'*, that can be computed by adding up adjacent d' values, as Equation 5.1 suggests. The value of cumulative d' obtained between both endpoint stimuli represents the total sensitivity of the observer to the stimulus set and is called *total d'*. Total d' is the basic measure of observer performance on the entire stimulus ensemble. It is important to realize that the use of cumulative and total d' depends cru-

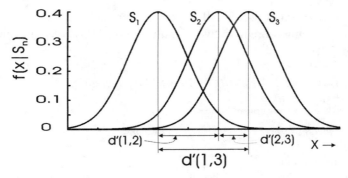

FIG. 5.1.   An additivity condition (Eq. 5.1) for perceptual one-dimensionality: $d'(1,3) = d'(1,2) + d'(2,3)$.

cially on the assumption of unidimensionality; when we consider multidimensional stimulus sets in Part II, we shall need other measures.

If there is reason to expect that a stimulus set will lead to a one-dimensional representation, Equation 5.1 can be used to infer sensitivities between remote stimulus pairs (like S_1 and S_3) from sensitivities to adjacent pairs (S_1 vs. S_2 and S_2 vs. S_3). Scales of sensory experienced magnitude have been constructed and verified by adding up or integrating d' values (Creelman, 1963b; Houtsma, Durlach, & Braida, 1980; Lim, Rabinowitz, Braida, & Durlach, 1977). This strategy can be used to measure quite large sensitivities.

Models of one-dimensional classification based on the normal distribution were first presented by Thurstone (1927a, 1927b) long before the advent of SDT, and the material in this chapter is largely drawn from the field of "Thurstonian scaling" (Bock & Jones, 1968; Torgerson, 1958). Similar classification models constructed from Choice Theory components (Luce & Galanter, 1963) are analogous to Thurstonian ones and are not described here. Nosofsky (1985) found, in fitting Kornbrot's data (1978), that the two approaches were about equally successful.

Two-Response Classification

Example 5a: Auditory Detection

In an experiment to determine the detectability of weak auditory stimuli, one of seven sound intensities is randomly chosen on each trial. The intensity of the weakest "sound," stimulus 1, is zero. The participant responds "yes" if the sound can be heard and "no" otherwise. This is a one-interval detection experiment, but it differs from other such experiments we have discussed in using more than two stimulus values. Possible data are shown in Table 5.1.

TABLE 5.1   *Results of an Auditory Detection Experiment*

Stimulus	Response	
	"Yes"	"No"
1 = no stimulus	2	98
2	6	94
3	15	85
4	40	60
5	80	20
6	92	8
7	96	4

Note. Values are percentages.

Measuring Sensitivity, Total Sensitivity, and Bias

The questions we can answer about classification data, like discrimination questions, fall into the two categories of sensitivity and bias. The perceptual spacing of each pair of stimuli, as well as cumulative and total sensitivity, can be calculated either directly or by using Equation 5.1. In two-response classification, a single bias parameter describes how the observer partitions the perceptual continuum to determine responses.

An SDT analysis of the data is presented in Fig. 5.2. The distances between the means of the distributions are, of course, values of d', calculated by considering the data in Table 5.1 two rows at a time. Thus, $d'(1,2)$, the perceptual distance between the first two stimuli, is $z(.06) - z(.02) = -1.555 - (-2.054) = 0.499$. As is apparent in the figure, the seven stimuli are not equally spaced, perceptually, but tend to cluster in two regions near the ends of the continuum.

Sensitivity for nonadjacent stimulus pairs can be found by two methods, one direct and one indirect. Directly, the distance between stimuli 1 and 3 equals $z(.15) - z(.02) = -1.036 - (-2.054) = 1.018$. The indirect approach uses Equation 5.1, which says that $d'(1,3) = d'(1,2) + d'(2,3)$. In fact the distance between Stimulus 1 and any other stimulus—cumulative d'—can be found by either technique, as shown in Table 5.2. Total d' is cumulative d' between Stimuli 1 and 7 and equals (directly or indirectly) 3.805 for these data.

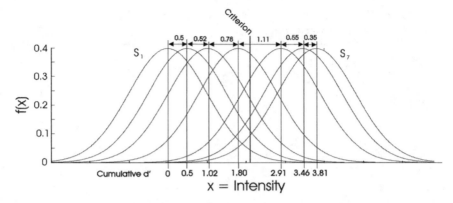

FIG. 5.2.   SDT analysis of auditory detection data (Example 5a) showing the differences between adjacent means (d') and the mean of each distribution relative to that of S_1 (cumulative d'). The criterion is located 2.05 units above the mean of S_1.

The participant decides whether to say "yes" or "no" by comparing observations with a criterion. The location of the criterion can be found from any row of Table 5.2. From the first row, for example, we see that the criterion must be 2.054 units above 0, the mean of Stimulus 1; or, from the fifth row, it is 0.853 units below 2.907, the mean of Stimulus 5, which is also 2.054. In this experiment, the observer tended to report hearing Stimuli 5 to 7, but not Stimuli 1 to 4. Distance along the internal continuum, our bias statistic here, is arbitrarily measured from the mean of an endpoint stimulus distribution. In chapter 2, the bias statistic c was referred to the zero-bias point, but with more than two stimuli this point is not unique. Another important measure of bias, likelihood ratio, is also defined in terms of two distributions: In this experiment, with its seven possible stimuli, there are 21 different likelihood ratios, and we therefore avoid this statistic.

One way in which the data might be plotted is shown in Fig. 5.3. In panel (a), cumulative d' is plotted against stimuli, which are equally spaced in physical units (say, decibels). The graph compares the physical and psychological spacing of the stimuli, and its slope tells us how rapidly the perceptual effect grows with stimulus value—that is, how sensitive the observer is to systematic stimulus changes. A straight line fitted to the data with standard least-squares methods (see Appendix 1) reveals that our data deviate somewhat from a straight line. Therefore, physical and psychological spacings are not exactly equivalent. The (cumulative) logistic distribution is also

TABLE 5.2   *Calculation of Cumulative and Total d'*
for the Data in Table 5.1

Stimulus	P("Yes")	z(p)	d'	Cumulative d'
1	.02	−2.054		
			0.499	0.499
2	.06	−1.555		
			0.519	1.018
3	.15	−1.036		
			0.783	1.801
4	.40	−0.253		
			1.106	2.907
5	.80	0.853		
			0.552	3.459
6	.92	1.405		
			0.346	3.805 = total d'
7	.96	1.751		

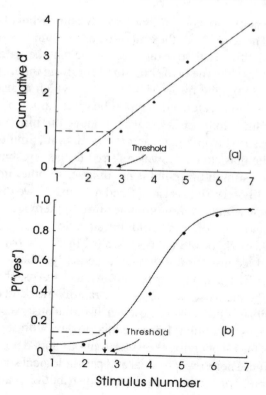

FIG. 5.3.   A psychometric function for the data of Example 5a in terms of (a) cumulative d', and (b) the proportion of "yes" responses.

sometimes used to fit data; as we have seen before, the normal and logistic curves are very similar.

Psychometric Functions and the Empirical Threshold

Each value of cumulative d' corresponds to a response proportion, and $P(\text{"yes"})$ is used as the ordinate in Fig. 5.3b. More precisely, cumulative d' is $z[P(\text{"yes"})]$ plus a constant equal to the criterion location (2.054 in this example).

When proportion of "yes" responses is plotted on a linear scale, as in panel b, the function follows an ogival shape; if the function is linear on z coordinates, it has the shape of the normal distribution function on linear coordinates. In a detection context, like Example 5a, a relation between stimulus intensity and proportion of "yes" responses is called a *psychometric function*. Such functions are often used to estimate the magnitude of the weakest detectable stimulus, conventionally termed the *threshold*. To distinguish this statistic from the theoretical "thresholds" discussed in chapter 4, we use the term *empirical threshold*.

The empirical threshold must be defined with reference to a particular performance level. That level is chosen arbitrarily, perhaps $d' = 1$. Examination of Fig. 5.3a reveals that the weakest stimulus value that can be discriminated from the Null stimulus with a d' of 1 is between 2 and 3. A threshold defined in terms of d', whatever the specific value, has the advantage of being uncolored by response bias.

A second (and probably more common) definition for threshold is the stimulus to which the observer responds "yes" on 50% of the trials. In Fig. 5.3, a value of about 4 is needed to reach this frequency of hearing. Because it depends only on the hit rate and not on the false-alarm rate, this measure is bias-contaminated. The shape of the curve in the lower panel of Fig. 5.3 depends on the response criterion, whereas that in the upper panel does not.

A third performance measure, proportion correct, does depend on both H and F, but is unbiased only when observers adopt a symmetric decision rule. When there are more than two possible stimuli, the criterion cannot be symmetric for all pairs, and $p(c)$ cannot be bias-free (Sperling & Dosher, 1986). Use of proportion correct in two-alternative forced-choice tasks (see chap. 7) is less problematic. If on each trial a Null stimulus and a non-Null stimulus are both presented in random order, then the same symmetric criterion can be applied to the difference between the observations for all stimulus pairs.

Example 5b: Length Classification With a Standard

Another application of classification has a long history. Consider a study in which a series of lines differing in length are presented for judgment, each line preceded by a standard from the middle of the range, and the observer must decide whether the comparison stimulus is "longer" or "shorter" than the standard. This *method of constant stimuli* dates back at least to Fechner (Jones, 1974). The method is used to measure discrimination between the standard and comparison stimuli, but is analytically a classification task. Modifying the two-response paradigm by adding a standard stimulus does not change the analysis, but experiments that use this design often have a different emphasis.

The PSE *and the* jnd

Figure 5.4 shows hypothetical psychometric functions for an experiment employing a standard, on both linear and z coordinates. The upper curve has a cumulative normal shape—as is often true, to a first approximation, in practice—so the lower curve is a straight line. Traditionally (see Luce & Galanter, 1963), two statistics are abstracted from such curves, a measure of slope and a measure of the intercept, both expressed in stimulus (x-axis) units. The intercept, the stimulus judged "longer" 50% of the time, is called the *point of subjective equality* (PSE). The usual measure of slope is half the stimulus distance between the 25th and 75th percentiles; this is termed the *just-noticeable difference* (jnd). We can see from Fig. 5.4 that the jnd is the stimulus difference yielding $d' = 0.675$; if the psychometric function is linear in z coordinates, the jnd is the same for any two points this distance apart.

Example 5c: False Memory

Roediger and McDermott (1995), following Deese (1959), conducted a recognition memory experiment with a twist: The study items were thematically related (e.g., *bed, night, dream, blanket*). The test included Old items, New items that were not related to the theme, and a "critical lure"—*sleep*, in this example—that was the core concept to which the study items were related. Of course there were many such sets of critical lures and related study items. Participants tended to recognize (incorrectly) the critical lures at a higher rate than other lures and sometimes at a higher rate than Old items. The paradigm (usually called *DRM* after its inventors) is of interest because it demonstrates, in a controlled situation, the phenomenon of "false mem-

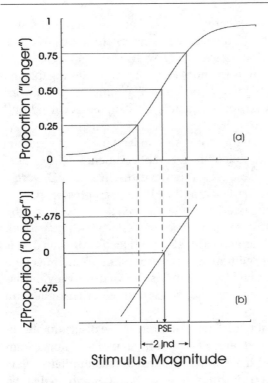

FIG. 5.4. A psychometric function for the data of Example 5b, with (a) linear and (b) z-score ordinates. Calculation of the PSE and jnd is shown.

ory." The possibility that such "memories" might occur in court testimony is a matter of evident concern.

The results of a typical experiment can be summarized in terms of the proportion of "yes" responses:

Type of Item	Proportion of "Yes" Responses	$z[P(\text{"Yes"})]$
Old words	.85	1.036
New words: Unrelated to Old words	.30	−0.524
New words: Critical lures	.80	0.842

A natural question about false memory (Miller & Wolford, 1999) is whether it is a sensitivity or a response-bias effect: Do participants "really" remember the critical lures as having been presented, or is the finding somehow due to a bias (that could, in principle, be manipulated)? The presence of three "yes" rates in a yes-no experiment raises the question of how sensitivity and bias are to be calculated. For example, is d' the distance between the Old and New/unrelated distributions or between the Old and New/critical distributions? If response bias is to be the distance from the criterion to a crossover between an Old and New distribution, which New distribution is relevant?

We have analyzed most previous memory examples by assuming that the underlying dimension being judged is *familiarity*. The familiarity of a word can be influenced by two factors: how frequently the item has occurred, and the number of associated words that have recently been presented. As Wixted and Stretch (2000) have pointed out, this understanding of familiarity leads to a straightforward account of the DRM phenomenon: The critical lures are highly familiar because of recently occurring associated words, and Old words are highly familiar because they have been presented.

If familiarity is the only characteristic of test words being evaluated by the participant, the appropriate detection theory model is one-dimensional, as sketched in Fig. 5.5. Old, New, and Critical stimuli lead to distinct distributions on the familiarity axis, with means corresponding to their average activation. Converting each yes rate to a z score, as shown in the previous table, reveals the values of these means; the location of the criterion is 0 on this scale.

There are three distinct sensory distances of interest: the discriminability of Old items from New/unrelated ones ($d' = 1.560$), that of Old items from critical lures ($d' = 0.194$), and that of critical lures versus unrelated New items ($d' = 1.366$). This last value is different from the other two, in that the correct response ("no") is the same for both stimulus classes. Dosher (1984) proposed the term *pseudo-d'* for a sensory distance estimated in this way. Pseudo-d' acts just like any other d': It is unaffected by criterion location, and it is additive in the sense of Equation 5.1.

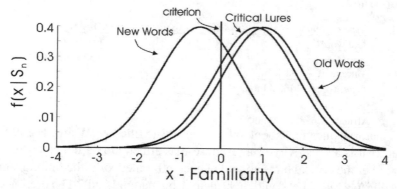

FIG. 5.5. Possible representation for a false-memory recognition experiment. Old items were presented in the study list, New items were not, and Critical lures are new items that are semantically related to Old items.

The analysis concludes that false memory is a sensitivity effect: The strength of critical lures is high due to associations, that of Old items through recent presentation, and these two sources cannot be distinguished by the participant. Response bias is fixed: There is only one criterion location. The representation could, of course, be wrong: Perhaps the experiment provides a separate source of information about whether test items were on the study list; this hypothesis suggests a multidimensional representation. The DRM design does not allow us to test alternative models in which such representations are required.

Example 5d: False Fame

Jacoby, Woloshyn, and Kelley (1989) introduced an influential variant of the recognition memory design. Their study list included both famous and nonfamous people (the latter being simply randomly chosen names). In the test condition, participants were asked not whether the names had occurred at study time, but rather whether the names were those of famous people. Some of the nonfamous names seen at study were judged famous by participants, who apparently could not always distinguish familiarity due to fame from familiarity due to recent exposure. In some ways, this design resembles the DRM study of the previous section.

We analyze here a similar experiment by Park and Banaji (2000) that differed in an important respect: There was no study trial. Participants were asked to judge whether names were those of basketball players; some of the names were players, and some were not. Among the players, some were African-American and some Euro-American, and among the nonplayers, some had names likely to be African-American (Lamont Turpin, Reggie Newton) and some likely to be Euro-American (Eric Griffin, John Merritt). The following table presents the results.

Type of Name	Proportion of "Yes" Responses	$z[P(\text{"Yes"})]$
Euro-American players	.932	1.49
African-American players	.893	1.24
Euro-American nonplayers	.334	−0.43
African-American nonplayers	.564	0.16

As in the earlier example, response bias is fixed, and the data are best interpreted as reflecting different levels of overall activation. Figure 5.6 illustrates the assumption that a single dimension, *basketball-playerness*, mediates judgments. Some of the sensitivities that can be estimated represent true d' (for distinguishing White players and nonplayers, or Black players and nonplayers). The key conclusions, however, are based on pseudo-d' (for distinguishing White and Black players, or White and Black nonplayers). On the playerness dimension, White players are higher than Black ones (perhaps being a minority they are more salient), whereas Black names are, on average, higher than White ones (apparent stereotyping).

Example 5e: Trading Relations in Speech Identification

In a common type of speech perception experiment, a set of synthetic stimuli is constructed along a continuum between two waveforms that correspond to different speech sounds. For example, a stimulus waveform perceived as /ga/[1] can be gradually converted into one perceived as /ka/ by lengthening voice-onset time (VOT), the amount of time between the beginning of the consonant and the onset of voicing. An apparently straightforward way to find out what a listener hears is to present a series of randomly chosen stimuli from this set and ask whether each sounds more like "ka" or "ga." Typically, the proportion of trials on which "ka" is the response increases as VOT increases (Lisker, 1975). Speech researchers term this design *identification*.

Table 5.3 presents data similar to those from Lisker's (1975) study. Lisker systematically varied VOT, and the second and third columns of the

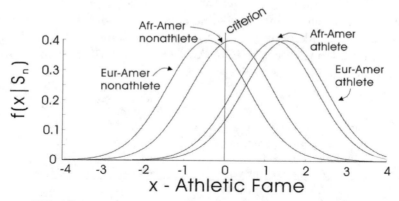

FIG. 5.6. Possible representation for the false-fame experiment of Park and Banaji (2000). Among real basketball players, Euro-American athletes were judged higher than African Americans, whereas nonplayers with African-American names were judged higher than those with Euro-American names.

---

[1]We use the common linguistic notation of enclosing phonetic utterances between slashes.

table show that as this parameter increased, the percentage of "ka" responses did also. The numbers in these columns are the same as in the auditory detection example (Table 5.1); the "identification function" (Fig. 5.7) is identical to the psychometric function shown in Fig. 5.3.

TABLE 5.3   *Results of a Speech Identification Experiment*

Stimulus		F1 Onset = 386 Hz		F1 Onset = 769 Hz	
Number	VOT (ms)	"ka"	"ga"	"ka"	"ga"
1 = /ga/	0	2	98	10	90
2	10	6	94	15	85
3	20	15	85	23	77
4	30	40	60	52	48
5	40	80	20	84	16
6	50	92	8	95	5
7 = /ka/	60	96	4	99	1

Note. Values are percent responses.

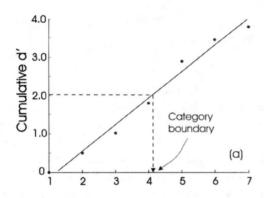

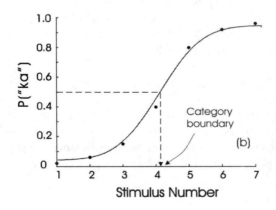

FIG. 5.7.   A function relating identification responses to stimulus levels for the data of Example 5e in terms of (a) cumulative d', and (b) proportion of "ka" responses. Although the experiment is different from Example 5a (which was detection rather than identification), the presentation of the data is identical.

Two features distinguish these formally identical examples. First, there is a correspondence function in auditory detection (the correct answer is "no" for Stimulus 1 and "yes" for the others), whereas there is none in speech identification: The *point* of the experiment is to find out how each sound is perceived. Second, the psychological interest in the detection experiment is in sensitivity (if the dependent measure used to estimate threshold is d'), whereas in the speech experiment the most popular dependent measure is the "category boundary"—the stimulus value at which each response is used on 50% of trials. Like the PSE, the category boundary is a bias statistic.

The final two columns in Table 5.3 approximate the results of a second condition in Lisker's study. Another feature of speech waveforms that influences the perception of voicing is the frequency at which F_1, the first formant (i.e., frequency band), begins. When the identification experiment was redone with a higher value of F_1 onset, the percentage of "ka" responses increased across the board. In effect, the two acoustic cues, VOT and F_1 onset, trade off against each other: To get 15% "ka" responses, either a 20-msec VOT and a 386-Hz F_1 onset or a 10-msec VOT and a 769-Hz F_1 onset will work. Such trading relations (Repp, 1982) reflect a kind of perceptual interaction between cues. Notice that the effect is one of response bias: In chapter 8, we introduce approaches to measuring such interactions that are sensitivity based.

The speech experiment described in Example 5e can be modified by including a standard. Now each trial contains two waveforms, the first always the /ga/ stimulus from the continuum endpoint (Stimulus 1), the second changing from trial to trial. The task is to say whether the second, comparison stimulus sounds like "ga" or "ka." In studies resembling Example 5e, the effect of the standard is frequently to change bias rather than sensitivity: If every presentation is preceded by Stimulus 1, the response criterion is farther to the left in Fig. 5.7 than if no standard is used. Listeners using the first /ga/ stimulus as a standard report hearing more intermediate stimuli as "ka" than those operating without a standard (Carney, Widin, & Viemeister, 1977; Diehl, 1981; Macmillan, Braida, & Goldberg, 1987).

Experiments with More Than Two Responses

The assignment of many stimuli to just two responses in the examples so far seems natural—all waveforms in the speech experiment, for example, resemble either /ga/ or /ka/, not a third utterance. But there are at least two reasons an experimenter might prefer a number of responses closer to the size of the stimulus set.

The first is familiar from our treatment of rating experiments. We found in chapter 3 that even with only two stimuli, the availability of a graded response provided us with a more detailed understanding of the discrimination process. This is no less true of classification experiments; if the underlying distributions have unequal variance, the use of multiple responses is essential.

The second reason is specific to classification tasks. In the hypothetical data of the examples so far, no response proportions of 0 or 1 occur—we have been protecting the reader from the complications such proportions cause for detection theory analysis. If total d' is more than about 3 (an unbiased proportion correct of about .93), the fact of binomial variability means that troublesome perfect proportions are quite likely. The availability of more responses solves the problem as long as *some* response proportion is not perfect for each stimulus pair.

Example 5f: Intensity Identification

Braida and Durlach (1972) conducted an auditory identification experiment with 15 pure-tone stimuli whose intensities ranged from 50 to 90 decibels. On each trial, one stimulus was presented, and listeners selected 1 of 15 responses. Some possible data for an analogous task in which $M = N = 4$, with four stimuli and four responses, are shown in Table 5.4.

Measuring Sensitivity, Total Sensitivity, and Bias

The data for an experiment with N stimuli and M responses fill an $N \times M$ stimulus–response matrix—in the two-response case, the matrix (e.g., Table 5.1) was $N \times 2$. Any two rows in such a table provide information about distinguishing two stimuli; when there are more than two responses, a pair of rows defines an ROC with more than one point. Considering only adjacent rows, an $N \times M$ experiment yields $N - 1$ ROCs, each with as many as $M - 1$ points.

The three 3-point ROCs for adjacent stimuli generated by Example 5f are shown (on normal coordinates) in Fig. 5.8, with straight lines of unit slope fitted to the points by eye. The values of d', shown in the figure, are approximately $d'(1,2) = 1.2$, $d'(2,3) = 0.4$, and $d'(3,4) = 0.8$.

Total d', a measure of total sensitivity in the experiment, is 2.4, the sum of the three d' values. The additivity condition (Eq. 5.1) implies that this is also $d'(1,4)$; when the data for Stimuli 1 and 4 are plotted, the distance between the ROC and the chance line is indeed 2.4.

As we saw in chapter 3, the best-fitting ROC can be found by maximum-likelihood estimation procedures (Dorfman & Alf, 1969). Addi-

TABLE 5.4   *Results of an Intensity Identification Experiment With Four Stimuli and Four Responses*

(a) Original Data

	Frequencies				Cumulative Proportions			
	Response				Response			
Stimulus	1	2	3	4	1	2	3	4
1	39	7	3	1	.78	.92	.98	1.0
2	17	12	10	11	.34	.58	.78	1.0
3	11	10	12	17	.22	.42	.66	1.0
4	3	5	9	33	.06	.16	.34	1.0

(b) Modified Data

	Frequencies				Cumulative Proportions			
	Response				Response			
Stimulus	1	2	3	4	1	2	3	4
1	39	11	0	0	.78	1.0	1.0	1.0
2	17	12	21	0	.34	.58	1.0	1.0
3	0	21	12	17	0	.42	.66	1.0
4	0	0	17	33	0	0	.34	1.0

(c) Further Modified Data

	Response			
Stimulus	1	2	3	4
1	0	39	7	4
2	0	17	12	21
3	0	11	10	29
4	0	3	5	42

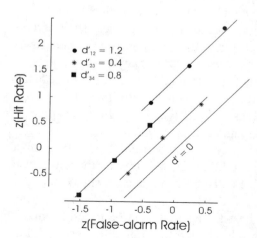

FIG. 5.8.   ROCs for the identification experiment of Example 5f. Each curve is for a separate pair of adjacent stimuli. Hit and false-alarm rates are estimated from successive rows in Table 5.4a. The data for each stimulus pair are analyzed using the methods described in chapter 3, with the higher number stimulus playing the role of S_2.

tional complexity is introduced when more than two stimuli are used, because the $M - 1$ criteria are assumed to fall in the same location for each stimulus pair. A program to fit classification data in this way has been published by Schönemann and Tucker (1967). In Braida and Durlach's (1972) modification of the program, equal variances for the underlying distributions are assumed.

Data Matrixes With Zeroes

In experiments with many responses, or generally high sensitivities, direct computation of total d' may not be possible because all responses may not have been used in response to all the stimuli. As long as each pair of successive stimuli has two adjacent response columns of overlap, d' can be computed, and the results can be summed to find total sensitivity. Consider the modified data matrix for the reduced Braida–Durlach experiment, presented in both raw frequency and cumulative proportion form in part (b) of Table 5.4.

For each adjacent stimulus pair, d' can be computed for one criterion placement. For example, $d'(1,2) = z(.78) - z(.34) = 1.184$, using the criterion to separate Response 1 from the other responses; for other response partitions, one or both proportions equal 1.0, and d' cannot be estimated. Similarly, $d'(2,3)$ and $d'(3,4)$ are found to be 0.404 and 0.824. All three values are approximately the same as for the original data matrix. We are still able to calculate d' for nonadjacent pairs, but only using the indirect, additivity-of-d' method. Thus, $d'(1,3) = 1.2 + 0.4 = 1.6$, and total $d' = 2.4$.

Biases and Feedback

In experiments that aim to measure sensitivity, it is common to provide trial-by-trial feedback, informing the observer, after each response, of the stimulus just presented. When the response used by the participant is interesting in its own right, feedback is usually not given. What effect does the feedback manipulation have?

The answer depends on both the spacing and presentation probabilities of the stimuli. Sometimes—again, most often in sensitivity-oriented experiments—the distribution of stimuli is uniform; in that case, feedback does not usually have much effect. But when stimulus distributions are not uniform, as (one might argue) in real life, the response biases of observers are given free rein. Such experimental situations have been theoretically influential. Helson's (1964) *adaptation-level theory* proposes that people make classification judgments by comparing their observations with a weighted average of stimulus effects, the *adaptation level*. For example, if Stimulus 1 in our speech study were presented more frequently than the others, the adaptation level, or neutral stimulus, would move toward Stimulus 1. Responses would therefore shift toward higher numbers. This approach has been extended in Parducci's (1974) *range-frequency model*, for which Parducci garnered support in experiments without feedback.

A participant who is informed about a nonuniform stimulus distribution might be expected to show an effect opposite to that just described: If one knew that Stimulus 1 would be presented half the time in an absolute identification experiment, surely it would be sensible to use Response 1 frequently. Results of this sort have been found by Chase, Bugnacki, Braida, and Durlach (1983) for auditory intensity and by Macmillan and Braida (1985) for a vowel continuum. Macmillan and Braida were replicating a study by Sawusch, Nusbaum, and Schwab (1980) that used the same continuum but no feedback and obtained results consistent with range-frequency theory. The pattern of results is similar to that found in discrimination experiments in which presentation probability is manipulated with and without feedback (see chap. 3). The moral is this: If one wants participants to mimic changes in presentation schedule in their responding, provide feedback; if no feedback is provided, they will act as if the distribution were uniform and show increased responding away from the most frequently presented stimuli. Exactly how to characterize these response changes is still a matter for study.

Nonparametric Measures

Pairwise Sensitivity: Mean Category Scale

Consider again the data of Example 5f, the intensity identification experiment. By finding d' for each pair of stimuli, we located the means of the four underlying distributions at 0, 1.2, 1.6, and 2.4 units on a psychological dimension. Another strategy for "scaling" these tone intensities is to compute the mean rating given to each stimulus. Stimulus 1 is assigned a mean rating of 1.32; Stimulus 2, 2.30; Stimulus 3, 2.70; and Stimulus 4, 3.44. These mean ratings constitute an alternative mapping of stimuli into psychological magnitude, called the *mean category scale*.

The mean rating assigned to a single stimulus is, of course, a measure of response bias, but the *difference* between two such ratings is not so obviously flawed as a measure of sensitivity. That it *is* flawed can be seen from an analysis of the simplest, 2×2 (one-interval discrimination) experiment.

The mean rating of Stimulus 1 equals $P(\text{"1"}|S_1) + 2P(\text{"2"}|S_1)$. In more usual notation, this is $(1 - F) + 2F$, or $1 + F$. The mean rating of Stimulus 2, similarly, is $1 + H$. The difference in mean rating equals $H - F$, which (when stimuli are equiprobable) equals $2p(c) - 1$. We saw in chapter 4 that threshold measures, such as those that depend on $p(c)$, are not bias-free; neither are the mean differences between category judgments.

The effects of stimulus range and frequency on mean category scales are substantial and have been studied extensively by Parducci (1974) and others. No one, to our knowledge, has succeeded in abstracting sensitivity measures from these impoverished scales. Once responses to a given stimulus have been summarized by a mean rating, the information needed to separate sensitivity and bias is lost. Quite useful analyses of mean rating data have been proposed (see also Anderson, 1974), but they are not detection theoretic and do not generate estimates of sensitivity.

Overall Sensitivity

There is a natural nonparametric response measure of overall sensitivity, namely, overall $p(c)$. How does it compare to our detection theoretic index, total d'?

A major accomplishment of detection theory is to abstract a measure of sensitivity, such as d', that does not depend on response criterion. The methods we have described in this chapter extend this accomplishment to designs with more than two stimuli. Whereas the bias parameter is usually viewed as a confounding influence in discrimination tasks, users of classification designs are often as interested in bias as in sensitivity.

Suppose the listeners in the auditory identification experiment (Example 5f) had been biased toward high-numbered responses and, in fact, *never* used Response 1. They might have generated the data matrix in part (c) of Table 5.4, in which Response 2 is assigned to Stimulus 1, Response 3 to Stimulus 2, and Response 4 to Stimulus 3 as accurately as correct assignments were made in the original matrix.

The ROC curves derived from this matrix consist of the two left-most points from each ROC in Fig. 5.8, so d' values are unchanged (leaving aside issues of sampling variability). Yet the participant has correctly identified only 69 of the 200 presentations correctly [$p(c) = .345$], whereas in the original matrix the proportion correct was 96/200 = .48. Is this a case in which a nonparametric measure, proportion correct, is more useful than a detection theoretic measure?

The answer is that this depends on the goal of the experiment. Is the experimenter interested in whether the participant can correctly name the stimulus or in the participant's resolution power? For the one-dimensional continua we have been considering, the answer is usually the latter; in any case, response criterion shifts of this sort are reflected in the estimated values of the bias parameters. We shall see in chapter 10 that correct naming often *is* important in investigations of multidimensional stimulus sets. For example, it seems natural to ignore response shifts of the sort just described for our /ga - ka/ continuum, but if the possible responses were all different words, a systematic mispairing of stimuli and responses would be viewed as a true decline in performance.

Information Theory

The ability of an observer to classify stimuli is often summarized by another nonparametric measure, *information transmitted*, a statistic proposed by information theory. We have elected not to describe this approach because information transmitted is not a true sensitivity measure. Information transmitted is unchanged by systematic reassignment of responses (e.g., "yes" for "no" and vice versa in yes-no discrimination), but *does* depend on presentation probability and response bias. Introductory treatments of information theory can be found in the work of Miller (1953, 1956); a more extended account has been provided by Garner (1962).

Comparing Classification and Discrimination

Using detection theory models, an investigator can measure values of d' for the same two stimuli in either classification or discrimination. Will both experiments (each of which has many possible variants) lead to the same result? If not, how will they differ?

Although classification and discrimination data both lead to estimates of sensitivity, they need not converge on the same truth. Detection theory and threshold theory offer ways to compare these two kinds of tasks.

SDT Models

Comparing classification and discrimination in detection theoretic terms is uncomplicated: One measures d' in one experiment of each type and examines the result to see whether sensitivity has changed. When this is done, sensitivity is almost always found to be better in discrimination.

In a few special cases, classification and discrimination d' are (theoretically or empirically) nearly equivalent. Empirically, Pynn, Braida, and Durlach (1972) compared identification and discrimination of pure-tone intensity on a very small (2.25-dB) range and found close agreement. Theoretically, an influential proposal about speech perception experiments of the sort described in Example 5e, the *categorical perception hypothesis*, says, in part, that discrimination is exactly as good as classification for some speech continua. This hypothesis has been presented in SDT language by Macmillan, Kaplan, and Creelman (1977); its original statement, to which we now turn, was in threshold terms.

Threshold Models

According to Liberman et al. (1957), the listener in a discrimination experiment (Liberman et al. used an "ABX" design; see chap. 9) covertly categorizes each sound of the three presented in a trial as S_1 or S_2 and makes a response consistent with these categorizations. The probabilities of these categorizations were to be directly measured in identification experiments, allowing discrimination to be predicted from identification.

Discrimination in ABX (see chap. 9 for details) can be predicted from two-response classification, according to these models, as follows:

$$p(c)_{\text{discrimination}} = \tfrac{1}{2}[1 + (p_1 - p_2)^2] , \qquad (5.2)$$

where p_1 and p_2 are the probabilities of classifying Stimuli 1 and 2 into a particular category, as estimated from a two-response classification experiment. We can use Equation 5.2 to predict discrimination results for the VOT continuum of Example 5e. Proportion correct in discriminating Stimuli 1 and 2, for example, is predicted to be $0.5[1 + (.98 - .94)^2] = .5008$, whereas for Stimuli 4 and 5 $p(c)$ should be .58.

In fact predictions made using threshold theory are always too low. Detection theory predictions tend to be higher, but still too low. The difference between observed and predicted results varies with a number of factors, including the type of continuum being studied, but there is always a difference. Why this is true is a long-standing puzzle in cognitive psychology.

Why Is Classification Harder Than Discrimination?

Psychologists were alerted to the discrepancy between classification and discrimination by George Miller in a famous 1956 paper. Miller summarized experiments showing that increases in the number of stimuli to classify led to corresponding increases in total sensitivity only up to about seven stimuli. When the range of stimuli was increased beyond that point, there were no further increases in classification performance, but total discrimination performance continued to improve.

A model that relates classification and discrimination has been offered by Durlach and Braida (1969). Although originally presented as a theory of intensity perception, the model applies to many other domains as well. The application of the model to speech perception in particular is discussed by Macmillan (1987) and Macmillan, Braida, and Goldberg (1987).

According to Durlach and Braida, fixed discrimination tasks (those using just two stimuli) measure only sensory resolution, whereas classification depends on both sensory and *context-coding*, or labeling processes. Both sensory and context-coding processes contribute to the variance of the underlying distributions, so if a is the distance between the two means, β^2 the sensory variance and C^2 the context-coding variance, then

$$d'_{discrimination} = \frac{a}{\beta}$$

$$d'_{classification} = \frac{a}{(\beta^2 + C^2)^{1/2}} .$$

(5.3, 5.4)

Clearly, the discrepancy between fixed discrimination and identification depends on the relative magnitude of the sensory and context variance components. The *relative context variance*—the size of the context variance in units of the sensory variance—can be estimated:

$$\frac{C^2}{\beta^2} = \left(\frac{d'_{discrimination}}{d'_{classification}} \right)^2 - 1 .$$

(5.5)

Equation 5.5 can be applied to total d' values as well as d' for particular stimulus pairs, and it provides a measure of the importance of context memory for a stimulus pair or continuum.

Context variance is a measure of memory, and the Durlach–Braida theory asserts that classification tasks are difficult because of the memory load they impose. What types of stimulus continua are hard to context code? In the theory, context variance is a function of stimulus range (measured, for intensity, in decibels). The greater the range, the greater the discrepancy between classification and discrimination. This accounts for the close agreement between the two tasks found by Pynn et al. (1972).

The range is not as well defined for most stimulus continua as it is for intensity, but the qualitative relation between range and sensitivity is nonetheless useful. Other proposals for relating performance in different classification tasks are discussed in chapter 10.

Summary

In a classification experiment, one stimulus from a set of possible stimuli is presented on each trial. This chapter has considered classification experiments for stimuli lying only on one perceptual dimension.

Two-response classification is a generalization of the yes-no design to many stimuli. Values of d' for any two adjacent stimuli can be found by subtracting z-transformed response proportions. Bias is measured by criterion location. Cumulative sensitivity is the perceptual d' distance between any stimulus and the endpoint stimulus, and total sensitivity is the sum of all adjacent values of d'.

When each stimulus to be judged is preceded by a standard, calculation of sensitivity and bias is unchanged. The function relating cumulative sensitivity to stimulus value is called the psychometric function. The midpoint of this function (the criterion location) is now called the (empirical) threshold or point of subjective equality; the (inverse of the) slope of the function is a sensitivity measure, the jnd.

Classification with more than two responses produces an ROC for each stimulus pair; measures of sensitivity and bias are otherwise the same as with two responses.

The mean category scale, another approach to analyzing classification data, leads to a sensitivity measure based on proportion correct and is not detection theoretic. As a measure of overall performance, $p(c)$ differs from total d' in assessing the ability to name the stimulus rather than sensitivity.

For the same stimulus set, discrimination is superior to classification. One model for this effect attributes the discrepancy to the need for context memory in classification and permits an estimate of the context variance caused by classification tasks relative to the unavoidable sensory variance inherent in all tasks.

Chart 8 of Appendix 3 provides guideposts to methods for analysis of classification data.

Problems

5.1. (a) If Stimuli A and B can be discriminated with a d' of 1 and Stimuli B and C with a d' of 2, what is the predicted d' for discriminating A and C (assuming that A, B, and C fall in that order on a single dimension)?
(b) If Stimuli A and B can be discriminated with a $p(c)$ of .69 and Stimuli B and C with a $p(c)$ of .84, what is the predicted $p(c)$ for discriminating A and C? Make the same assumptions as in (a), plus unbiased responding by the participants. Is it possible to state an additivity condition like Equation 5.1 for $p(c)$?
(c) Repeat part b, but with a $p(c)$ of .84 for (A, B) and .93 for (B, C).

5.2. Is it possible to state an additivity condition like Equation 5.1 for d'_1 (mean difference in units of the S_1 distribution)? Redo Problem 5.1a, letting the values be for d'_1 rather than for d', and let $s = 0.5$ in all cases.

5.3. In a recognition memory experiment using faces, some stimuli are presented once, some twice, and some four times. The test sequence contains some faces from each condition, plus New faces. The proportions of "yes" responses are .92, .76, and .60 for 4, 2, and 1 presentations, and .18 for New faces.
(a) Assuming that all judgments are mediated only by familiarity, find cumulative d' for all categories of faces. Where is the criterion? Is this location closer to the average Old face or the average New face?
(b) Suppose the participants in the experiment are presented, in a separate condition, with faces they have seen either once or four times, and asked to say which. Predict d' and $p(c)$ for this "frequency discrimination" task assuming an unbiased criterion setting.

5.4. In an auditory frequency-discrimination experiment, a 1000-Hz tone is used as the standard, and the observer responds "higher" or "lower" to five other frequencies as follows.

Frequency (Hz)	Number of "Higher" Responses	Number of "Lower" Responses
998	10	40
999	20	30
1000	30	20
1001	40	10
1002	45	5

Find the PSE and the jnd.

5.5.   The jnd depends on the 75% and 25% points on the psychometric function and, as Fig. 5.4 shows, yields two stimuli that are 0.675 z units apart. But these percentages are arbitrary: We could use 80% and 20%, 65% and 35%, or any other pair that add to 100%. What percentages should be used to find two stimuli that are 1 z unit apart?

5.6.   In a modification of the method of constant stimuli (as applied to line length), participants are allowed to respond "longer," "shorter," or "same." The results of such an experiment are as follows (the standard length was 28 cm).

Length of Comparison Stimulus (cm)	"Longer"	"Same"	"Shorter"
10	0	2	18
15	0	4	16
20	0	7	13
25	3	9	8
30	5	10	5
35	10	10	0
40	14	6	0

Find d' for each adjacent stimulus pair (if two estimates are available, average them). Calculate cumulative d' for each stimulus, and total d'. Assuming that the observer uses two criteria to partition the decision dimension, find their locations relative to the endpoint distributions. To what approximate stimuli (lengths) do the criteria correspond?

5.7.   Reduce the data matrix of Problem 5.6 into two response categories by dividing the "same" responses equally between "longer" and "shorter." Reanalyze the data. What is the effect of this reduction on total d'?

5.8.   Using the data from Problem 5.7, draw a straight-line psychometric function by eye to the data on z coordinates. Use the function to estimate the jnd and PSE.

5.9.   In a visual "categorical perception" experiment (similar to that of Yasuhara and Kuklinski, 1978), the letter E is modified by decreasing the length of its lowest horizontal segment in four steps until it becomes the letter F. The five resulting stimuli are flashed briefly, many times each, to the observer, who must identify them as "E" or "F." The proportion of "E" judgments for Stimuli 1 through 5 is, respectively, .97, .9, .6, .2, and .1. Find d' for each adjacent stimulus pair, cumulative d' for each stimulus, and total d'. Also find the location of the criterion relative to the means of the stimulus distributions. To approximately what stimulus or stimuli does the criterion correspond?

5.10. Apply the categorical perception hypothesis to the data of Problem 5.9 assuming (a) an SDT model, and (b) a threshold model. Predict the proportion correct in discriminating each adjacent pair and each pair two steps apart (like Stimuli 1 and 3) in a yes-no discrimination experiment.

5.11. Suppose that the actual total (fixed) discrimination d' in the *E/F* experiment (Problem 5.9) is 4.0. Estimate the relative context variance for this continuum according to Durlach and Braida's theory. Reestimate it assuming that total discrimination d' is 6.0 and assuming that it is 8.0.

5.12. Observers are asked to identify four sucrose solutions that differ in saturation. There are four solutions and four responses, and the data (numbers of responses) are as follows.

Saturation	*Response*			
	1	2	3	4
1	6	2	2	0
2	0	6	4	0
3	0	3	5	2
4	0	0	3	7

Find each pairwise d', total d', and the location of the three criteria (letting the mean of the #1 solution equal 0).

II

Multidimensional Detection Theory and Multi-Interval Discrimination Designs

---

The one-interval experiment has now been analyzed in some detail. For two-stimulus experiments, we have learned how to estimate sensitivity and bias from yes-no data and how to plot ROCs from rating data. The analysis generalizes easily to experiments with more than two stimuli. Our models also provide us with a picture of the decision space and the manner in which decisions are made. What is common to all the situations we have so far considered is the assumption that observers base their decisions on a single variable or axis and determine their responses by dividing this continuum into segments using one or more criteria.

In Part II, we consider some of the many situations in which this assumption fails. Most obviously, more than one variable is needed to describe many perceptual and cognitive representations: Changes in tone intensity and light intensity, to take a simple example, have distinct neural and psychological outcomes. To model this additional complexity, we take the obvious step of increasing the dimensionality of the representation. For the most part, we use two-dimensional geometric analyses.

We progress from the simplest cases toward (but not *to*) the most complex on two parallel tracks. Chapter 6 considers the detection and discrimination of two stimuli whose representation is two-dimensional (such as simultaneous tone-light pairs). Because there are only two stimuli, it turns out that the optimal strategy of relying on a single dimension—the sum of perceived brightness and loudness—is sufficient to analyze such experiments, but that two dimensions are required to allow for reasonable but nonoptimal decision rules. In chapter 8, we examine classification designs, in which collapsing the two dimensions is not possible. The primary substantive questions addressed by such designs are the abilities to separate (selective attention) or combine (perceptual interaction) distinct dimensions. Chapter 10 considers identification, as in chapter 5, but without the restriction to a single dimension.

The second track in Part II concerns more complex designs for studying discrimination. In these tasks, each trial contains two or more intervals, the "intervals" being arranged in either time or space. These experimental designs serve the same goal as yes-no discrimination: to estimate the observer's ability to tell two stimuli apart. In multi- and one-interval designs alike, therefore, the stimuli are of only two types. Investigators choose to measure discrimination with these apparently more complicated designs for practical reasons, some of which we discuss.

Multi-interval designs can be described by SDT models in which the "dimensions" are the intervals that compose the task. In chapter 7, we use the tools required to model compound detection (chap. 6) to examine two-alternative forced-choice and the "reminder" experiment. In chapter 9, we use the tools developed for studying attention and interaction (chap. 8) to model the same-different, matching-to-sample, and oddity experiments. In chapter 10, we find that multi-alternative forced-choice is a special case of multidimensional identification, so that the same models apply to both.

6

Detection and Discrimination of Compound Stimuli: Tools for Multidimensional Detection Theory

In *Flatland*, Abbott's (1991/1884) classic mathematical fantasy, a two-dimensional world is visited by someone from the third dimension who shows an eager acolyte the splendors of 3D. So far we have described a *one*-dimensional psychological world that even flatlanders would disdain: Sensation, familiarity, and other such dimensions have been the single subjective variables involved. For the initial applications of detection theory to auditory and visual detection, the idea that a single variable—subjective intensity—characterized the decision process was quite reasonable. We saw in chapter 5 that some apparently more complex problems such as false memory and social judgment can be interpreted unidimensionally as well.

The problems we consider in this chapter are the detection and discrimination of "compound" stimuli, that is, those with two or more perceptually distinct components. The key questions are whether these "cues" are combined by the observer and, if so, in what way. Treisman (1998) offers some nonlaboratory examples: To decide whether there is an aircraft in the sky (a detection task) or whether the aircraft is a plane or a helicopter (discrimination), one may rely on visual appearance or the quality of sound it produces. In assessing the degree of impairment of a particular patient, a clinician's judgment may be based both on a deficiency in movement control and signs of disordered thought. The question may be whether impairment exists (detection) or what type of impairment it is (discrimination). Cues may be in conflict or in agreement, and how they are best combined is a complex problem. Should the nature of combination change, in the plane-spotting example, if clouds limit the view or traffic noise masks the auditory signal?

The experimental illustration for most of the chapter is a simplified version of the aircraft example, the detection of the simultaneous presentation of a tone burst and a light flash. Compound detection of this type was explored early by Tulving and Lindsay (1967). To analyze the problem requires multidimensional tools, and the first part of the chapter provides them. We begin by reviewing the general one-dimensional model with which we have worked so far, and then we expand its application into perceptual spaces of two or more dimensions. After discussing the principles needed to analyze multidimensional problems in general terms, we finally apply them to the compound detection problem.

The attentive reader will note that we have considered "multidimensional" stimuli, such as faces, words, and X-rays, without venturing beyond one-dimensional representations. The single decision axis has been "strength of evidence," even if that evidence had multiple components. Formally, the decision axis in a single two-response task can always be considered as likelihood ratio. Why is this familiar flatland mode no longer adequate?

In many applications, the interest in the detection of multidimensional objects lies in its relation to the detection of the objects' components. How much more accurate is an observer who listens to, as well as looks at, an aircraft? A clinician who considers two categories of behavior rather than one? Providing a quantitative framework for such comparisons of tasks is one of detection theory's significant contributions, and the possibility of such comparisons is much greater in multidimensional domains. The only such case considered in unidimensional Part I was the relation between discrimination and identification at the end of chapter 5. In Part II, connections of this type will be far more salient.

Distributions in One- and Two-Dimensional Spaces

One-Dimensional Review

Figure 6.1 shows two familiar one-dimensional normal distributions, each with a standard deviation of 1; the upper distribution arises from S_1 (Noise) trials and the lower one from S_2 (Signal) trials. The mean of each distribution is labeled 0 so that locations on both axes correspond to z scores. The criterion in this example is one standard deviation below the mean of the S_2 distribution and one standard deviation above the mean of the S_1 distribution. The hit and false-alarm rates are given by the areas under the two curves to the right of the criterion.

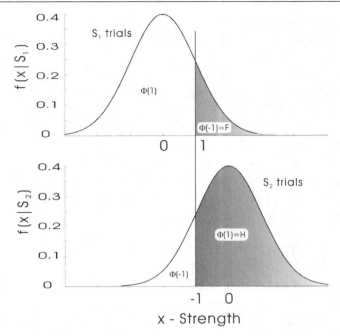

FIG. 6.1.   The one-dimensional equal-variance SDT model for $d' = 2$ and an un-biased criterion placement. Hit, miss (lower distribution), false-alarm, and correct rejection (upper distribution) probabilities can be written in terms of the normal distribution function Φ.

To calculate these proportions, we need information about the normal distribution function, and for this purpose Table A5.2 is the more convenient of our two tables of this curve (see chap. 2). For each positive z score, the table gives $\Phi(z)$, the area from the left tail of the distribution to the criterion.[1] The general rule is that areas in Table A5.2 are from one tail of the distribution to a z score on the opposite side of the mean. For z scores on the same side, the areas in the table must be subtracted from 1.

The four basic SDT probabilities are easily found from the table:

- *Correct rejections.* In the upper panel of Fig. 6.1, the area below the criterion is the probability of a correct rejection. The criterion $c = 1$, so this probability is $\Phi(c) = \Phi(1) = .84$.

---

[1] Remember that (uppercase) Φ is not the same as (lowercase) ϕ, the height of the normal curve, which is given in Table A5.1.

- *False alarms*. Still in the upper panel, the area *above* the criterion is the false-alarm rate. The total area under the curve is 1, so this probability is $1 - \Phi(1) = .16$.
- *Hits*. In the lower panel, the value of z at the criterion is negative, specifically -1. The table does not contain negative numbers, and the symmetry of the normal distribution must be used. The area to the right of a negative z score equals the area to the left of the corresponding positive z score, so the hit rate is $\Phi(1) = .84$.
- *Misses*. Still in the lower panel, the area *below* the criterion is the miss rate. The total area under the curve is 1, so this probability is $1 - \Phi(1) = .16$. Because this is an area below the criterion, it is also a value of Φ itself, namely, $\Phi(-1)$.

Two-Dimensional Distributions That Can Be Analyzed One-Dimensionally

Multidimensional distributions build on the familiar one-dimensional variety, but there are several steps in the generalization. Our goal in this chapter is to describe the compound detection problem, in which two compound stimuli are discriminated, but to simplify things we temporarily consider the unrealistic case in which only one (two-dimensional) distribution rather than the usual pair is possible. We still refer to the observer as making a decision, although a well-informed decision maker would simply produce the same response on every trial.

Figure 6.2 shows two ways to draw the joint distribution of two variables, produced when a light and a tone are presented simultaneously. One strategy is to add a third dimension to the graph: A two-dimensional graph (like Fig. 6.1) was needed to display one variable and its likelihood distribution, and a three-dimensional graph (Fig. 6.2a) can show two variables plus the likelihood of the combination. The overall distribution is a hill situated on a surface defined by loudness and brightness dimensions. A particular value of loudness and brightness is a point on the surface, and the likelihood of that value is the height of the hill over that point. The highest point, which represents the greatest likelihood, lies over the means of both variables, the point (μ_x, μ_y).

As decision problems grow in complexity, three-dimensional pictures of perceptual spaces quickly lose their charm. We instead use cross-sections to represent distributions, omitting the likelihood dimension. The circles in Fig. 6.2b connect (x, y) points of equal likelihood from Fig. 6.2a. They can be thought of as paths of constant height around the hill in Fig. 6.2a or pla-

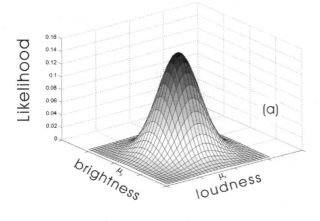

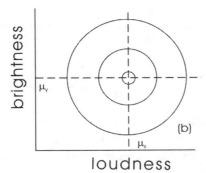

FIG. 6.2. Two representations of a two-dimensional distribution of the brightness and loudness of the light/tone pair: (a) The likelihood of each (x, y) point is a value in the third dimension, and (b) the likelihood dimension is suppressed, and each circle is the locus of points having the same likelihood.

teaus obtained by slicing off the top of the hill at a constant height. The center of the circles is still (μ_x, μ_y), the means on the two axes, and the diameters represent the standard deviation or a multiple of the standard deviation. Notice that, compared with Fig. 6.1, in which the psychological space is one-dimensional, the cross-section picture in Fig. 6.2b portrays a two-dimensional space, each dimension representing a psychological variable. Likelihood is not shown, but the equal-likelihood contours do convey useful information as we shall see.

Now what about the observer's criterion? In one-dimensional problems, this was just a point ($z = +1$ or -1 in Fig. 6.1), but with two internal dimensions we need a curve or line, called a *decision boundary*, that gives values of y for all possible values of x. The example in the lower half of Fig. 6.1—a criterion one standard deviation below the mean—is rendered as a two-dimensional plot in Fig. 6.3. The problem of finding the "yes" rate looks much more difficult in this representation: Instead of finding the area to the

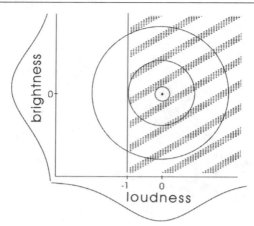

FIG. 6.3. A two-dimensional distribution (in the style of Fig. 6.2b) and a decision boundary. Points to the right of the boundary represent above-criterion values of loudness, and the value of brightness is ignored. The marginal distributions of brightness and loudness are shown along their respective axes.

right of $z = -1$, we are interested in the *volume* to the right of the line $z_x = -1$. Fortunately, there is a shortcut.

To understand the shortcut, called *projection*, we need to know a little more about the *joint distributions* (i.e., those that depend on more than one variable). If the distributions are normal, they can be described by five values: the means on both variables, the standard deviations on both variables, and the correlation between them. For calculations involving only one dimension, we can use the *marginal distribution* on that axis, that is, the distribution of x (ignoring y) or y (ignoring x). One way to think about marginal distributions is to imagine that the three-dimensional joint distribution is tipped on its side so that all the mass piles up on one axis. The height at each value of x in the marginal distribution of x corresponds to the summed heights of the joint distribution at every point for that value of x for *any* value of y. The marginal distributions, shown along the axes of Fig. 6.3, are also normal, and the mean and standard deviation of the marginal distribution on x are the same as the x-mean and x-standard deviation of the joint distribution. The joint distribution is said to be *projected* onto the x-axis.

Now we can calculate the "yes" rate for an observer with the representation shown in Fig. 6.3 for repeated presentations of a tone-light pair. The vertical criterion line means that the decision is based solely on the loudness of the tone—as if the judgment was made with the eyes closed. The probability of an observation to the right of the decision boundary is the volume to the right of that boundary in the joint distribution, but this is the same as the *area* to the right of the criterion in the marginal distribution. If $z = -1$, this area is $1 - \Phi(-1) = .84$—the same as in Fig. 6.1b. This is exactly what one should expect: The probability of detecting the tone is the same when only the tone is presented as when both a tone and light are presented, but the light is ignored.

In Fig. 6.3, the joint distribution is drawn as *circular*, with equal standard deviations on the two dimensions. For many stimuli (including the simultaneous tone and light presentation we have been discussing), there is no reason to expect equal standard deviations. When variability is unequal on the two dimensions, equal-likelihood contours are *elliptical* rather than circular, as in Fig. 6.4. Computations in which the joint distribution of x and y is projected onto either x or y are unchanged. For example, in Fig. 6.4, the standard deviation on x is 2 and on y is 1. The area to the right of a vertical line at $x = -1$ is $\Phi(0.5) = .69$, but above the line $y = -1$ it is $\Phi(1) = .84$. If the distribution were circular, these two numbers would be equal.

Two-Dimensional Decision Rules
That Can Be Analyzed One-Dimensionally

We have now succeeded in finding the "yes" rate in a two-dimensional perceptual space by projecting a joint distribution onto a single-dimensional one. This simplification always works when the decision boundary is a straight line, and the line need not be perpendicular to one of the axes.

Suppose the standard deviations are the same on the two axes, so that the likelihood contours are circles. An observer might reasonably decide to *add* the values of loudness and brightness and use the sum as the basis for a deci-

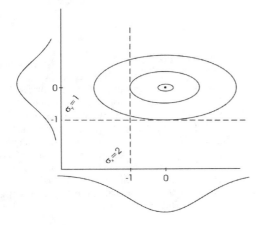

FIG. 6.4. A two-dimensional distribution in which the standard deviation of x is greater than the standard deviation of y. The outer ellipse represents points one standard deviation from the mean, and the inner ellipse represents points 0.5 standard deviations from the mean.

sion about whether the tone-light pair was presented. This observer's decision axis, shown in Fig. 6.5a, is a line at a 45-degree angle to both axes. Values increase as we move up and to the right along (or parallel to) this axis: The point $(-1,-1)$ has a sum of -2, $(0,0)$ has a sum of 0, $(1,1)$ has a sum of 2, and so on. The decision boundary, as in earlier examples, is perpendicular to the decision axis. In the figure, the boundary is set so that any sum of loudness and brightness greater than -2 leads to a "yes" response. Thus, an observation of $(0, -1)$ produces a "yes" and $(-2,-1)$ a "no."

For this boundary, what is the probability of a "yes" response? The projection strategy is appropriate, but the projection must be onto the *decision* axis (the sum of loudness and brightness), not the x-axis (loudness) or the y-axis (brightness). In the figure, the marginal distribution is drawn on an axis parallel to the decision axis; notice that all points on the decision boundary project onto the same point on the decision axis, as is necessary for projection to work.

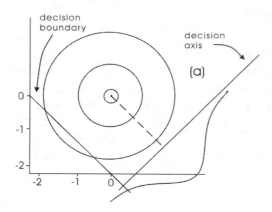

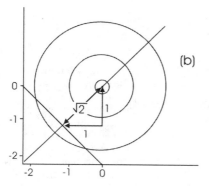

FIG. 6.5. (a) A two-dimensional distribution, a decision axis for increasing values of the sum of x and y, and a decision boundary that is the locus of points with a fixed sum of x and y. The marginal distribution of $x + y$ is shown parallel to the decision axis. (b) Demonstration that the distance from the decision boundary to the mean of the distribution is the Pythagorean sum of the distances along the x- and y-axes.

How far is the projected boundary—the criterion—from the mean? At the critical point $(-1,-1)$, the distance to the mean of $(0,0)$ is, by the Pythagorean Theorem, $\sqrt{2}$ or 1.41 units (see Fig. 6.5b). The area to the right of the boundary is therefore $\Phi(1.41) = .92$. The observer who uses both loudness and brightness in deciding whether the tone-light pair occurred has a higher hit rate (92% detections) than the one who is detecting only one or the other (84%) because the former has two useful pieces of information, the latter only one. (Keep in mind, however, that this number is just a hit rate, not an index of sensitivity; a true detection theory analysis is yet to come.)

Some Characteristics of Two-Dimensional Spaces

So far the analysis of two-dimensional perceptual spaces has been only a matter of properly reducing them to one-dimensional problems. When this is not possible, it is because the distributions, the decision rule, or both require the second dimension to be taken seriously. Before we return to the compound detection problem, a brief tour of two-dimensional-space geography is necessary.

Perceptual Independence and Dependence of Distributions

The essential simplicity of distributions like those in Fig. 6.2 through 6.5, and the possibility of analyzing either component dimension separately, arises from the lack of correlation between the dimensions. Normal bivariate distributions with zero correlation result from statistically independent variables and are said to be *perceptually independent* (Ashby & Townsend, 1986). The opposite case is called *perceptual dependence*; this condition arises in vision, for example, because increasing the brightness of a patch of light tends to increase its yellowness, and it arises in hearing because increasing the loudness of a pure tone slightly increases its pitch.

There are two equivalent ways to represent perceptual dependence. In Fig. 6.6a, the x and y axes are nonorthogonal; in Fig. 6.6b, the axes are orthogonal, but the distribution is elliptical. In this figure, the marginal as well as the joint distributions are displayed, and one way to see that the elliptical distribution is not perceptually independent is to compare it with the distribution shown in dashed lines. This distribution is the result of multiplying the marginals together, and it is circular.

Perceptual dependence always refers both to the shape of the bivariate distribution and its orientation in the perceptual space, or equivalently the

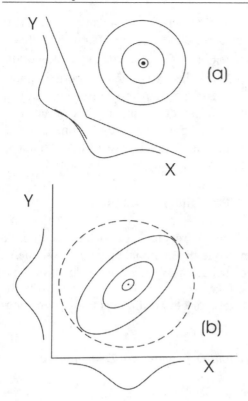

FIG. 6.6.   Two equivalent representations of perceptual dependence: (a) the x- and y-axes meet at a nonright angle, and the distribution is circular (correlation equals 0); (b) the axes are orthogonal, but the distribution is elliptical (correlation is not equal to 0). The dashed lines represent a perceptually independent distribution constructed from the two marginal distributions.

angle between the underlying axes. There is a simple quantitative relation between the two depictions: The correlation of the bivariate distribution in panel b equals minus the cosine of the angle between the axes in panel a.

Two-Dimensional Decision Boundaries: The Product Rule

Figure 6.7 illustrates another way to divide up the three-dimensional perceptual space, one that explicitly makes use of the two distinct dimensions by placing a separate criterion on each of them. The distribution still describes the internal effect of a tone-light pair, and it exhibits perceptual independence. The space (and the distribution) is divided by the criteria into four regions according to whether the observation is above or below each.

Suppose a cautious observer requires a combined observation to be above *both* of the criteria in order to respond "yes." We are interested in the volume that looms over the area shaded in Fig. 6.7a. The proportion of the distribution's volume to the right of the x criterion (when it is located one standard deviation below the mean, as in the figure) is $\Phi(1) = .84$, as we

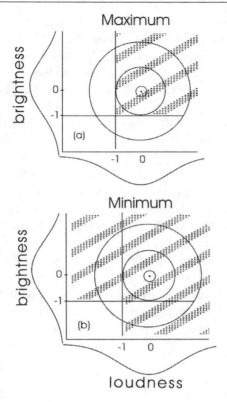

FIG. 6.7. The maximum and minimum rules. In the maximum rule (a), the observer responds "yes" only if both x and y exceed their respective criteria. In the minimum rule (b), the observer responds "yes" if either x or y exceeds its respective criterion.

found earlier (see Fig. 6.3), but we are interested in just some of that volume, the part that is higher than the y criterion. What fraction would that be? For the whole distribution, the proportion of volume above the line is .84, and a convenient consequence of perceptual independence is that this same proportion applies to any fraction of the distribution to the left or right of the x criterion. The fraction of the marginal distribution that is to the right of the criterion on x is .84, but not all of that area can be counted because of the criterion on y. Thus, the volume over the shaded area is $.84 \times .84$, or .71. We call this principle, naturally enough, the *product rule*: The volume under a distribution that is above a horizontal criterion z_y *and* to the right of a vertical one z_x equals the proportion above the horizontal criterion [$\Phi(-z_y)$, found from the y marginal distribution] times the proportion to the right of the vertical criterion [$\Phi(-z_x)$, found from the x marginal]. As an equation:

$$\begin{aligned} &\text{volume over an infinite "rectangle" above } z_y \\ &\text{and to the right of } z_x = \Phi(-z_y)\,\Phi(-z_x)\,. \end{aligned} \qquad (6.1)$$

   With this result in hand, we can easily find the volume beneath the joint distribution over the unshaded area in Fig. 6.7a. This represents the likelihood that the compound stimulus would lead to a "no" response—that is, the miss rate. This area is L-shaped rather than rectangular, so the product rule cannot be used directly, but the likelihood is the complement of the hit rate, $1 - .71 = .29$.

   Now consider an alternative decision rule. The shaded area in Fig. 6.7b corresponds to all observations that are *either* above the y criterion or to the right of the x criterion, and it reflects the "yes" rate of an incautious observer whose decision rule is to say "yes, the compound is present" if either tone or light yields a sufficiently large input, regardless of the value of the other. This time it is the miss rate that can be calculated directly from the product rule; it equals $(.16) \times (.16) = .026$. The hit rate, the volume over the shaded area, is the complement of this value, or .974.

   Even if the standard deviations for two dimensions differ, as in Fig. 6.4, the procedure for finding the volume over (infinite) rectangular area is the same. For example, the area in the upper right-hand quadrant of Fig. 6.4 is the volume above the y criterion, which is $\Phi(-z_y) = .84$, times the volume to the right of the x criterion, which is $\Phi(-z_x) = .69$, for a product of .58.

Compound Detection

We are at last ready to tackle the problem with which the chapter began, the detection of compound stimuli such as a simultaneous tone burst and light flash. An important part of any solution must concern the comparisons likely to interest the experimenter, in particular detection using the same components of the stimulus, but alone rather than in combination. For example, the research question might be how combined detection of the tone-light combination compares with detection of the tone or the light when either is presented separately. This focus on relative performance in more than one task with the same stimuli is a strength of multidimensional detection theory; it allows for theoretical "converging operations," relating performance in separate tasks (Garner, Hake, & Eriksen, 1956).

Equal-Variance Uncorrelated Representation
for Compound Detection

Half of the representation for compound detection—the distribution due to the compound stimulus—has been displayed in previous figures. The missing half, in detection, is the distribution due to no stimulus. In Fig. 6.8, two circular unit normal distributions arise—one with a mean of (0, 0) for the no-stimulus distribution, the other with a mean of (d_x', d_y') for the stimulus distribution. We develop equations for this general case, but also track the specific example in which d_x' and d_y' both equal 1.

Decision Rules

A characteristic of multidimensional tasks is that observers may plausibly adopt any of a number of response strategies, as spelled out in earlier sections of this chapter. In the one-interval design, variations in performance could be produced by changing the location of the criterion, and by some degree of in-attention, criterion fluctuation, and so on. A criterion shift, according to detection theory, represents a change in the likelihood of using one response rather than the other, and it does not affect sensitivity as measured by d'. Inattention and criterion fluctuation produce lower performance, arising because the observer acts nonoptimally. The alternate rules adopted in multidimensional tasks provide an additional level of complexity.

How can we compare different strategies for dealing with the same multidimensional decision problem? In analyzing the one-interval design, we stressed the bias-free measure d', but d' is a characteristic of the task or problem, not of the decision rule. To understand decision strategies for a representation in which d' is the same for the strategies to be compared, we are forced to depend on some other index. One possibility is $p(c)$, although we have seen that this depends on the criterion when d' is fixed. A natural

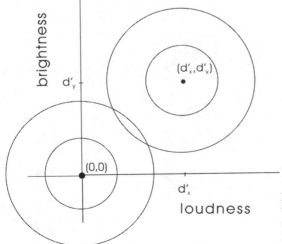

FIG. 6.8.   The compound detection problem: A compound stimulus must be discriminated from a null stimulus.

criterion-free measure is $p(c)_{max}$, the value of $p(c)$ when responding is unbiased, and we sometimes adopt this measure. For the one-dimensional yes-no task, best performance is with a criterion halfway between the distributions; we saw in chapter 2 (Eq. 2.10) that in that case,

$$p(c)_{max,\,yes-no} = \Phi(d'/2) .$$

(6.2)

Decisional Separability.   The first decision rule to be considered is the simplest and the most obviously inadequate: Attempting to detect a stimulus that has two components, the observer ignores one of them. We considered this strategy early in the chapter; it makes use of the marginal distribution of one component. As shown in Fig. 6.9, the decision boundary is a straight line parallel to one of the axes, a condition called *decisional separability.* For example, in the tone-light detection example, the observer considers only the amount of activation on the loudness dimension. The effective sensitivity to the combination is d'_x, so $p(c)$, assuming equal presentation probabilities, is the average of the hit and correct rejection rates, $0.5[\Phi(d'_x - k_x) + \Phi(k_x)]$. The maximum value is for a criterion halfway between the distributions, and $p(c)_{max} = \Phi(d'_x/2)$. If $d'_x = 1$, as in the running example, $p(c)_{max} = .69$.

Maximum, Minimum: Nonparametric Solution.   Two similar but opposite rules are shown in Fig. 6.10. The observer sets criteria on both axes and responds "yes" if the observation exceeds *both* criteria (the *maximum*

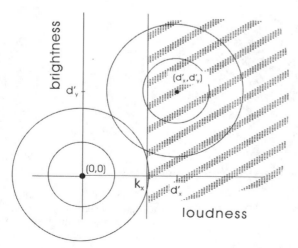

FIG. 6.9.   A decisionally separable rule for the compound detection problem.

rule) or *either* criterion (the *minimum* rule). These rules resemble somewhat the conservative and liberal settings of the criterion in the simple one-dimensional case, in that the maximum rule leads to a low "yes" rate and the minimum rule to a high one.

To relate performance in the compound task to that for detecting single components, a nonparametric approach is possible. For the maximum rule, the observer says "yes" to a compound signal whenever an above-criterion event occurs on both dimensions. Such events are exactly those that would generate "yes" responses in the single-component task, so the yes-rate for compound detection is the product of the yes-rates for single-feature detection. This argument applies to both the hit and false-alarm rates. We denote the hit rates for the x and y dimensions as H_x and H_y and the false-alarm rates as F_x and F_y. Then the hit and false-alarm rates for compound detection, H_2 and F_2, are

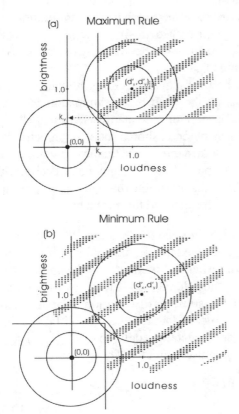

FIG. 6.10. (a) Maximum and (b) minimum rules for the compound detection problem.

$$H_2 = H_x H_y$$
$$F_2 = F_x F_y \ .$$

(6.3)

For the running example, if the criteria are $k_x = k_y = 0.5$, then $H_x = H_y = .69$ and $H_2 = .69^2 = .48$. Similarly, $F_x = F_y = .31$ and $F_2 = .31^2 = .10$. To find $p(c)$, we average the hit and correct rejection rates to get $0.5(.48 + .90) = .69$, perhaps surprisingly the same as using only one dimension.

In the minimum rule, the compound yes-rate is broken into two parts—one for an above-criterion observation on the x-axis, the other if there is no visual detection but an above-criterion event on the y-axis. Thus,

$$H_2 = H_x + (1 - H_x) H_y$$
$$F_2 = F_x + (1 - F_x) F_y \ .$$

(6.4)

For the example, $H_2 = .90$, $F_2 = .52$, and $p(c)$ is again .69. The maximum and minimum rules produce the same accuracy, but only when the criteria are halfway between the distribution means as they are here. The minimum rule is often called *probability summation* because any increase in two-dimensional accuracy over one dimension results only from having two chances to (probabilistically) detect the stimulus.

The weak part of the analysis is the assumption that criteria remain the same between conditions. (Applications of the probability summation idea are sometimes even weaker, in that no account of false alarms is taken at all; see Treisman, 1998.) To maximize $p(c)$, an observer should adopt a symmetric criterion, and in the example we have assumed this choice for the one-component tasks. The result, however, is strong response bias in compound detection, with an overall yes-rate $[(H + F)/2]$ of .29 for the maximum rule and .71 for the minimum rule. A detection theory approach can avoid this problem.

Maximum, Minimum: SDT Solution. To evaluate the maximum and minimum rules in SDT terms, we need to find the volume over an infinite rectangle, calculated as shown earlier in this chapter by the product rule (see Figs. 6.7a and 6.7b). The only additional step is to apply the product rule twice, to both the stimulus and no-stimulus distributions. For the maximum rule, the easiest proportions to calculate are the hit and false-alarm rates because they are volumes over the rectangle defined by the upper right quadrant, and the product rule can be used directly. These are:

$$H_2 = \Phi(d'_x - k_x) \, \Phi(d'_y - k_y)$$
$$F_2 = \Phi(-k_x) \, \Phi(-k_y) \ .$$

(6.5)

Notice that Equation 6.5 is just a special case of Equation 6.3.

For the minimum rule, the easiest values to calculate are the miss and correct rejection rates, which fall in the lower left quadrant. These rates, which we write as the complements of H_2 and F_2, are

$$1 - H_2 = \Phi(-d'_x + k_x)\, \Phi(-d'_y + k_y)$$

$$1 - F_2 = \Phi(k_x)\, \Phi(k_y) . \tag{6.6}$$

Rearrangement of Equation 6.6 reveals that it is a special case of Equation 6.4.

Notice that an observer using either rule can adjust the yes-rate by moving one of the decision boundaries, or both together. That is, the observer can generate an ROC. What is the shape of these curves? We consider the simplest cases, in which sensitivity is equal for the two dimensions ($d'_x = d'_y = d'$), and the decision boundaries are moved at the same time and in the same direction along the axes. For the maximum rule, the hit and false-alarm rates are

$$H_2 = [\Phi(d' - k)]^2$$

$$F_2 = [\Phi(-k)]^2 \tag{6.7}$$

and for the minimum rule they are

$$H_2 = 1 - [\Phi(-d' + k)]^2$$

$$F_2 = 1 - [\Phi(k)]^2 . \tag{6.8}$$

Figure 6.11 shows these curves together with the ROC for decisional separability (which is the same as that for single-component detection). The maximum and minimum rules are virtually indistinguishable, and both are slightly better than decisional separability.

This latter conclusion may seem to be in conflict with our analysis of the running example, in which the maximum rule ($H_2 = .48, F_2 = .10$), minimum rule ($H_2 = .90, F_2 = .52$), and decisionally separable rule ($H_2 = .69, F_2 = .31$) all produced $p(c) = .69$. Clearly, however, these three ROC points differ in bias. We found in chapter 1 that the same value of $p(c)$ represents greater true sensitivity for biased performance (as in our calculations for the maximum and minimum rules) than for unbiased performance (as in the decisionally separable example). The criterion $k = 0.5$ for decisional separability, but to obtain the highest value of $p(c)$ [$p(c)_{max} = .73$], we need $k = 0$ for the minimum rule and $k = 1$ for the maximum rule.

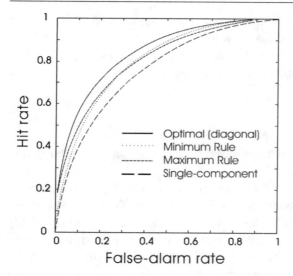

FIG. 6.11. ROCs for decisional separability, maximum, minimum, and optimal rules in the compound detection problem.

The Optimal Rule: Diagonal Decision Boundary. The optimal rule has been illustrated (considering only the S_2 distribution) in Fig. 6.5a: The decision axis runs between the means of the alternative distributions, and the criterion is a line perpendicular to the decision axis, dividing the space in two. Reasonably, both components contribute to the decision, and strong evidence for one component of the compound can compensate for weaker evidence for the other. The decision rule is to use the sum of the values from a stimulus on the two axes and to say "yes" if the sum is above a set value. A large sum corresponds to a strict decision boundary, a small number to a lax one.

A consequence of reducing the problem to one dimension is that it makes sense to ask: What is the effective d' value for this pair of distributions? The answer comes immediately from the Pythagorean Theorem:

$$d'_{compound} = (d'^2_x + d'^2_y)^{1/2} . \tag{6.9}$$

In the running example, if the one-modality d' values are each 1.0, d' for the compound is 1.41. Accuracy as measured by $p(c)_{max}$ is .76, as compared with .69 on each dimension alone.

An alternative way to reach this conclusion is to calculate H_2 and F_2 from the one-dimensional projections. The value of d' is $\sqrt{2}$ greater than for the individual components, so:

$$H_2 = \Phi(\sqrt{2}d' - k) \tag{6.10}$$

$$F_2 = \Phi(k) \ . \tag{6.11}$$

Taking the unbiased case in which $k = \sqrt{2}(d'/2)$, we find again that $H_2 = 1 - F_2 = .76$. The ROC for the diagonal rule can be calculated from these equations and is shown in Fig. 6.11. Clearly this rule yields the best performance of any decision strategy we have considered.

The optimal rule allows for integration of information before, rather than after making decisions about the individual components, and it is this better use of the available input that produces the improvement in accuracy. Distinguishing the optimal rule from the maximum or minimum rule is difficult when only one data point is available—performance levels are not *that* different—but when multiple comparisons are available the predictions are more discrepant. Treisman (1998) pointed out that in a method-of-constant-stimuli context, the rules predict psychometric functions with both different slopes and different intercepts.

Inferring the Representation From Data

If the representation and decision rule in a compound detection task are known, the tools of the previous section allow us to calculate hit and false-alarm rates, and thus ROC curves. Being able to deduce predictions from models in this way is valuable, but what about the opposite, inductive problem of deriving representations and decision rules from the data? This latter sequence dominated Part I, and it is important to extend it to multidimensional situations.

An immediate hurdle is that more parameters are needed to describe two-dimensional representations than one-dimensional ones. For detection of a single feature, the hit and false-alarm rates each depend on a sensitivity value and a criterion, which can therefore be uniquely determined. In two dimensions, these same two observables depend on two values of sensitivity and one or two values of criterion, even with assumptions about independence, the form of the decision bound, and the nature of the decision rule. There are two ways to attack this problem.

First, simplifying assumptions can make any problem more tractable. Consider the nonparametric approach to predicting H_2 and F_2 in compound detection from the same statistics in single-component detection. There are no sensitivity or bias parameters—the issue is whether accuracy in the single-component conditions can be predicted from accuracy in the compound case. Equations 6.3 (maximum rule) and 6.4 (minimum rule) express compound H_2 and F_2 (two values) in terms of single-component H_x, H_y, F_x, and

F_y (four values). Sometimes it is natural to assume that both components are equally detectable, so that $H_x = H_y = H_1$ and $F_x = F_y = F_1$.

Consider, for example, a "compound" recognition memory experiment. Following a study trial in which a list of words is presented, observers are shown pairs of words with one of two instructions: say "yes" only if *both* words are Old (maximum rule) or if *either* word is Old (minimum rule). Both components of the stimulus pair are equally detectable, and we may hope that the instructions determine the decision rule. In that case, the hit and false-alarm rates for the one-component tasks, H_1 and F_1, can be related to the values for compound performance, H_2 and F_2, by solving Equations 6.3 and 6.4. For the "both" condition,

$$H_1 = (H_2)^{\frac{1}{2}}$$
$$F_1 = (F_2)^{\frac{1}{2}} , \qquad (6.12)$$

and for the "either" condition,

$$(1 - H_1) = (1 - H_2)^{\frac{1}{2}}$$
$$(1 - F_1) = (1 - F_2)^{\frac{1}{2}} . \qquad (6.13)$$

The shortcoming of the nonparametric model, noted earlier, is that it assumes an unrealistic kind of bias constancy. The SDT model overcomes this problem by postulating an underlying representation on which all tasks draw. The assumption needed—plausible in the compound memory experiment—is that both sensitivity and bias are the same for the two words in each stimulus. Then Equations 6.7 and 6.8 can be solved for d'; for the "both" condition,

$$d' = z(H_2^{\frac{1}{2}}) - z(F_2^{\frac{1}{2}}) , \qquad (6.14)$$

and for the "either" condition,

$$d' = z[(1 - F_2)^{\frac{1}{2}}] - z[(1 - H_2)^{\frac{1}{2}}] . \qquad (6.15)$$

To be concrete, suppose we observe $H_2 = .8$ and $F_2 = .2$ in each of the two compound conditions. Because $H_2 = 1 - F_2$, both the nonparametric and SDT models make the same prediction for the two conditions. Nonparametrically, $H_1 = \sqrt{8} = .89$ and $F_1 = \sqrt{2} = .45$. The SDT model uses these same values to find d', which equals $z(.89) - z(.45) = 1.35$. Values of $H_1 = .89$ and $F_1 = .45$ satisfy this prediction, of course, but so do $H_1 = .55$ and

$F_1 = .11$ and many other (F_2, H_2) values on the ROC defined by Equations 6.14 and 6.15.

The diagonal rule, which might be used in the "both" or "either" recognition task, makes a different prediction. Solving Equations 6.8 and 6.9 yields

$$d' = (1/\sqrt{2})[z(H_2) - z(F_2)] . \tag{6.16}$$

In the example, $d' = (1/\sqrt{2})[(0.842 - (-0.842)] = 1.19$. Notice that the inferred value of d' is smaller in this case: Because the decision rule is optimal, a lower level of sensitivity is required to reach the same level of performance.

Summary

Some stimulus sets are best represented using a perceptual space with more than one dimension. In the two-dimensional case, each stimulus leads to a distribution taking on values of likelihood for pairs of observations. In an experimental context, the observer establishes a decision boundary that divides the space into regions corresponding to each possible response.

In compound detection, the task is to detect the presence of a compound stimulus, saying "yes" to presentation of a simultaneous tone-light pair and "no" to noise alone on both channels. A variety of decision rules can be applied to this problem: (a) decisional separability, in which only one dimension is considered; (b) the maximum rule, in which an above-criterion event is required on both channels; (c) the minimum rule (probability summation), in which either a visual or an auditory detection is required; and (d) an integration rule, in which the effective decision variable is the sum of the two channels. The latter rule is optimal.

Problems

6.1. On a decision axis x, a value of 2 is observed. What is the area below this point if (a) the mean of the distribution of observations is 0 and the standard deviation is 1, (b) the mean is 2 and the standard deviation is 1, (c) the mean is 3 and the standard deviation is 1, and (d) the mean is 3 and the standard deviation is 2?

6.2. (a) For the representation in Fig. 6.3, suppose the decision boundary on the x-axis is -1.41 standard deviations. How does this affect the volume to the right of it?

(b) Same question, except that the standard deviation on the y-axis is 1.41. How does this effect the volume to the right of the criterion line?

(c) Now suppose the criterion line is *horizontal* at $z = -2$. Describe this decision rule in words. What proportion of the volume of the distribution is above this line?

6.3.   (a) Suppose an observer in the situation of Fig. 6.5a uses a decision boundary that is parallel to the one shown, but goes through the point (0,0). What is the implied decision axis? What percent of the distribution is to the right of the boundary?

(b) Suppose the observer decides that loudness is more important than brightness in making a detection decision. How should this affect the decision axis and criterion line? (*Hint*: An observer who ignores one dimension completely uses an extreme version of this strategy.)

6.4.   Assume that the underlying distributions are normal.

(a) In the tone-light example, what is the probability that a particular stimulus will be louder than average? brighter than average? both? neither?

(b) Do the answers to (a) depend on (i) the x and y standard deviations, (ii) the distributions being normal, (iii) the distributions being perceptually independent?

(c) In the tone-light example, what is the probability that a particular stimulus will be at least 1.5 standard deviations louder than average? at least 1.5 standard deviations brighter? both? neither?

(d) Do the answers to (c) depend on (i) the x and y standard deviations, (ii) the distributions being normal, (iii) the distributions being perceptually independent?

6.5.   In the tone-light example, suppose that the distribution for the Null stimulus (S_1) and means of the distribution for the Signal (S_2) are as shown in Figs. 6.8 to 6.10, but the standard deviation of S_2 equals 1 for the loudness dimension and 2 for the brightness dimension. What is $p(c)$ for the maximum rule if the criteria are (a) $k_x = 1, k_y = 1$, (b) $k_x = 1, k_y = 0.5$, (c) $k_x = 0.5, k_y = 1$?

6.6.   Generalize the tone-light example to three stimuli: the Null stimulus (S_1), a Weak Signal (S_2), and a Strong Signal (S_3). Suppose all standard deviations equal 1, and the means are located at (0,0), (1,1), and (2,2). (a) Does the additivity condition of Equation 5.1 hold for an observer using the optimal rule? (b) Does additivity hold for a minimum-rule observer? (Set the criteria to $k_x = 1, k_y = 1$. For each pair of stimuli, find H and F, then calculate d' from Eq. 1.5 and $p(c)_{max}$ from Eq. 1.7.)

6.7.   In Fig. 6.10a, the maximum rule for compound detection is illustrated, with $d' = 1$ for both dimensions. Plot ROCs for three ways of changing the criterion:

(a) Move the vertical segment of the decision boundary to the left or right, but leave the horizontal one where it is.

(b) Move the horizontal segment of the decision boundary up or down, but leave the vertical one where it is.

(c) Move both segments together.

In each case, you can choose any locations you want, perhaps 1 z unit apart. Plot the ROCs on z coordinates as well as linear coordinates.

6.8.   In Fig. 6.10a, what do the "isobias" curves for the maximum rule look like? To find out, change the location of the tone-light distribution along the major diagonal decision axis [i.e., put it at (1,1), (2,2), etc.], but leave the noise-noise distribution where it is. Two cases:

(a) Keep the "corner" of the horizontal and vertical segments of the decision boundary halfway between the two distributions.

(b) Keep the "corner" fixed at (0,0), the mean of the noise-noise distribution.

6.9.   The decisional separability, maximum, minimum, and diagonal decision rules for compound detection make different predictions about how performance in that task should be related to simple (tone-alone or light-alone) detection.

(a) Design an experiment that would allow you to tell which rule a naive observer is using.

(b) What instructions or payoffs could you give participants to encourage them to adopt these strategies?

7

Comparison (Two-Distribution) Designs for Discrimination

So far we have discussed only situations in which one stimulus at a time is evaluated. In psychophysics, even these designs are sometimes termed *discrimination* because they permit estimates of the ability to distinguish two stimuli or stimulus classes, and in common psychological usage discrimination means telling two things apart. In this chapter, we introduce paradigms in which the process of discrimination is more salient because two or more stimuli are explicitly compared on each trial.

There are two types of such paradigms, which we term comparison designs and classification designs. *Comparison designs* (considered in this chapter) resemble one-interval designs in that the observer makes a binary decision based on an underlying representation containing only two distributions. These paradigms require the observer to make a direct comparison between two stimulus presentations. In *classification designs* (chap. 9), the observer again compares (two or more) stimuli, but the world of possible stimuli and their representation contains more than two distributions. Because multiple distributions are mapped onto each response, the task facing the observer is more complex than a simple comparison.

There are only two comparison designs: two-alternative forced-choice and the reminder design. It is possible to analyze both with only the one-dimensional, flatland tools of Part I. The advantage of considering two-dimensional representations is that each design offers multiple alternative decision rules, and the perspective of a multidimensional spaceland view renders the relations among these rules more visible. A representation containing a pair of bivariate distributions in a two-dimensional perceptual space was illuminating in analyzing the compound detection problem in chapter 6, and we construct similar decision spaces to describe two-alternative forced-choice and the reminder design. The decision boundaries turn out to be straight lines in the space, allowing us to use the projection strategy to find a one-dimensional decision axis that is the basis for the observer's decision.

Two-Alternative Forced Choice (2AFC)

In using the traditional name for this design, we continue an unfortunate historical precedent. The choices made by observers in two-alternative forced-choice (2AFC) studies are no more constrained than in other correspondence experiments. As in the one-interval design, the possible stimuli come from two categories (Old or New, Loud or Soft), and the experimenter is interested in the correspondence between the correct response and the observer's "forced choice."

The new feature of the 2AFC design is that *both* alternatives are presented on every trial, in a random spatial or temporal order. The observer reports not which stimulus occurred—both did—but in which order. A better name for this paradigm might be "order discrimination," but other designs also manipulate the order of stimuli.

There are forced-choice designs that, while also using two stimulus classes, have more than two intervals in which the stimuli may occur and a response for each. Three-, four-, and larger-number alternative variants are identification rather than comparison designs, and these are described in chapter 10.

Example 7: Recognition Memory for Words

Glanzer and Bowles (1976) presented English words to participants, first for study and later for recognition. In the recognition test, two words were presented on each trial, one above the other. One of the two (an Old word) had been presented during the study phase of the experiment, and the other (the New word) was a "lure." Whether the Old or New word was on top varied randomly from trial to trial. The participants indicated which of the two words was more likely to have been the Old one, the one on the top or the bottom one. The words to be remembered and the lures with which they were paired were selected to be either high or low in frequency of occurrence in English (see Glanzer & Adams [1985] for a summary of this and other experiments on the effects of frequency of occurrence in recognition memory).

The experimenters elected to use the apparently more cumbersome 2AFC procedure to avoid possible contamination of yes-no results by effects of the decision criterion. If participants make judgments in a memory experiment on the basis of familiarity, as suggested in the discussion of previous recognition memory examples, we might expect them to be more willing to call frequent or familiar words "old" even if they had not appeared in the study phase of the experiment. The yes-no paradigm, in presenting one word at a time and asking whether it has been seen earlier, confounds familiarity in general with familiarity due to recent exposure. By looking separately at the four possible combinations of test item frequency (high and low in English) and lure item frequency (also high and low), the experimenters hoped to disentangle the effects of recent and remote history on judgments of familiarity.

Experimental Design and Sensitivity Estimates

Data for one condition of such an experiment can be represented in a 2×2 stimulus–response table:

	Responses		
Stimulus Sequences	*"Old on Top"*	*"Old on Bottom"*	*N*
<Old, New>	16	9	25
<New, Old>	7	18	25

The rows of the table correspond to stimulus sequences, denoted by angle-bracketed lists, rather than to individual stimuli. In this example, the stimuli are listed spatially from top to bottom in the stimulus presentation so that, for instance, <Old, New> represents the presentation of an Old word on top and a New word below. A temporal sequence rather than a spatial one is used in auditory (and many visual) 2AFC experiments, but the analysis is the same.

The two possible responses are "old on top" and "old on bottom." The designations "hit" and "false alarm" in this case are arbitrary; we define them as:

$$H = P(\text{"old on top"} | \text{<Old, New>})$$

$$F = P(\text{"old on top"} | \text{<New, Old>}) \ . \tag{7.1}$$

Two questions we can ask of these data are parallel to those posed for the one-interval design: How sensitive are the observers? How biased are they? A third question concerns the relation between 2AFC and yes-no. As many readers will surmise, 2AFC is the easier task. Models for 2AFC must build on those for yes-no and give an account of this discrepancy in performance.

The first and third questions can be answered together: To compute sensitivity for these data, we first subtract the transformed hit and false-alarm rates, as we did for one-interval data in chapter 1. To take account of the difference in difficulty between 2AFC and yes-no, this difference must be adjusted downward by a factor of $\sqrt{2}$ as follows:

$$d' = \frac{1}{\sqrt{2}}[z(H) - z(F)] \ . \tag{7.2}$$

For the example at hand, $H = .64$ and $F = .28$, the transformed difference $z(H) - z(F) = 0.358 - (-0.583) = 0.941$ (from Table A5.1), and $d' = 0.665$.

Choice Theory leads to exactly the same prediction about the 2AFC/yes-no relation. From the hit and false-alarm rates in 2AFC, $\ln(\alpha)$ can be found from

$$\ln(\alpha) = \frac{1}{2\sqrt{2}}\left[\ln\left(\frac{H}{1-H}\right) - \ln\left(\frac{F}{1-F}\right)\right] \ , \tag{7.3}$$

which is a factor of $\sqrt{2}$ less than if the data had arisen from a yes-no experiment.

Representation and Analysis

To understand why forced choice should be easier for the observer than yes-no (and why the discrepancy should be $\sqrt{2}$), we must derive the characteristics of the decision space underlying a forced-choice task. Each trial involves two stimuli, the top word and the bottom one. We assume that the observer estimates the familiarity of each word independently, which means that we can treat each spatial location as a separate dimension in the decision space. Geometrically, independence is interpreted as orthogonality, so the two stimulus locations in Fig. 7.1 are drawn at right angles. The internal effect of a single experimental trial is a point in the two-dimensional space: The top word has a familiarity value on the vertical axis, the bottom word on the horizontal axis. The underlying distributions are surfaces above a plane, but they are indicated in the figure as circles of equal likelihood, as in chap. 6.

The mean of the New distribution on both axes is, arbitrarily, chosen to equal zero. A 2AFC trial involves one observation that is most likely to be near 0 and another near the original d', so points will center around coordi-

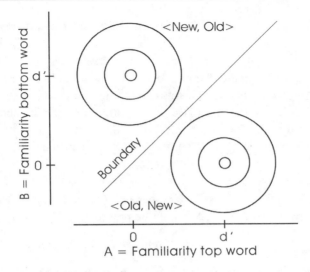

FIG. 7.1. A two-dimensional interpretation of the 2AFC task. The decision axes are the observation strengths for the two intervals. Each distribution is represented by a set of concentric circles defining contours of equal likelihood for one possible stimulus sequence. The decision boundary is perpendicular to the line connecting the two means. The observer responds "old on top" in the region to the right of the boundary

nates $(0, d')$ on <New, Old> trials and around $(d', 0)$ when the stimulus sequence is <Old, New>. The observer's task is to decide whether a two-interval observation is drawn from one or the other distribution on the decision plane, that is, to decide which axis is nearer to the observation. Application of Pythagoras' theorem to the triangle in Fig. 7.1—formed by the two means and the origin—shows that the means of the two probability density distributions are separated by a distance $\sqrt{2}$ as great as d', confirming Equation 7.2. The logic is the same as for compound detection in chapter 6, but notice that the distances along the two axes between the means are now guaranteed to be the same because each represents the psychological distance between Old and New stimulus classes.

Also as in chapter 6, it is possible to reduce the decision space to one dimension by projecting the bivariate distributions onto a decision axis. The appropriate axis for the optimal rule runs through the means of the two distributions, the decision boundary perpendicular to this orientation. In the resulting one-dimensional picture, shown in Fig. 7.2, the means are separated by $\sqrt{2}d'$, and this is now the value that is estimated by calculating

$z(H) - z(F)$. To find d', therefore, we must divide this result by $\sqrt{2}$, as in Equation 7.2.

One nonoptimal rule is worth mentioning: decisional separability. Suppose, for example, that the decision bound is a vertical line in Fig. 7.1. The strategy implied by this bound is to use only the top word in making a decision. The effective sensitivity in this case is simply d', the distance between the marginal distributions on the *x*-axis. The observer has converted the 2AFC task to yes-no by ignoring one of the two pieces of useful information on each trial.

For measuring response bias, the methods of chapter 2 are entirely adequate. No $\sqrt{2}$ adjustment is necessary, because (as the reader may not be surprised to learn) bias in one task cannot be predicted from bias in the other.

Sensitivity Measures Based on Proportion Correct

Proportion correct is a poor measure of sensitivity for the yes-no experiment because as bias varies the ROC implied by $p(c)$ has a threshold shape (chaps. 1 and 4). This flaw is not remedied by moving to a new design, but its impact is greatly reduced by an empirical discovery about 2AFC data: Observers tend not to display extreme biases. If we could be sure that all responding is unbiased, so that ROC points lie on the minor diagonal, then $p(c)$ would be as good a measure as any for simply saying which of two conditions yields the higher accuracy.

The 2AFC Area Theorem.   In fact, Green and Swets (1966, pp. 45ff) showed that the proportion correct in 2AFC by an unbiased observer

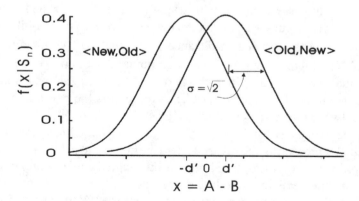

FIG. 7.2.   A one-dimensional interpretation of the 2AFC task. The decision axis is the difference between observations in the two intervals *A* and *B*.

equals the area under the yes-no isosensitivity curve. This important result, which we call the *2AFC area theorem*, is sometimes cited as justification for using (a) proportion correct, or (b) area under the yes-no ROC as a "non-parametric" measure of discrimination performance. These complementary arguments require some attention.

First, the area theorem says nothing about proportion-correct scores that *do* exhibit bias. Although extreme response strategies are rare, the forced-choice design does not guarantee the complete absence of bias. To measure sensitivity, bias must be eliminated; this can be done either by calculating a statistic such as d' or by correcting $p(c)$ for bias, as discussed later.

Second, although the theorem does indeed offer justification for the use of area under the yes-no ROC, it does not abet all area-like measures equally. We saw in chapter 4 that A', Pollack and Norman's (1964) statistic, has threshold properties at high levels that make it undesirable. The area we want is the *whole* area under the ROC, not simply that marked off by a point on it. When more points are estimated, as in rating experiments, the area A_g found by "connecting the dots" is closer to the desired value. Alternatively, measured points can be used to estimate a continuous curve consistent with a pair of distributions of known shape and the area under that continuous curve calculated. An SDT statistic of this type is A_z, which was introduced in chapter 3.

Maximum Proportion Correct [$p(c)_{max}$]. Proportion correct is a desirable measure, we have seen, to the extent that responding is unbiased. Why not "adjust" observed values of $p(c)$ to find the value that would have been obtained by a truly unbiased participant? Would this not solve all our problems without recourse to detection theory?

Such an adjustment is possible, but is not innocent of theory. We begin with the equal-variance SDT model. For a constant value of d', the largest value of $p(c)$ arises at the minor diagonal, where responding is unbiased. The proportion correct at this point, $p(c)_{max}$, can be found from

$$p(c)_{max} = \Phi\{[z(H) - z(F)] / 2\} . \qquad (7.4)$$

The difference between the transformed hit and false-alarm rates does not depend on bias, according to the equal-variance assumption; so $p(c)_{max}$ is a true sensitivity statistic.

Equation 7.4 can be applied to either yes-no or 2AFC data. In the yes-no case, it reduces to

$$p(c)_{max, \text{yes-no}} = \Phi(d'/2) , \qquad (7.5)$$

which was presented earlier (Eqs. 2.10 and 6.2) as an expression for unbiased $p(c)$. In 2AFC, combining Equations 7.2 and 7.4 leads to

$$p(c)_{\text{max, 2AFC}} = \Phi(d'/\sqrt{2}) \, . \tag{7.6}$$

For the equal-variance case, $p(c)_{\text{max, 2AFC}}$ is identical with A_z, the area under the yes-no isosensitivity curve (see Eq. 3.8). This conclusion follows from the area theorem, which says that $p(c)_{\text{max, 2AFC}}$ equals the yes-no area. We shall see presently that Equation 7.6 can be easily generalized to the unequal-variance case.

Equation 7.6 can be solved for d', with the result:

$$d' = \sqrt{2} \, z[p(c)_{\text{max, 2AFC}}] \, . \tag{7.7}$$

The resulting values of d' are listed in Table A5.3. Given the simplicity of Equation 7.7, the table is not really necessary for 2AFC, but it *is* necessary for multiple-interval forced-choice designs, which are also included (and which we discuss in chap. 10). It is important to recognize that the table gives correct results only for unbiased performance. For 2AFC data, the table is most useful in reanalyzing published data that may not include separate hit and false-alarm rates.

The adjustment provided by $p(c)_{\text{max}}$ is consequential with moderately biased data. Suppose that in a yes-no experiment $H = .4$ and $F = .04$, indicating a strong "no" bias. Observed $p(c) = .68$, but $d' = 1.498$ and $p(c)_{\text{max, yes-no}} = .77$, which is a noticeably different value from .68.

Proportion Correct Corrected for Guessing.   Proportion correct can also be adjusted according to the assumptions of threshold theory. In chapter 4, we presented a simple (but ineffective) method for correcting the hit rate for guessing; the corrected hit rate q was found to be

$$q = \frac{H - F}{1 - F} \, . \tag{7.8}$$

One could apply exactly the same formula to 2AFC, but a modification is usually made. Because observers often display little bias, H is set equal to proportion correct $[p(c)]$ and F to the proportion of trials on which one could be successful by guessing, namely, ½. In 2AFC, then

$$q_{\text{2AFC}} = \frac{p(c)_{\text{2AFC}} - \frac{1}{2}}{1 - \frac{1}{2}} = 2 \, p(c)_{\text{2AFC}} - 1 \, . \tag{7.9}$$

The hit rate adjusted in this way is the level that could be obtained if F were 0—the leftmost point on a high-threshold ROC. Of all the points on that ROC, this is the one with the highest value of $p(c)$—a point of resemblance between q_{2AFC} and $p(c)_{max}$.

Proportion Measures Versus Distance Measures.   Equations 7.5 and 7.6 describe a monotonic relation between two sensitivity measures, a proportion measure $[p(c)_{max}]$ and a distance measure (d'). The two are equivalent in the sense that they imply the same ROC. Is either type of measure preferable on other grounds?

The greatest advantage of proportion measures is their familiarity–many people who do not know whether a d' value of 4.0 represents good or poor performance appreciate immediately that a proportion correct of .98 is excellent. A second advantage concerns perfect accuracy, which corresponds to a finite value of $p(c)$, but an infinite value of d' and α. The shadow side of this finiteness is the disguising of ceiling effects; occasional proportions of 1.0 can be adjusted downward to obtain finite values of d', as discussed in chapter 1.

Distance measures have three important advantages. First, they can be compared across different experiments. For example, an observer who obtains $d' = 1$ in both yes-no and 2AFC has displayed the same level of performance, whereas someone for whom $p(c) = .75$ in both tasks has not. Second, they can be compared across different values of physical variables. Which is better, $p(c) = .9$ for stimuli two units apart or $p(c) = .74$ for stimuli differing by one unit? Conversion to d' shows that the two results are the same in d' per unit.

Finally, distance measures can be added and subtracted. In chapter 5, we took advantage of this fact in several ways. To recall one, the sum of d' values across a one-dimensional set of stimuli can be used as a measure of total sensitivity to the set. No summary statistic based on proportions is as useful.

Distributions With Unequal Variance

Form of the 2AFC Isosensitivity Curve.   Although it is not common, the forced-choice rating design (first explored by Schulman & Mitchell, 1966) deserves attention. In this experiment, the participant chooses a response ranging from high confidence that $<S_1 S_2>$ was presented to high confidence in the occurrence of $<S_2 S_1>$. As in chapter 3, the data are plotted as ROCs.

The interesting aspect of this experiment is the shape of the ROC. Figure 7.3a shows the standard model for the yes-no task, assuming that S_1 and S_2

lead to distributions with unequal variances s^2 and 1. Figure 7.3b is the one-dimensional representation for 2AFC under the same assumption. The decision variable in 2AFC is the difference in strength between Intervals 1 and 2, which we denote $A - B$ (see Fig. 7.2). The variance of this difference is the sum of the variances, $1 + s^2$, for either possible stimulus order. Thus, the representation for 2AFC is equal-variance even if that for yes-no is not, and the ROC in 2AFC should be a straight line with unit slope on z coordinates in all cases. The unit slope reflects an important theoretical advantage of forced choice over yes-no: No matter what the criterion is, apparent sensitivity—the difference between the transformed hit and false-alarm rates—is the same. In a one-interval experiment, this is true only if the underlying distributions have the same variance. It is ironic that an advantage of 2AFC should be the robustness of its sensitivity measure in the face of extreme biases that do not normally arise.

Implications for One-Interval ROC Analysis.   The relation between the yes-no and 2AFC isosensitivity curves provides a theoretical rationale for the use of d_a in the one-interval task, as proposed in chapter 3

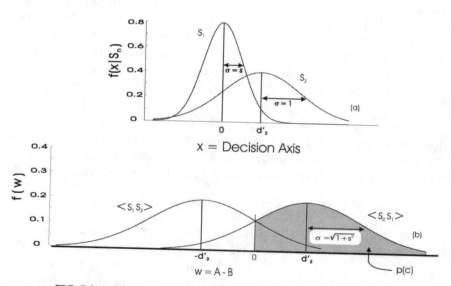

FIG. 7.3.   (a) Decision space for yes-no when the variances of S_1 and S_2 are unequal. (b) Decision space (in the style of Fig. 7.2) for 2AFC, according to SDT, when the variances of S_1 and S_2 are unequal and the observer uses an unbiased cut-point decision rule. The area under the $<S_2S_1>$ distribution to the right of the criterion (and the area under the $<S_1S_2>$ distribution to its left) equals $p(c)$, which by the area theorem equals the area under the unequal-variance yes-no ROC.

(Schulman & Mitchell, 1966). The two distributions in Fig. 7.3b each have variance $(1 + s^2)$ and differ in mean by $2\,d'_2$. The mean difference divided by the common standard deviation can be estimated by subtracting the z-transformed hit and false-alarm rates:

$$[z(H) - z(F)]_{2\text{AFC}} = \frac{2d'_2}{\left(1 + s^2\right)^{\frac{1}{2}}} \; . \tag{7.10}$$

The right side equals $\sqrt{2}\,d_a$ (see Eq. 3.4), so

$$d_a = [z(H) - z(F)]_{2\text{AFC}}/\sqrt{2} \; . \tag{7.11}$$

For the case in which $s = 1$, we recommended earlier that d' be estimated from 2AFC by dividing $z(H) - z(F)$ by $\sqrt{2}$ (Eq. 7.2). It now appears that when the unit-slope assumption is unwarranted, this method is still desirable and yields an estimate of d_a.

Finally, Fig. 7.3b can be used to illustrate the area theorem for this normal unequal-variance case. Maximum proportion correct is the same for either stimulus sequence and equals

$$p(c)_{\text{max, 2AFC}} = P(A - B > 0 \mid <S_2 S_1>)$$

$$= \Phi\left[\frac{d'_2}{\left(1 + s^2\right)^{\frac{1}{2}}}\right] = \Phi\left(\frac{d_a}{\sqrt{2}}\right) . \tag{7.12}$$

This expression equals A_z, the area under the yes-no ROC, in the SDT case (Eq. 3.8), confirming the area theorem. The equivalence of a distance measure d_a to the area under the yes-no ROC is a strong argument for preferring it to other possible distance measures of sensitivity in the one-interval experiment (Simpson & Fitter, 1973).

Some Empirical Findings and Their Implications for Theory

Although 2AFC appears to be a simple extension of the one-interval design, a number of experimental results using this paradigm have been fodder for perceptual theory. In particular, the data force us to think seriously about the limitations imposed on discrimination by imperfect memory.

Empirical Comparisons Between 2AFC and Yes-No.   Signal detection theory and Choice Theory agree exactly on the relation between

2AFC and yes-no data. It seems almost impolite to ask whether the data respect this unanimity.

Much of the early work that introduced SDT established that different tasks yielded constant estimates of d'. The results of most early experiments using simple auditory and visual *detection* tasks (see Green & Swets, 1966, ch. 4; and Luce, 1963a, for summaries) supported detection theory in this respect. Extending the theory to *discrimination* tasks uncovered a systematic failure. Jesteadt and Bilger (1974) found that 2AFC performance was a factor of 2, rather than $\sqrt{2}$, better than yes-no, both in their own frequency-discrimination experiments and others they surveyed. Creelman and Macmillan (1979) found the same result for discrimination of both auditory frequency and monaural phase.

What accounts for the confirmation of the predicted yes-no/2AFC relation originally found by SDT advocates in detection experiments? Wickelgren (1968) enumerated the many processing assumptions underlying the $\sqrt{2}$ prediction and concluded:

> When one considers all the ways in which the $\sqrt{2}d'$ prediction might fail for reasons having nothing to do with the essential validity of strength theory [detection theory] for both absolute [yes-no] and comparative [2AFC] judgments, it is truly amazing that it has not failed thus far. However, the present analysis makes it clear that, if the $\sqrt{2}d'$ prediction fails in some future application of strength theory, one cannot reject strength theory without a detailed study of the reason for the failure. (p. 117)

Subsequent data, as we have seen, justified Wickelgren's suspicions. Two variables in particular—time between intervals and stimulus range—are known to affect performance in 2AFC, and thus its relation to yes-no. There has been some progress in interpreting the effects of these variables theoretically without, as Wickelgren also foresaw, abandoning the basic detection theory approach.

Effects of Interstimulus Interval.   In temporal 2AFC, the two stimuli are separated by time rather than space. How much time should elapse between the two stimuli? Our analysis has assumed that the particular order of the stimuli, and the time between them, makes no difference, but it turns out that the interstimulus interval (ISI) does affect both sensitivity and response bias.

The response-bias findings are classic. When the two stimuli on a trial differ in intensity, the second interval is commonly called "larger" more often than the first, an effect called *time order error*. The sequence <Small, Large> is, accordingly, correctly reported more often than <Large, Small>. Further, the greater the ISI, the greater the bias. These data have been interpreted to show decay of a central representation of the stimulus over time (Köhler, 1923; discussed in Osgood, 1958).

Increases in ISI also lead to decreases in sensitivity. Berliner and Durlach (1973), Kinchla and Smyzer (1967, in a same-different task), and Tanner (1961) systematically varied ISI, and all found sensitivity to be a decreasing function of time. In Tanner's auditory experiment, a very short ISI (less than 0.8 seconds) also led to decreased discrimination, a result Tanner interpreted as evidence for short-term auditory interference.

Effects of Stimulus Range: The Roving Discrimination Design. Two-alternative forced choice permits manipulation of another experimental variable: the range of stimulus values. In our word-recognition experiment, each stimulus class contains many words, but there are only two values of recency—words tested have either been seen recently, in the study phase, or not at all in the experiment. However, we could allow a wider range of recency among words tested for recognition. Suppose participants learn one list on Monday, a second on Tuesday, and so on for a week. In the test phase, they are presented with two words, one from day n, the other from day $n + 1$, and must choose the more recent. The stimulus difference being discriminated—one day's difference in presentation time—is a constant. Notice that, unlike other 2AFC designs we have discussed, *roving discrimination* does not have an analogous yes-no task. An attempt to generate one leads to absurdities: "Here is a word that you saw last week. Did you see it before or after another word, which I will not show you?"

Although this roving discrimination task is more difficult than the corresponding *fixed discrimination* experiment, a common decision strategy applies to both: Each word leads to an estimate of recency, and the decision variable is the difference between the two values. The observer compares this difference with a criterion to make a response.

Roving and fixed 2AFC discrimination have been compared for auditory amplitude and frequency by Jesteadt and Bilger (1974). The fixed task used one pair of tones differing in (say) amplitude; the roving design used a constant amplitude difference, but the two stimuli ranged together over many amplitudes from trial to trial. A 40-dB range of amplitudes yielded a 27% drop in intensity discrimination d', and a 465-Hz range in frequency led to a 37% drop in frequency discrimination d'. Berliner and

Durlach (1973) found that the decline in intensity discrimination performance depended systematically on the intensity range, reaching 58% for the largest (60-dB) range.

Trace-Context Theory.   The same $\sqrt{2}$ relation between 2AFC and yes-no clearly cannot hold for both roving and fixed discrimination, and a model of how decisions are made in roving discrimination tasks is needed to relate the two types of tasks. Durlach and Braida's (1969) trace-context theory addresses this problem and unifies the perceptual phenomena we have been discussing.

Durlach and Braida's proposal about the one-dimensional classification experiment, described in chapter 5, is that both sensory noise (β^2) and range-dependent context noise (C^2) limit performance, and that these sources of variance add. Context noise is proportional to the square of the range R so that $C^2 = G^2R^2$ (G is a constant). Sensitivity is the mean difference a divided by the standard deviation, or

$$d'_{\text{classification}} = \frac{a}{\left(\beta^2 + G^2R^2\right)^{\frac{1}{2}}} . \tag{7.13}$$

Discrimination performance in 2AFC depends on both sensory and context variance, and also on *trace variance*—noise that increases with the interstimulus interval T. How do these limitations combine? Durlach and Braida suggested that they do so optimally, the result being that whichever memory process is more accurate—has smaller variance—dominates. In a roving experiment, each pair of stimuli is discriminated according to the following relation:

$$d'_{\text{roving discrimination}} = \frac{a}{\left[\beta^2 + \dfrac{1}{(AT)^{-1} + (G^2R^2)^{-1}}\right]^{\frac{1}{2}}} . \tag{7.14}$$

Although it may not be obvious, this form of combining variances has the properties we want. First, what if the range R is small, as in fixed discrimination? Then the right-hand variance term is small as well, and

$$d'_{\text{fixed discrimination}} \approx a/\beta . \tag{7.15}$$

Reducing T also improves performance so that if the two stimulus intervals are adjacent in time, range does not matter and Equation 7.15 again holds.

Trace-context theory has been extensively tested for sets of tones differing in intensity and describes many regularities of the data. One systematic violation is the prediction that d' ratios across tasks will be the same throughout the range. What is found instead is that at particular points, like the edges of the range, the advantage of fixed d' over classification and roving d' is reduced. In the current version of the theory (Braida & Durlach, 1988), this effect is attributed to *perceptual anchors* that narrow the effective value of R in certain parts of the range.

Two Reasons for Using Two Alternatives

Two-alternative forced choice has been a very popular procedure, for two excellent reasons. First, the procedure discourages bias. The assumption of symmetric bias is often a good first approximation, and in any case bias can be easily evaluated using the methods of chapter 2. Detection theoretic measures are preferable to "nonparametric" ones in 2AFC, as in yes-no, but small amounts of bias reduce the experimenter's theory dependence, because most measures are equivalent at the ROC's minor diagonal. Low expected bias makes 2AFC a convenient task for use with adaptive procedures, in which stimulus differences are changed depending on the current level of performance (see chap. 11).

Second, performance levels in 2AFC, as measured by $p(c)$, are high. The predicted $\sqrt{2}$ difference between yes-no and 2AFC permits measurement of sensitivity to smaller stimulus differences than would be practical with yes-no, and we have seen that, for many possible reasons, the disparity observed in practice may be even greater. The relative ease of 2AFC has an impact on some aspects of subjective experience: Observers often report surprise that they can perform above chance with small stimulus differences, which they might be unwilling to report as above a yes-no criterion.

In a 2AFC experiment, two stimuli are presented on each trial. The design is occasionally confused with other paradigms that also happen to use two intervals. The defining characteristics of 2AFC are that both S_1 and S_2 occur on each trial, and that the order of the stimuli determines the corresponding response. One or the other of these properties is violated by other, similar designs. In the task we discuss next, one of the two stimuli is merely a *reminder* or *standard* that may improve performance (an empirical question), but is not essential to the judgment process.

Reminder Paradigm

Design

Consider again the lowly yes-no experiment, in which one of two stimuli is presented on each trial. In a detection task, the two stimuli are Signal and Noise. If the discrimination is difficult, an observer sometimes has the sense of not being able to remember what the signal looks, sounds, or feels like. As we have seen, the data support the idea that "memory" for the stimuli to be detected is fragile, and in the *reminder design* the experimenter attempts to jog the observer's memory. Each trial contains two intervals, the first of which always contains the same stimulus. The observer's task is to determine whether the second stimulus matches the first. If the reminder is S_1, then the presentations are $<S_1S_1>$ and $<S_1S_2>$, and the participant in effect decides "same" or "different" rather than "1" or "2." A variant of the reminder experiment is the method of constant stimuli, in which the comparison stimulus varies from trial to trial, but the reminder in the first interval is always the same. We considered this design in chapter 5, but without incorporating the reminder into our theoretical analysis.

Analysis

Figure 7.4 portrays the observer's problem in the usual two-dimensional space. There are two distributions corresponding to the two stimulus possibilities $<S_1S_1>$ and $<S_1S_2>$. As for 2AFC, we consider both decisionally separable and differencing strategies.

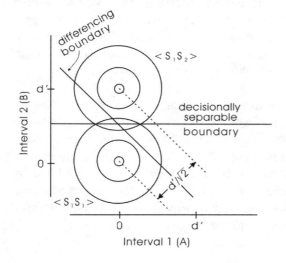

FIG. 7.4. Decision space for the reminder experiment, with decision boundaries for both the decisionally separable and differencing models. Each model postulates that the observer responds "different" in the region above the appropriate decision boundary.

In the decisionally separable strategy, the observer places a boundary perpendicular to the line connecting the two distributions. In effect the space is viewed from the vantage of the vertical axis, and the distributions are projected on this axis. Because the distance between the two distributions from this perspective is d', performance is the same as in yes-no. Indeed this is quite sensible: The boundary line is independent of the observation from Interval 1, implying that the decision maker is ignoring the "reminder" stimulus.

In the *differencing strategy*, the observer bases a decision on the difference between the two observations, $A - B$. In the space of Fig. 7.4, this is accomplished by using a boundary line of the form $A + B$ = constant. Along the $A - B$ axis, the difference between the means of the two distributions is not d', but only $d'/\sqrt{2}$, so observed $z(H) - z(F)$ will be *poorer* than yes-no d' by a factor of $\sqrt{2}$. Thus, counterintuitively, observers who use the reminder stimulus as a decision aid will suffer a decline in performance. The reminder stimulus has as much variance as the stimulus to be judged, so the variable $A - B$ has twice the variance of A alone.

Data

We may thus hope that data will help us decide between the intuitive but deleterious differencing strategy and the decisionally separable rule of ignoring the reminder. Some empirical comparisons of the reminder, yes-no, and 2AFC designs are summarized in Table 7.1. All of these experiments were essentially fixed discrimination, although in all except the line-orientation study of Vogels and Orban (1986) both stimuli roved slightly from trial to trial, the difference (on trials where there was a difference) being constant. Such "jittering" had little effect in the experiments of Jesteadt and Bilger (1974), who also measured unjittered performance.

Table 7.1 reports ratios of d' measures. If our SDT models were all correct, each entry in the table would equal 1.0. This prediction is most nearly fulfilled for the 2AFC/reminder comparison, for which the geometric mean ratio is 1.15. (Some individual ratios are nearer $\sqrt{2}$ than 1 and have been interpreted as supporting the differencing model for both tasks; see Vogels & Orban, 1986.) The two comparisons with yes-no show, once again, the relative difficulty of that task. Comparison of yes-no to reminder data shows that reminders *aid* rather than harm performance. Relative to the optimal performance described by our models, 2AFC yields the best performance and yes-no the worst, with the reminder task intermediate between them.

TABLE 7.1   *Experiments Comparing Yes-No, 2AFC,*
and Reminder Performance

Reference	Continuum	Relative Performance (d' Ratio)		
		Reminder/ Yes-No	2AFC/ Yes-No	2AFC/ Reminder
Jesteadt & Bilger (1974)	Intensity	1.26	1.51	1.20
	Frequency	1.00	1.48	1.48
Jesteadt & Sims (1975)	Frequency	1.37	1.74	1.27
	Frequency modulation	1.06	0.86	0.81
Creelman & Macmillan (1979)	Frequency	1.23	1.37	1.11
	Phase	1.33	1.33	1.00
Vogels & Orban (1986)	Line orientation	–	–	1.31
Geometric mean		1.20	1.35	1.15

It is possible to force the use of a differencing strategy in reminder experiments by employing a real roving design, in which the standard varies across a substantial range of stimuli from trial to trial. Jesteadt and Bilger (1974) conducted such an experiment; for both intensity and frequency discrimination, and for both 2AFC and the reminder task, d' declined by about the predicted $\sqrt{2}$ compared to the fixed design. A similar result was obtained for intensity discrimination by Long (1973).

Essay: Psychophysical Comparisons and Comparison Designs

The basic psychophysical process, we believe, is comparison. All psychophysical judgments are of one stimulus relative to another; designs differ in the nature and difficulty of the comparison to be made. In the one-interval experiment (or any of our two-interval designs if decisional separability is in use), comparison is made to events remembered from previous trials. We have seen evidence that this is a challenging task: Yes-no performance is not as good as it should be relative to both 2AFC and the reminder experiment.

A comparison task of great importance in psychophysics, but one we have slighted here, is the *matching* procedure. To judge the subjective magnitude of a stimulus, a participant selects a value on some other continuum that seems to "match" the standard. For example, the brightness of a light might be matched by the intensity of white noise or the brightness of a light of a different color. What can we say about the reliability of such judgments?

When the two stimuli being matched are from the same continuum (e.g., both are pure tones and differ only on the dimension being studied), adjustment is *more* accurate than fixed methods. At least that was the finding of Wier, Jesteadt, and Green (1976) for frequency discrimination. But when the comparison is across continua, the need to compare disparate stimuli harms performance: Lim, Rabinowitz, Braida, and Durlach (1977) and Uchanski, Braida, and Durlach (1981) measured roving intensity discrimination of pure tones (or noises) that differed in frequency (or spectrum). They found that comparing stimuli from different continua contributed additional, additive variance to the decision process.

One assumption we have been making that may be incorrect concerns the independence of the intervals being compared. The variance of the difference between two variables is the sum of their variances only if the two variables are independent; if they are positively correlated, the standard could effectively *increase* performance. This effect may account for the small advantage of the reminder and 2AFC designs seen in Table 7.1. The matching procedure, by providing the observer with control, may allow strategies that maximize this correlation.

Some researchers have combined the 2AFC and reminder designs. In experiments with binaural noise samples, Trahiotis and Bernstein (1990; also Heller & Trahiotis, 1995) preceded and followed each 2AFC presentation with an example of the standard, so that the possible stimulus sequences were $<S_1 S_2 S_1 S_1>$ and $<S_1 S_1 S_2 S_1>$. The analysis is the same as for 2AFC, but the instructions no longer require discussion of stimulus order. Instead the listeners are asked to say whether it is the second or third stimulus that is different from all the others. Trahiotis and colleagues found superior results with this design, but Gerrits and Schouten (2004) found that it lowered performance with their speech syllables.

Gerrits and Schouten invoked perceptual memory to account for their results, and a more detailed understanding of memory may be required to unify findings in this field. McFadden and Callaway (1999) conducted reminder experiments in which the standard was a "commonly encountered" stimulus or, in other conditions, "less commonly encountered." For example, in musical chord discrimination, the standard was either an in-tune chord, so that the comparison was out of tune, or an out-of-tune chord, with a comparison that was in tune. The result was that performance was much better (a factor of about 2 in the chord experiment) for the commonly encountered standard. McFadden and Callaway suggested that such stimuli have stable memory representations and may allow a more efficient form of processing. Whatever the explanation turns out to be, it will necessarily require an understanding of the stimulus domain being studied, not just general processing principles.

Summary

In comparison designs for discrimination, each trial contains two stimuli, and the decision problem can be represented by two bivariate distributions that can be projected onto a single dimension. In two-alternative forced choice (2AFC), S_1 and S_2 are presented in either of two orders, and performance is expected to be better than in yes-no. In the reminder design, a yes-no interval containing either S_1 or S_2 is preceded by a constant "standard" (say S_1), and performance is expected to be worse than yes-no if the observer compares the two intervals.

Accuracy in 2AFC (and thus its relation to yes-no) depends on two aspects of the two-interval design: the interstimulus interval and the range of stimuli. Long intervals and wide ranges lower performance, and models that are explicit about perceptual memory can account for the pattern of results in some domains. Yes-no accuracy tends to be lower, relative to both 2AFC and reminder performance, than detection theory predicts; a likely culprit is the need to make comparisons across trials rather than across the shorter intervals within a trial.

Chart 6 in Appendix 3 provides pointers to calculations of sensitivity in 2AFC.

Problems

7.1.   For the following stimulus–response matrixes, calculate d', the criterion c, $p(c)$, and $p(c)^*$ assuming that the data arose from (a) a 2AFC experiment, and (b) a yes-no experiment.

Matrix A		Matrix B		Matrix C		Matrix D	
12	8	18	2	4	16	9	6
8	12	14	6	1	19	2	1

7.2.   Find $p(c)_{max, 2AFC}$ for each matrix of Problem 7.1.

7.3. Marsh and Hicks (1998) conducted both yes-no and 2AFC experiments on source monitoring. In the study phase, participants saw some words and generated others by rearranging anagrams.

(a) In one yes-no task, words of both types were presented, and the possible responses were "seen" and "not seen"; the results were $H = .66$, $F = .18$. In a second yes-no task, the possible responses were "generated" and "not generated"; the results were $H = .84$, $F = .37$. Find d' and c for both conditions. How do the two tasks differ?

(b) In a 2AFC task, two words were presented on each test trial, one Seen and one Generated. In one version, participants were asked to choose the one that was generated, and $p(c)$ was .83 (separate hit and false-alarm rates are not reported). How does this compare with detection theory predictions?

(c) In a second version of 2AFC, participants were asked to choose the word that was seen. This time $p(c)$ equaled .7. How would you account for the discrepancy between the two 2AFC tasks?

7.4. In a 2AFC recognition memory experiment, the participants correctly identify both Old and New items at the same rate, .8. (a) Predict d_a in a yes-no experiment with the same stimuli, assuming $s = 0.5$; assuming $s = 2$. (b) Predict d'_2 in a yes-no experiment with the same stimuli, assuming $s = 0.5$; assuming $s = 2$.

7.5. You conduct three intensity-discrimination experiments with the same observer using the same stimulus pair for each. The first experiment uses a 2AFC paradigm, the second a reminder paradigm, and the third a yes-no task to figure out what strategy the observer is using in the reminder task. What would you expect the data to be if

(a) the observer is using a differencing strategy in each condition;

(b) the observer is using a decisionally separable strategy in each;

(c) the observer is using an optimal strategy in each.

8

Classification Designs:
Attention and Interaction

In a *classification* design, a number of stimuli are sorted into a smaller or equal number of categories. When introducing this type of experiment in chapter 5, we restricted the discussion to sets of stimuli that differed on a single internal dimension, but we now abandon that limitation and examine paradigms in which the stimuli lead to representations that differ multidimensionally. Proceeding gently, we consider apparently simple problems in which just three or four stimuli must be classified into only two categories. This project turns out to be sufficiently challenging for one chapter.

This set of problems has both methodological and substantive applications. Methodologically, there is a set of discrimination paradigms that can be thought of as classification tasks. Recall that the comparison designs of chapter 7 always lead to a representation with only two distributions. As a result, although they *can* be modeled in two dimensions, they can also be analyzed by projecting the bivariate distributions onto a single axis and conducting a unidimensional calculation. As long as there are only two stimulus classes, and thus only two distributions, the projection strategy always works. This simplification cannot be made for classification paradigms, and in chapter 9 we use the tools developed here to analyze them.

Substantively, classification designs are extensively used to study two important topics: (a) independence versus interaction between two aspects of a stimulus, and (b) attention. The independence question was the first, historically, to which multidimensional detection theory was applied (Tanner, 1956), but the idea of independence turns out to be multifaceted. In chapter 6, we encountered the concept of perceptual independence—a characteristic of the representation of a single stimulus or stimulus class. An analogous concept applies to stimulus *sets*; this was Tanner's focus, and many other paradigms have been introduced more recently to explore the presence or absence of interaction in this sense.

Classification experiments are more complex than discrimination designs in that they require grouping multiple stimuli together (i.e., assigning them the same response). Does this structural complexity have a corresponding cognitive cost? We discuss three classification paradigms in which *attention* has been invoked by theorists. If several distinct stimuli occur that require the same response, we refer to the design as one of *uncertainty* about which of these stimuli will occur. If the response partition is such that some aspects of the stimulus set must be appreciated and others ignored, attention is *selective*; if all aspects are relevant, attention must be *divided* among dimensions or features.

A critical distinction is that between *extrinsic* and *intrinsic* attentional limitations (Graham, 1989). Extrinsic uncertainty is inherent in the situation, whereas intrinsic uncertainty is internal to the observer. It is essential to find the extrinsic difficulty of a classification design so that poor performance that is in fact inevitable is not blamed on the experimental participant's inefficiency. Most of this chapter concerns models for extrinsic uncertainty, which are useful for establishing a performance baseline.

One-Dimensional Representations and Uncertainty

Multiple distributions *may* lie on a single axis, of course. We explored such examples in chapter 5, and we begin this chapter with some unidimensional problems that are special cases of true multidimensional designs.

Inferring a One-Dimensional Representation

There are no one-dimensional stimuli: Every perceptual and cognitive object has multiple characteristics. But stimulus classes can be represented one-dimensionally if they differ from each other in only one way that is relevant to judgment. The projection strategy of chapters 6 and 7 is a method for mapping two complex stimuli onto a single decision axis. How can we know if this is appropriate for more than two stimuli?

As discussed in chapter 5, the additivity of d' permits a simple test of one-dimensionality: If three stimuli lead to a one-dimensional representation, as in Fig. 8.1a, then

$$d'(1,3) = d'(1,2) + d'(2,3) \ . \tag{8.1}$$

This prediction can be tested by estimating the three d' values. In some cases, the nature of the stimuli makes the hypothesis of one-dimensionality plausible; this is true of amplitude-discrimination experiments, which provide our first two examples.

Example 8a: Detection of Signals of Different Amplitudes

Suppose that Fig. 8.1a is the representation of three auditory tones differing only in amplitude: S_1 is a constant background tone, and S_2 and S_3 are small and large increments in the background. In a simple detection experiment, the listener may be presented with any of the stimuli and must say whether an increment (S_2 or S_3) has occurred or not (S_1). This is an uncertain detection experiment, and the question is whether the listener can do as well in this roving, three-stimulus situation as in the fixed, two-stimulus variants (S_1 vs. S_2 and S_1 vs. S_3).

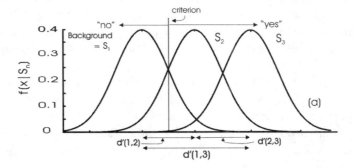

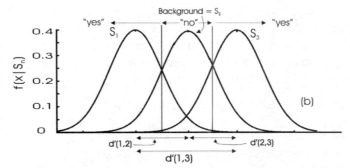

FIG. 8.1. One-dimensional representations of three auditory stimuli differing in intensity; the listeners' task is to say "yes" to changes in the background, "no" to background alone. (a) S_1 is background, S_2 is a small increment in the background, and S_3 is a larger increment. A single criterion is used. (b) S_1 is a decrement in the background, S_2 is background, and S_3 is an increment. Two criteria are required.

The appropriate decision strategy—using a single criterion to divide the stimulus set into two regions—is just a minimalist version of the method of the rule we used to construct psychometric functions from the method of constant stimuli in chapter 5. The model predicts that the d' values obtained in the certain and uncertain situations will be the same, that there is no extrinsic loss due to uncertainty about the magnitude of change. Although there have been few direct tests, most experimenters assume that this is true in practice as well, that is, listeners display no intrinsic loss. The experiment is technically one of uncertainty, but there is no reason to worry about it.

Example 8b: Detection of Increments and Decrements in a Background

In an apparently minor modification of this experiment, Macmillan (1971) studied the detection and discrimination of increments (S_3) and decrements (S_1) in a background (S_2). In one condition, listeners were obliged to distinguish changes (S_1 or S_3) from nonchanges (S_2), as in Fig. 8.1b. Is *this* partition of the stimulus set just as demanding as the fixed discrimination tasks, or does it lead to an extrinsic loss?

A reasonable decision rule (if the representation is still unidimensional) is to say "yes" (there was a change in amplitude) for observations sufficiently large or small and "no" for those that are not. As Fig. 8.1b shows, two criteria are needed to implement this rule. To take a simple example, let us suppose that the means are 1 d' unit apart and the criteria halfway between the means.

For the fixed conditions, in which there are just two stimuli and one criterion (i.e., S_1 vs. S_2 and S_2 vs. S_3), $d' = 1$ and $p(c) = .69$. To evaluate the uncertain condition S_2 versus (S_1 or S_3), we find the proportion of correct responses for each of the three stimulus types. Each "yes" probability has two components, one for observations below the lower criterion and the other for observations above the higher one.

$$P(\text{"yes"}|S_3) = \Phi(0.5) + \Phi(-1.5) = .69 + .07 = .76 .$$

$P(\text{"yes"}|S_1)$ is the same as $P(\text{"yes"}|S_3)$.

$$P(\text{"yes"}|S_2) = 2\Phi(-0.5) = .62, \text{ so } P(\text{"no"}|S_2) = .38 . \tag{8.2}$$

The overall $p(c)$ depends on the proportion of each type of trial. Macmillan used 50% no change (S_2) and 25% of each of the others, so predicted $p(c) = (.5)(.38) + (.25)(.76) + (.25)(.76) = .57$.

Thus, uncertainty about the *direction* of the change produces a 12-point deficit in performance. One might interpret the drop as an attentional effect (i.e., the participant does better by knowing the direction of change to attend to), but really it is the situation that forces the decline. This is our first example of extrinsic uncertainty: The 12-point loss is the *best* the listener can do, not an indication that central capacity is limited or attention is wandering. To justify inferences about such effects (which are types of intrinsic uncertainty), the data would have to show a greater-than-12-point discrepancy between the certain and uncertain tasks.

Two-Dimensional Representations

We have just seen that an understanding of uncertainty for stimuli varying along one dimension requires knowledge of the (unidimensional) representation of these stimuli. Before we can advance to uncertainty in other attention experiments that are analyzed using multidimensional representations, we must consider how such representations can be derived.

Example 8c: Tanner's "Theory of Recognition"

Detection theory was still young when it was first extended to multiple dimensions. In a 1956 article, Tanner devised no new experimental designs, but instead proposed measuring the discriminability of each pair in a multidimensional set and then examining the relations among these indexes. He analyzed the simple but important case of three stimuli, one of which is Noise. In his experiments, tones of different frequencies were the other two stimuli.

These three stimuli can be thought of as varying on two dimensions, as in Fig. 8.2. The notation in the figure generalizes that introduced in chapter 7: Each distribution is labeled as an ordered pair, the elements of which are the stimuli on the different dimensions. Thus, $<S_1N_2>$ is the distribution due to Signal on Dimension 1 and Noise on Dimension 2; the notation is more explicit than referring to this distribution simply as S_1 (which we continue to do when no ambiguity is possible).

Three two-stimulus discrimination experiments can be constructed from the three stimuli: detection of S_1 (i.e., discrimination of S_1 vs. the Noise stimulus), detection of S_2, and recognition of S_1 versus S_2. The results of these tests can be used to decide whether the S_1 and S_2 dimensions are orthogonal. Nonorthogonality implies (for normal distributions) a correlation between the dimensions; if the dimensions intersect at an angle θ, the correlation equals $\cos(\theta)$. The results of the three experiments can be used to esti-

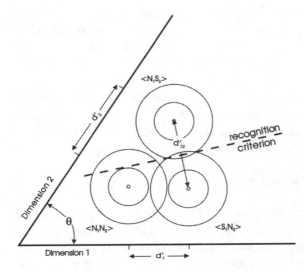

FIG. 8.2. Decision space showing distributions for the Null stimulus and two stimuli differing from it, each along a different dimension. The angle between the axes measures the dependence between the two dimensions.

mate θ. The relation between recognition performance ($d'_{1,2}$) and detection sensitivity (d'_1 and d'_2), measured in separate sessions, can be calculated from the geometry of Fig. 8.2:

$$(d'_{1,2})^2 = d'^2_1 + d'^2_2 - 2d'_1 d'_2 \cos(\theta) \ . \tag{8.3}$$

The criterion for deciding which response to give in recognition divides the line joining the means of the S_1 and S_2 distributions. The particular criterion shown in Fig. 8.2 is halfway between the two distributions. Because the values for d'_1 and d'_2 are not equal in the figure, the criterion line does not pass through the origin, the mean of the Noise distribution.

Equation 8.3 covers all possible relations between pairs of imperfectly detectable stimuli. In one important special case, the alternative stimuli produce independent effects, which are said to require independent sensory *channels*, a metaphor introduced by Broadbent (1958). In that case, the axes are orthogonal so that $\theta = 90°$, $\cos(\theta) = 0$, and

$$(d'_{1,2})^2 = d'^2_1 + d'^2_2 \ . \tag{8.4}$$

We derived the same equation for compound detection of orthogonal stimuli in chapter 6 (Eq. 6.9).

Values of θ less than 90° arise from overlap between the channels' regions of sensitivity—a signal that activates one maximally also activates the other to some extent. Angles of θ greater than 90° might arise from inhibition between the separate perceptual or sensory channels (Graham, Kramer, & Haber, 1985; Klein, 1985).

When $\theta = 0°$, we are back in the unidimensional world of the previous examples, where pairwise d' values add: $\cos(\theta) = 1.0$, so $d'_{1,2} = d'_1 - d'_2$. When $\theta = 180°$, another one-dimensional case, the distance between the two Signals in the recognition task is the sum of the individual detectability values. This is the well-known *city-block* metric, first described by Shepard (1964) for the scaling of similarity judgments. For a discussion of the range of application of this metric, see Nosofsky (1984).

In his own experiments, Tanner found that dimensional orthogonality held when tones were sufficiently different in frequency, but that θ was less than 90° when they were similar. The result is consistent with the "critical-band" hypothesis, according to which auditory inputs are divided into channels according to frequency. Tanner's approach offers a convenient summary of the data in geometric terms, but it has a shortcoming: The three experiments result in three values of d'. This is just enough data to determine the internal angles of the triangle in Fig. 8.2 (by the side-side-side theorem of geometry, which underlies Eq. 8.3), but does not provide any internal test of validity (Ashby & Townsend, 1986). Later in the chapter, we shall see how the addition of just one more stimulus can give us more confidence in the representations inferred from data like these.

Example 8d: Item and Source Recognition for Words

In a typical recognition memory experiment, participants are asked whether test items were on a study list they saw earlier. In real life, a question just as important as whether an item can be recognized is whether its source can be identified: Did I see this face yesterday at work or yesterday on TV? at the scene of the crime or in the police station? To simulate this problem in the laboratory, two lists are presented for study, and tests can be of two kinds: *item recognition* (was this presented on a study list?) and *source identification* (which list was it on?).

Banks (2000) pointed out that these two tasks are analogous to the detection and recognition problems in Tanner's (1956) model. The question Tanner posed is important in this application: Do the three distributions fall on a single dimension or are two dimensions required? A single dimension would be appropriate if all judgments were based on "familiarity," a variable commonly thought to underlie item-recognition decisions. Banks presented his participants with a visual list of words on a computer monitor and an auditory list via loudspeakers. ROC curves were collected, and d_a in item recognition was found to be 1.55 for the visual list and 1.63 for the auditory list. Clearly if the same dimension were also responsible for source identification, we would expect $d_a = 1.63 - 1.55 = 0.08$. In fact d_a was 1.59, implying a representation like Fig. 8.2, with $\theta = 59°$. Furthermore, as Banks pointed out, the implied decision axes for item recognition (with list membership "uncertain") and source recognition are perpendicular to each other (as is approximately true in Fig. 8.2). This result is roughly consistent with the idea that source identification depends on a (probably conscious) recollection process, whereas item recognition depends primarily on a (probably unconscious) familiarity judgment.

Perceptual Separability and Integrality

Tanner's model captures the idea of interaction, and the degree of interaction (as measured by θ) maps naturally onto concepts in psychoacoustics, recognition memory, and other fields. As noted earlier, however, an important limitation of the model is the use of only three stimulus classes and thus three distributions. The data consist of three values of d', and the representation is a triangle, the length of each leg equal to a d'. Except in extreme cases, for which the triangle inequality is violated (one value is greater than the sum of the other two), the model is guaranteed to work.

Expanding the stimulus set overcomes this technical limitation and allows for the study of interesting substantive questions. Many stimulus sets can be constructed by varying two or more dimensions: height and width to make rectangles, the first and second formants to make vowels, contrast and spatial frequency to make gratings, and so on. In the resulting stimulus sets, every value of one dimension can (in principle) combine with every value of the other. The smallest set of stimuli that has this property is built from two values on each of two dimensions; the stimuli can be denoted S_{11} (value 1 on both x and y), S_{12} (value 1 on x and 2 on y), S_{21}, and S_{22}.

Such sets have been studied extensively by Garner (1974) and his colleagues, with the intent of distinguishing "integral" pairs of dimensions (which interact) from "separable" ones (which do not). Garner proposed a series of classification tests to distinguish these possibilities operationally; all of them are discussed in this chapter, and we return to an evaluation of the

"Garner paradigm" in a later section. For now we simply adapt his terminology for use in our psychophysical context. To avoid confusion, we follow General Recognition Theory (Ashby & Townsend, 1986) in using the terms *perceptual integrality* and *perceptual separability* for characteristics of representations, and the Garner terms *integrality* and *separability* in his original operational senses.

Perceptual separability (Fig. 8.3a) is defined by a rectangular arrangement of distribution means; in this case, a change on one dimension has no effect on the value of the other. In perceptually integral cases (Fig. 8.3b), the two dimensions are correlated so that a change on one is at least partly confusable with a change on the other. Such representations display *mean-shift* (or just *mean-*) *integrality* (Kingston & Macmillan, 1995; Maddox, 1992) because the means of the distributions are shifted compared to the perceptually separable case. In Fig. 8.3b, lines connecting the means are drawn, and the angle θ is a measure of how integral the two dimensions really are. Notice that these

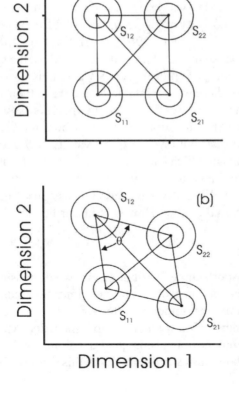

FIG. 8.3. (a) A perceptually separable arrangement of distributions due to four stimuli. (b) A mean-integral arrangement of the four distributions; the angle θ measures the degree of mean integrality.

concepts refer to *sets* of distributions, whereas the idea of perceptual (in)dependence introduced in chapter 6 refers to a single distribution.

The use of four stimuli produces constraints among the fixed, two-stimulus tasks needed to generate the representation: There are six values of d' constrained by only five degrees of freedom. The four outside segments in Fig. 8.3, plus one diagonal, force the value of the other diagonal. Predictions about classification tasks can also be made from such a representation, as we shall see shortly.

Two-Dimensional Models
for Extrinsic Uncertain Detection

Example 8e: The Uncertainty Design in Multimodal Detection

Bonnel and Miller (1994) asked observers to detect a change in background that, on different trials, was unpredictably an increment in either the luminance of a spot or the intensity of a tone. The research question was whether uncertainty would lower performance compared with control conditions in which the modality to be attended to was known in advance. This basic design was earlier used within a modality (e.g., using tones of different temporal frequencies; Creelman, 1960; Green, 1961) or gratings of different spatial frequencies (Davis & Graham, 1981). In all these studies, the uncertainty design was used as a tool for exploring sensory channels.

Bonnel and Miller assumed there was no interaction between their visual and auditory stimuli, and that the representation was thus perceptually separable, as illustrated in Fig. 8.4. The locations of the distribution means for visual (S_1) and auditory (S_2) distributions are the d' values (1.5 and 2.0) found in the control conditions in which each increment was discriminated from no change (N). The uncertainty task requires that observers establish a decision boundary in the space of Fig. 8.4 that accurately assigns stimuli S_1 and S_2 to one response and N to the other. How should this be done?

Summation Rule

Although the representation sports three distributions in two dimensions, it is still possible to reduce the decision problem to one dimension using the projection technique. Just as in the compound detection case of chapter 6, the observer might base a decision on total subjective intensity, which is greater for points farther out into the upper right quadrant along the decision axis $y = x$. Possible decision boundaries consistent with this rule are all per-

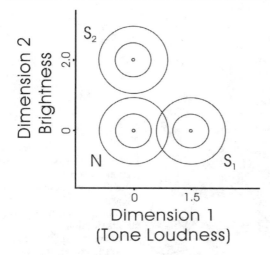

FIG. 8.4. A representation of an uncertain detection experiment in which an auditory stimulus S_1, a visual stimulus S_2, or no stimulus (N) may occur, and the auditory and visual dimensions are independent.

pendicular to this line, as shown in Fig. 8.5. When the S_1 and S_2 distributions are projected onto the decision axis, the means are no longer 2 and 1.5 units from the N mean, but instead (using the Pythagorean theorem as usual) $2/\sqrt{2} = 1.41$ and $1.5/\sqrt{2} = 1.06$ units away. Choosing a criterion location halfway between the means of N and S_2 places it at .71 units from the N mean. The hit rates are therefore $\Phi(0.71) = .761$ for S_2 and $\Phi(0.36) = .641$ for S_1; the false-alarm rate is $\Phi(-0.71) = .239$. If Noise is presented on half the trials and each Signal on one quarter, then $p(c) = (.5)(.761) + (.25)(.641) + (.25)(.761) = .73$. The model predicts a drop due to uncertainty in visual performance from $p(c) = .84$ in visual detection and $p(c) = .77$ in auditory detection to an overall level of $p(c) = .73$.

The summation rule is a natural one.[1] Furthermore, it resembles the optimal strategy for detecting compound stimuli that we developed in chapter 6. It is clear, however, that Bonnel and Miller's observers did not use this rule because their performance turned out to be better than the rule predicts. Recall that models of extrinsic uncertainty give the best performance possible, and intrinsic uncertainty can only lower observer accuracy. When extrinsic models are outpaced in practice, they are wrong.

Independent-Observation Rule

Bonnel and Miller's observers were not using a straight-line boundary, but perhaps they employed another relatively simple rule: Compare the obser-

---

[1] A slightly different summation rule, in which the decision boundary is parallel to a line passing through the S_1 and S_2 distribution means, yields slightly better performance.

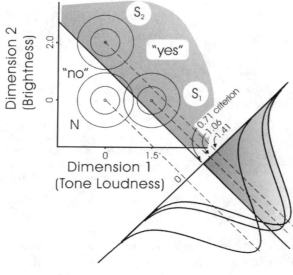

FIG. 8.5. The uncertain detection experiment of Fig. 8.4 and an integrative (total intensity) decision rule. The decision boundary permits all distributions to be projected onto a single dimension, shown at the lower right, and the observer responds "yes" if the total intensity exceeds a criterion.

vation to criteria on each dimension independently, and say "yes" if either criterion is exceeded. We also encountered this "minimum" rule in the compound detection problem of chapter 6. Continuing the present example, suppose the observers placed criteria perpendicular to the x and y axes at unbiased locations, 0.75 units along x and 1 unit along y, and responded "yes" if their joint observation exceeded either criterion. This leads to the two-segment rectilinear decision boundary shown in Fig. 8.6a.

It is easiest to calculate $P(\text{"no"}|\text{each possibility})$ because a "no" response can only be made if the observation is below *both* the criteria and the product rule introduced in chapter 6 applies. For the N distribution, the probability of a correct rejection, $P(\text{"no"}|N)$, therefore equals $\Phi(1)\Phi(0.75) = (.841)(.773) = .650$. Applying the same logic to the non-null stimuli gives us the "miss" rates for each modality: For visual stimuli, $P(\text{"no"}|S_1) = \Phi(-0.75)\Phi(1) = (.229)(.841) = .193$; for auditory stimuli, $P(\text{"no"}|S_2) = \Phi(-1)\Phi(0.75) = (.159)(.771) = .123$. Subtracting the latter two values from 1 gives us the hit rates .807 and .877. The overall proportion correct is a weighted average of the correct rejection and hit rates. Bonnel and Miller presented each signal on one quarter of the trials, no signal on half, so $p(c)$ is $(.5)(.647) + .(.25)(.807) + (.25)(.877) = .744$. Fixed (no uncertainty) $p(c)$ was $(.841 + .771)/2 = .806$ so there is a 6.2 percentage drop due to uncer-

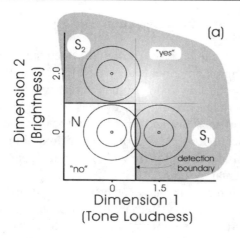

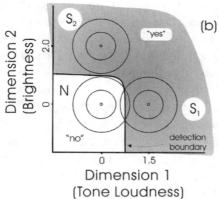

FIG. 8.6. The uncertain detection experiment of Fig. 8.4. (a) An independent-observation decision rule. The observer responds "yes" if either a Dimension 1 criterion or a Dimension 2 criterion is exceeded. (b) The optimal decision rule.

tainty according to this model. This *independent-observation* rule is always superior to the summation rule, although in this case the difference is small (6.2 vs. 7.5 percentage points).[2]

Is this corner rule the best strategy? Almost. The optimal boundary, shown in Fig. 8.6b is (as always) the locus of points for which the likelihood ratio is the same. This boundary is only slightly discrepant from the rectilinear boundary, and the predictions of the corner rule are much easier to calculate (Irwin & Hautus, 1997). For more than two dimensions, the discrepancy between the corner and optimal rules is even smaller.

---

[2]The calculation here is not precise because the locations of the criteria for best performance are not always midway between the means. The difference between the strategies tends to be larger for lower values of d'.

Uncertain Simple and Compound Detection

Now consider a variant on the uncertain detection design in which the compound stimulus as well as the single components are possible signals. For example, a radiologist may be attempting to determine the presence of tumors without knowing how many are present; the dimensions might be strength of evidence in two areas of an X-ray, and tumors could be present in neither, one, or both regions. Or an observer—perhaps an airline pilot—may be engaging in "bimodal" signal detection, in which a warning signal (if it occurs) may be auditory, visual, or both.

Like other multidimensional problems, this task can be solved with an integration or independent-observation strategy. Because of the richness of the data—in two dimensions, there are four possible stimuli, rather than two or three—more powerful tests are available to distinguish the two possibilities than in other designs we have considered. Our analysis follows that of Shaw (1982), who presented these models and others in more detail and generality.

Figure 8.7 shows a two-dimensional decision space for such an experiment, assuming orthogonal channels. The first panel illustrates the independent-observation model: The observer responds "no" in the shaded area. The decision rule is the same as in the uncertain detection task.

It is convenient to use SDT terminology in writing the predictions of the model, but we shall soon see that no distributional assumptions are necessary. A "no" response results from observations from the relevant distribution that are below both the channel-1 criterion k_1 and the channel-2 criterion k_2. Because we are assuming perceptual independence, the product rule applies, and the probability of this event is the product of subcriterion observations on the two axes, and

$$P_n = P(\text{"no"}|<N_1 N_2>) = \Phi(k_1)\,\Phi(k_2)$$

$$P_1 = P(\text{"no"}|<S_1 N_2>) = \Phi(k_1)\,\Phi(k_2 - d'_2) \tag{8.5}$$

$$P_2 = P(\text{"no"}|<N_1 S_2>) = \Phi(k_1 - d'_1)\,\Phi(k_2)$$

$$P_{12} = P(\text{"no"}|<S_1 S_2>) = \Phi(k_1 - d'_1)\,\Phi(k_2 - d'_2)\,.$$

Shaw (1982) pointed out that the four equations imply two simple interrelations among the response probabilities:

$$P_1 P_2 = P_n P_{12} \tag{8.6}$$

$$P_1 + P_2 < P_n + P_{12}\,.$$

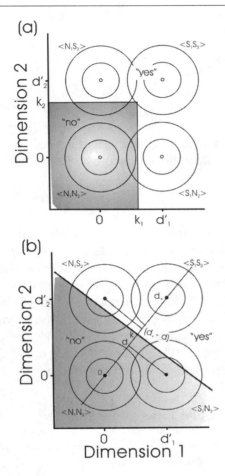

FIG. 8.7. Decision space for the simultaneous simple and compound detection task: (a) independent-observation rule, and (b) integration rule. Observers respond "no" in the shaded regions.

These relations depend on the perceptual independence and orthogonality assumptions, but they do *not* require that Φ be the normal distribution or even that the marginal distributions on the two channels have the same form.

Figure 8.7b shows the same decision space, but with an integration-rule decision criterion. The observer's criterion line is perpendicular to a line connecting the mean of the $<S_1S_2>$ and $<N_1N_2>$ distributions. To find the response probabilities, we need to know only the distances along this line corresponding to the means of the various distributions. In the figure, the distance from the mean of $<N_1N_2>$ to the criterion is k and that to the mean of $<S_1S_2>$ is d_+. By geometry, if the distance to the mean of $<N_1S_2>$ is d, the distance to the mean of $<S_1N_2>$ is $d_+ - d$. Thus,

$$P_n = \Phi(k)$$

$$P_1 = \Phi(k - d_+ + d) \tag{8.7}$$

$$P_2 = \Phi(k - d)$$

$$P_{12} = \Phi(k - d_+) .$$

The interrelation that follows from these equations *does* depend on the assumption of normal distributions. Denoting the z score corresponding to P_i by z_i, it is

$$z_1 + z_2 = z_n + z_{12} . \tag{8.8}$$

Mulligan and Shaw (1980) applied this approach to the problem of bimodal (auditory and visual) detection and found the independent-observation predictions (Eqs. 8.6) supported over the integration prediction (Eq. 8.8). Shaw reached the same conclusion in her analyses of experiments on visual detection and Bayesian decision making. The relatively firm preference for one type of model over the other does not depend simply on a comparison of d' values or other performance measures, but on finer, structure-revealing aspects of the data (Fidell, 1982; Shaw & Mulligan, 1982). That the predictions are to some degree nonparametric is another advantage of Shaw's approach.

Selective and Divided Attention Tasks

Is the uncertain detection task a "selective"or "divided" attention design? Recall that in selective tasks the goal is to attend to one dimension and ignore others, whereas in divided tasks attention to both dimensions is necessary. The uncertain-detection task can be viewed either way depending on the model assumed: The one-dimensional "intensity" model (Fig. 8.5) treats attention as selective, in that the observer must attend to subjective intensity and ignore characteristics, like modality, that distinguish stimuli S_1 and S_2. The corner and optimal models (Fig. 8.6), however, appear to be strategies for dividing attention.

Selective and divided attention are easier to distinguish operationally with four-stimulus sets. There are three ways in which four elements can be partitioned into two equal parts, two of these being examples of selective attention and one of divided. We consider these in turn, following an analysis presented by Kingston and Macmillan (1995) for speech discrimination experiments.

Selective Attention

Figure 8.8a displays a perceptually separable representation, as in Fig. 8.3a. In one selective attention task, observers are instructed to respond strictly on the basis of the x variable, assigning one response to S_{11} and S_{12}, the other to S_{21} and S_{22}. A decisionally separable boundary—the vertical line in the figure—is optimal, and the distributions project onto a single (horizontal) axis. Performance is just as good as if only the two distributions S_{11} and S_{21} were being discriminated, so the model predicts that for separable dimensions there is no performance deficit due to *filtering*, as the selective task is sometimes called. An analogous task for selective attention to the vertical dimension is analyzed in the same way.

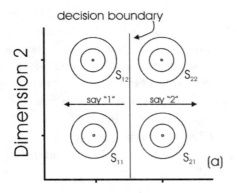

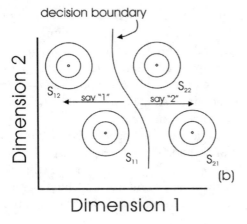

FIG. 8.8. Representation for selective attention task in which S_{11} and S_{12} are assigned to one response and S_{21} and S_{22} to the other. (a) Perceptual separability, and (b) mean integrality.

The mean-integral arrangement in Fig. 8.8b requires a different boundary. This is the kind of problem for which the likelihood ratio analysis of response bias (see chaps. 2 and 4) is essential. When two different distributions correspond to the same response, the likelihood of an observation due to either of them is the sum of their likelihoods—an example of the additive rule for combining probabilities. For the representation in Fig. 8.8b, this is true of both stimulus subsets. The boundary shown in the figure connects all points for which the likelihood of either S_{11} or S_{12} is the same as the likelihood of either S_{21} or S_{22}—that is, for which the likelihood ratio is 1. It may seem surprising that the optimal boundary has this curved shape, rather than being parallel to lines connecting the means, but some insight can be gained by considering points far up above S_{12} and S_{22}. In this region, the S_{11} and S_{21} distributions matter little, so the boundary must be perpendicular to the Dimension 1 axis.

The attention question is how performance in the task sketched in Fig. 8.8b compares to performance with just stimuli S_{11} and S_{12}. Can an observer do as well as in the baseline two-stimulus control condition, or is there a "filtering loss," that is, a deficit due to the additional stimuli. Our simple methods of calculating proportion correct fail us here—numerical integration is needed—but performance is indeed lower than for the baseline task. The magnitude of the drop depends on θ, the degree of integrality (see Fig. 8.3b). Larger declines arise as θ nears $0°$ or $180°$. For example if $d' = 2$ for all one-dimensional comparisons, so that baseline $p(c) = .84$, then predicted $p(c) = .82$ if $\theta = 60°$ and $.78$ if $\theta = 30°$.

Divided Attention

To force attention to both dimensions, the observer is required to assign stimuli S_{11} and S_{22} to one response, S_{12} and S_{21} to the other. An optimal strategy for doing this in a perceptually separable representation is shown in Fig. 8.9. The observer divides the decision space into four quadrants and gives one response for the NE and SW regions, the other for NW and SE. It is clear that this strategy has no equivalent one-dimensional model, but is the optimal strategy good enough to prevent a performance decline?

To analyze this perceptually separable case, we denote the discriminability of S_{11} and S_{21} by d'_x and that of S_{11} and S_{12} by d'_y. Because of the assumed symmetric criteria, proportion correct is the same for all four stimuli, so we need to consider only one of them, say stimulus S_{12}. The observer makes a correct response to this stimulus if the observation falls in either the

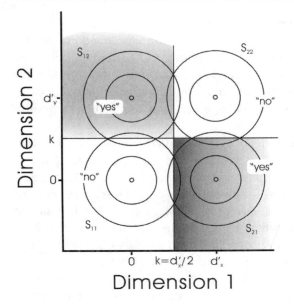

FIG. 8.9. Representation for divided attention task in which S_{11} and S_{22} are assigned to one response and S_{12} and S_{21} to the other.

upper left or lower right quadrant, and we can calculate the probabilities of each of these events using the product rule from chapter 6:

$$P(\text{upper left} \mid S_{12}) = \Phi(d'_x/2)\,\Phi(d'_y/2) \qquad (8.9)$$

$$P(\text{lower right} \mid S_{12}) = \Phi(-d'_x/2)\,\Phi(-d'_y/2)\ .$$

If $d' = 2$ on both dimensions, so that baseline $p(c) = .84$, these terms are $(.84)^2 = .706$ and $(.16)^2 = .026$, for a sum of .732. For $d'= 1$, the decline is from .69 to .572. Clearly the divided attention task is, extrinsically, quite a difficult one.

We do not discuss the mean-integral case in detail. The optimal decision boundary is constructed by combining two curves like the one in Fig. 8.7b. The interesting result is that performance is relatively unaffected by θ over its entire range.

The Garner Paradigm for Assessing Interaction

We are now in a position to consider the complete Garner paradigm. Garner (1974) argued that determining whether two dimensions interact should not rely on a single test, but on "converging operations." In typical experiments by Garner and his colleagues, the two dimensions are sampled at two points each, as in the last few examples. Separability is defined by no filtering loss, that is, selective attention equal to baseline performance; and no "redundancy gain," for example, the ability to distinguish S_{11} and S_{22} being the same as the ability to distinguish S_{11} and S_{12}. Integrality is the opposite pattern, both a filtering loss and a redundancy gain. Divided attention is not always included and is not considered diagnostic in distinguishing integrality and separability.

Does the perceptual-space model agree with Garner's definitions? Both approaches agree that integrality is associated with filtering loss, separability with no loss. As for redundancy gain, the parallelogram model predicts this effect for *all* arrangements if optimal decision rules are used, but can predict no gain in the separable case if decisional separability is assumed (see chap. 6). In many experiments using the Garner paradigm, participants are instructed to attend to one dimension even in the redundant case, so it is perhaps not surprising when redundancy gains are not found.

The analyses in this chapter provide a theoretical convergence of operations that allows for quantitative predictions of the relations among these tasks, but there are two important limitations. The first we have just seen: Predicted performance depends on the particular decision strategy used by the observer. Second, detection theory applies to imperfectly discriminable stimulus sets and the measurement of accuracy. Most Garner-paradigm studies have instead used response time, and explicit modeling of this measure is required if quantitative predictions are to be made (Ashby & Maddox, 1994).

Attention Operating Characteristics (AOCs)

Extrinsic models of attentional paradigms provide a useful baseline for performance: Even a substantial drop due to divided attention, for example, can be consistent with no real limit on intrinsic attention allocation. We now consider a detection-theoretic approach to an intrinsic concept, "paying attention." To begin, it helps to return to the problem of compound detection introduced in chapter 6.

"Multiple-Look" Experiments

Remember that the detectability of a "compound" stimulus, for example, a simultaneous tone and light flash, is the Pythagorean sum of each component's d'. If the stimuli are equally detectable, the improvement (or redundancy gain) is a factor of $\sqrt{2}$. Now imagine a slight modification in which the observer gets multiple "looks" at the *same* stimulus, say a light flash. The argument still applies, so that the detectability of a double look is $\sqrt{2}$ times the d' for a single one. In fact the argument can be extended to any number of looks, so that 10 looks should improve d' by $\sqrt{10}$. Early research (Swets, Shipley, McKee, & Green, 1959) roughly supported this way of modeling multiple presentations, although observers were not completely efficient.

This same relation can be derived in a different way with reference to a single decision axis. Assume that the decision variable is the sum of observations (on a single dimension). Then n stimuli produce a mean difference of nd' and a variance of n (because the variance for one observation is 1), so the effective normalized mean difference is $nd'/\sqrt{n} = \sqrt{n}d'$. This one-dimensional perspective allows us to easily go beyond two samples, whereas visualizing six-dimensional spaces is hard.

Capacity and the Sample-Size Model

What has this to do with attention? Suppose that (as is postulated by many models of attention) a person has a fixed "capacity" to allocate among whatever tasks are at hand.[3] For convenience, let us call this capacity T (for "total") units. As in the previous discussion, assume that as each unit is allocated it adds a fixed amount to both the mean and variance. Hypothetical performance using one unit of capacity is denoted by d'.

Consider now the uncertain detection experiment with which we began the chapter. If all attention is allocated to dimension x, performance will be $\sqrt{T}d'$ on that dimension, but 0 on dimension y. The reverse is true if all attention is allocated to y. But what if P of the T units are allocated to x and $T-P$ to y? Then performance on x, denoted d'_x, is $\sqrt{P}d'$ and d'_y is $\sqrt{(T-P)}d'$.

The model says that capacity can be allocated to one dimension only at the cost of the other, and so it describes a tradeoff between accuracy on the two tasks. When P is large, the observer will do well on Dimension x and poorly on Dimension y, whereas when P is small (so that $T-P$ is large) the opposite will be true. The relation between x and y performance is an "operating characteristic," analogous to the receiver operating characteristic (ROC), which describes a tradeoff between hits and correct rejections.

---

[3]Most such models distinguish "controlled" tasks, which require attentional capacity, from "automatic" tasks, which do not.

To find the form of the attention (or performance) operating characteristic between d'_x and d'_y, we need to solve the prior expressions derived from the sample-size model for one in terms of the other. This can be done most easily in terms of the squares of the sensitivities:

$$d'^2_y = (T - P)d'^2 = Td'^2 - Pd'^2 = Td'^2 - d'^2_x .$$

(8.10)

This is a circle (the usual equation is $y^2 = r^2 - x^2$) as shown in Fig. 8.10. Rearranging the terms provides another perspective:

$$d'^2_x + d'^2_y = Td'^2 = \text{constant} .$$

(8.11)

The idea that *squared* sensitivities are added to estimate overall capacity is an old one, dating to Lindsay, Taylor, and Forbes (1968).

What would happen if participants were asked to give, say, 80% attention to x and 20% to y? They should allocate 80% of their *capacity* to x and operate at the point labeled (80%, 20%) on the diagram. Experiments of this type have often shown that participants not only follow a circular tradeoff function, but are also accurate at assigning the requested percentage of capacity.

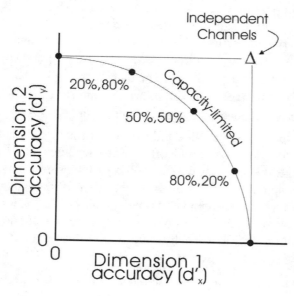

FIG. 8.10. Schematic representation of a hypothetical attention operating characteristic (AOC) showing joint performance (or sensitivity) in the dual-task paradigm. Solid symbols depict resource limitation in which a fixed capacity is allocated to each task alone in the single task, but is divided according to instructions in the dual task. The open triangle represents a case of independence in which neither of the dual-task components affects the other.

For some pairs of stimuli, however, no tradeoff is found. For example, Graham and Nachmias (1971) found that attention could be simultaneously paid to gratings of two different frequencies so that the AOC consisted of two straight line segments, as also illustrated in Fig. 8.10. This result is strong quantitative evidence that separate perceptual "channels" are used in processing the two gratings.

Summary

Classification experiments, in which a number of stimuli are partitioned by the observer into a smaller number of categories, can be used to study perceptual independence versus interaction and a variety of paradigms for measuring attention. Uncertainty about which element of a stimulus subset is to be presented can force performance to be lower than in a corresponding fixed-discrimination condition. Uncertainty effects occur even for unidimensional stimulus sets if multiple criteria are used in the decision process.

Stimuli that differ perceptually in more than one way can be represented as distributions in a multidimensional space. Sensitivity measures, such as d', are distances in such a space, and multiple experimental conditions can allow the geometric arrangement of the distributions to be determined. To determine whether two stimulus dimensions are represented independently, a set of stimuli in which both dimensions vary must be used. As few as three stimuli lead to an answer to the independence question, but a 2×2 set permits stronger conclusions.

Many multidimensional tasks are susceptible to either integration or independent-observation decision strategies. These can sometimes be distinguished on the basis of predicted accuracy, but more powerful methods examine more detailed aspects of the data.

Selective and especially divided attention are usually intrinsically more difficult than the corresponding baseline tasks. The loss due to attention depends on whether the dimensions on which the stimuli vary are independent or interacting.

Detection theory can be used to quantify the idea of a limited attentional capacity that must be allocated among various tasks. Data from experiments in which observers are instructed to allocate attention differently can be used to determine whether different stimulus dimensions are processed by a single channel or separate ones.

Problems

8.1.   In detection experiments for two audio frequencies, $H = .78$, $F = .24$ for a weak 1000-Hz tone and $H = .72$, $F = .31$ for a weak 1200-Hz tone. Find detection d' for both frequencies. Predict identification d' assuming that the tones are analyzed by independent channels.

8.2.   What would a 2AFC identification experiment yield for $p(c)$ if tones of frequencies 1000 Hz and 1000.5 Hz were each detectable at $d' = 1.5$ and (a) $\theta = 60°$ or (b) $\theta = 30°$ in Fig. 8.2?

8.3.   (a) In Example 8b, on increment-decrement uncertainty, recalculate the hit and false-alarm rates assuming the criteria are located at $+/- 0.75$ rather than $+/- 0.5$ SDs from the mean of S_2. Do the stricter criteria lead to better performance in the uncertain task, and thus a smaller decline due to extrinsic uncertainty?

(b) Extend the calculation to plot an ROC for this task.

8.4.   In the representation of Tanner's detection/recognition experiment shown in Fig. 8.2, suppose that $d'_1 = d'_2 = d'_{12} = 2$. These are the lengths of the three legs of an equilateral triangle in which all the interior angles are 60°. Therefore, participants could get 84% correct in any of the three tasks (S_1 detection, S_2 detection, and recognition). But what if they only consider Dimension 1 in making their decisions? That is, they have decisionally separable boundaries perpendicular to Dimension 1 or (equivalently) project all three distributions onto Dimension 1? What is percent correct for the three tasks in that case?

8.5.   In the illustration of mean integrality (Fig. 8.3b), what would happen if $\theta = 0°$ or $180°$?

8.6.   Redo the uncertain-detection example (Figs. 8.4–8.6) assuming $d' = 1$ on both dimensions. (a) For both the summation and independent-observation rules, find $p(c)$ for the uncertain condition and compare it with the fixed condition. (b) This is not the best possible performance for the independent-observation rule. How well can the participant do if the two criteria go through the *means* of the S_{12} and S_{22} distributions rather than halfway between those means and the origin?

8.7.   On each trial of a detection experiment, an auditory signal can be presented to the listener's left earphone (S_L), right earphone (S_R), both, or neither. Two observers produce the following data:

	Observer 1		Observer 2	
Signal	*"Yes"*	*"No"*	*"Yes"*	*"No"*
Null	8	32	17	23
S_L	32	8	32	8
S_R	24	16	24	16
Both	36	4	36	4

For each listener, determine whether an integration or independent-observation strategy is being used.

8.8. Redo the example in Figs. 8.8a and 8.9 assuming $d'_x = d'_y = 1$. That is, predict performance in selective and divided attention assuming that this is the accuracy level in the fixed, baseline tasks.

8.9. In Fig. 8.8b, the optimal decision rule cannot be interpreted as a projection on a single decision axis, but there is a simple non-optimal rule that can: The decision axis could be parallel to a line connecting the means of S_{11} and S_{12}. Suppose all d' values for one-dimensional comparisons equal 2, $\theta = 45°$, and the decision criterion goes through the middle of the parallelogram. Analyze the problem in this one dimension and calculate $p(c)$.

8.10. The divided attention problem (Fig. 8.9) can also be reduced to one dimension. Assume the same representation as in Problem 8.9, but with a decision axis parallel to the line connecting the means of S_{12} and S_{21} and three response regions (as in Fig. 8.1b). What is $p(c)$?

8.11. Participants study a list of words. There are two test conditions. One is standard: Single words are presented, and the participant says "yes" or "no." In the "expanded" condition, four words are presented on each trial, either all Old words or all New words. If $p(c)$ is .75 in the one-word condition, what do you predict it will be in the four-word condition? (Assume unbiased responding.)

9

Classification Designs for Discrimination

We return again to designs for studying discrimination. The tasks described to this point—yes-no with or without a rating response or a reminder, and 2AFC—provided the experimental cornerstone for detection theory in psychology. They are natural paradigms for studying the detection of weak signals and, as we have seen, are simply related to each other on theoretical grounds.

Each design, however, has shortcomings. The failure of the predicted relation between yes-no and 2AFC (Eq. 7.2) led us to the suspicion that participants are limited in the one-interval task by imperfect memory. Two-alternative forced choice, which survives this criticism, is subject to another: In some applications, the task is difficult to describe to participants. Observers in 2AFC are instructed to "choose the picture you think you have seen before" or "choose the interval that contained a tone added to the noise background." The dimension of judgment—recency and "tone-ness," in these examples—is made explicit. But observers may not share the experimenter's definition of the dimension being judged, and may even be able to distinguish the stimuli without having names for them at all. Many listeners in simple auditory tone-detection experiments, for example, discover that "tone-ness" is not, introspectively, the basis for judgment: The experience of a very weak stimulus is not a small version of a more intense one, but participants usually learn to respond appropriately with training.

Frequently, the problem of describing the dimension on which the stimuli differ is not so readily solved. Sometimes the physical dimension is difficult to characterize for participants; the experimental design precludes training; or participants are unsophisticated, and forced-choice instructions are difficult to convey. We now discuss three participant-friendly designs that seem well suited to such situations: same-different, ABX, and oddity. These tasks have been used in experiments with animals, unsophisticated participants by most standards, and in human studies in which the differences among stimuli are difficult to describe. Our examples illustrate these applications: We consider people categorizing visual objects, animals discriminating visual shapes, and people discerning subtle differences among wines.

The cost of using these accessible designs is borne by the experimenter, for they are not psychophysicist-friendly. The "comparison" tasks discussed in chapter 7—2AFC and reminder—assumed two distributions, one for each of the possible stimuli (or stimulus sequences). These distributions were represented in a two-dimensional perceptual space, but the optimal strategy could be displayed in one dimension by an appropriate projection. The discrimination designs in this chapter require classification—that is, there are more possible stimulus sequences than responses. The attention designs analyzed in chapter 8 can be adapted with only minor modification to describe same-different and ABX (matching to sample). Oddity requires a slightly different classification analysis.

Same-Different

Example 9a: Semantic Judgments of Pictures

Irwin and Francis (1995a) explored the perception of line drawings of objects that were either *natural* (e.g., alligator, leaf) or *manufactured* (e.g., various tools). Pairs of such objects were briefly presented, and the observers had to say whether they belonged to the same or different categories. Thus, the correct response for the pair <hammer, leaf> was "different," whereas for <leaf, alligator> it was "same."

Letting S_1 and S_2 denote the natural and manufactured stimuli, there are four possible pair types: $<S_1S_1>$, $<S_2S_2>$, $<S_1S_2>$, and $<S_2S_1>$. The participant has only to respond "same" or "different" and need not know or be able to articulate the ways in which the stimuli actually differ. The results can be summarized in a 2 × 2 table as in earlier chapters, but with new labels for the rows and columns. Here are some possible data:

Stimulus Pair	Response	
	"Different"	"Same"
$<S_1S_2>$ or $<S_2S_1>$	30	20
$<S_1S_1>$ or $<S_2S_2>$	10	40

Hit and false-alarm rates can be defined in a natural way:

$$H = P(\text{“different”} | <S_1S_2> \text{ or } <S_2S_1>) \qquad (9.1)$$

$$F = P(\text{“different”} | <S_1S_1> \text{ or } <S_2S_2>) \, .$$

We assume that presentations of the two kinds of Same trials and the two kinds of Different trials are equally likely. How can we estimate d' for data of this sort?[1]

Representation

To appreciate the peculiarity of the same-different task, consider its underlying distributions shown in Fig. 9.1. As with 2AFC, the two dimensions are the two intervals of the task, and every point in the two-dimensional space represents a possible outcome of a trial. For each interval, the mean given S_1 is 0 and the mean given S_2 is d', so that d' is the distance between the means of any two distributions differing along just one axis. The four possible stimulus sequences generate four probability distributions in the space. If the stimulus sequence is $<S_2S_1>$, for example, the observer's observation is drawn from the distribution at the lower right. Our task is to estimate d', the original normalized distance between the means of the S_1 and S_2 distributions, a sensitivity statistic that characterizes only the stimulus pair, not the method.

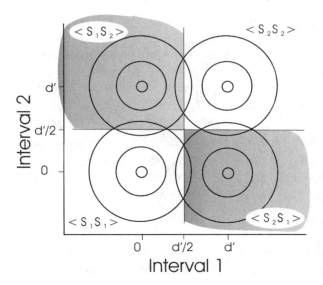

FIG. 9.1. Decision space for the same-different experiment. The effects of the two observations are combined independently. The unbiased decision rule is to respond "different" in the shaded area.

---

[1]Because we know of no Choice Theory models for the tasks described in this chapter, we consider only SDT models.

We explore two decision rules based on this representation: an independent-observation and a differencing rule. Both are special cases of rules we developed for divided attention in chapter 8, and the independent-observation rule is again the optimal one. However, we shall see that some experimental designs conspire against any decision maker's attempts to use this strategy, and for those designs differencing is the best available analysis.

Independent-Observation Decision Rule

Statement of Rule. The optimal decision rule is like that used for the divided attention task: To determine which points on the plane lead to which response, a pair of criterion lines is used to partition the space of Fig. 9.1. If an observation falls either to the right of the vertical criterion line and below the horizontal one (in the lower right quadrant) or to the left and above (upper left quadrant), the response is "different"; otherwise the observer responds "same." For the $<S_1 S_2>$ distribution, the proportion in this region is the hit rate; because the decision rule is symmetric, this is also the proportion correct for all other trials and for the task as a whole. The calculation of proportion correct in same-different using independent observations $[p(c)_{SD,IO}]$ is the special case of divided attention (Eq. 8.9) in which d'_x equals d'_y :

$$p(c)_{SD,IO} = [\Phi(d'/2)]^2 + [\Phi(-d'/2)]^2 \ . \tag{9.2}$$

Solving for d' yields[2] (see Computational Appendix to this chapter)

$$d' = 2z\left(\frac{1}{2}\left\{1 + \left[2\,p(c)_{SD,IO} - 1\right]^{1/2}\right\}\right) \ . \tag{9.3}$$

Comparison With Yes-No. How difficult is this task compared with yes-no? To relate the two tasks, recall that $\Phi(d'/2)$ is the proportion correct for an unbiased participant in yes-no. Combining this identity with Equation 9.2 reveals that

$$p(c)_{SD,IO} = p(c)_{yes\text{-}no}{}^2 + [1 - p(c)_{yes\text{-}no}]^2 \ . \tag{9.4}$$

Sample predictions from this equation, given in Table 9.1, clearly show that observers are expected to find same-different more difficult than the corre-

---

[2]Equation 9.3 assumes that $p(c) \geq .5$. If not, the equation cannot be used. A heuristic solution is to replace $p(c)$ with $1 - p(c)$ and treat the result as a *negative* value of d'.

TABLE 9.1 *Comparison of Yes-No Performance With Two Decision Strategies in Same-Different*

| | | *p(c)* | |
| | | *Same-Different* | |
d'	*Yes-No*	*Independent-Observation*	*Differencing*
1	.69	.57	.55
2	.84	.73	.68
3	.93	.88	.80
4	.98	.96	.89
5	.994	.987	.95
6	.999	>.999	.98

sponding yes-no task, just as they find the divided attention task quite challenging compared with baseline.

Equation 9.4 contains no explicit reference to d'; does that mean it is a nonparametric result? The requirement is that the underlying distributions be perceptually independent and that the arrangement be perceptually separable. These assumptions may or may not be correct in general, but in this application the two dimensions are the two observation intervals of a single trial. It is common to assume independence and separability in this case (although remember that non-independence was one of the reasons conjectured to account for the superiority of 2AFC and reminder over the level predicted from yes-no).

Threshold Analysis. Same-different data are often summarized by proportion correct, but this measure turns out to imply a threshold model in which the participant covertly classifies each stimulus into one of two categories. Let p_1 and p_2 be the probabilities that S_1 and S_2, respectively, are classified covertly by the participant as stimulus S_2. The observer responds "same" whenever the two classifications agree, "different" otherwise. Then as Pollack and Pisoni (1971) have shown,

$$p(c)_{\text{same-different}} = \tfrac{1}{2}[1 + (p_2 - p_1)^2] \ . \tag{9.5}$$

A similar analysis (see Creelman & Macmillan, 1979) reveals that proportion correct by an unbiased observer in both yes-no and 2AFC is

$$p(c)_{\text{yes-no}} = \tfrac{1}{2}[1 + (p_2 - p_1)] \ . \tag{9.6}$$

Combining Equations 9.5 and 9.6 leads to a prediction about the relation between same-different and yes-no performance, for an unbiased observer, and it is again Equation 9.4 (see Computational Appendix). Apparently, for an unbiased observer, the covert-classification and independent-observation models are the same. Discrepancies arise when observers display bias, because the ROC implied by proportion correct has the wrong shape. This familiar shortcoming of proportion correct is of even greater significance for same-different than for other paradigms we have discussed, because participants seem to naturally adopt strong response biases in same-different experiments. In particular, a preference for "same" is commonly observed for hard-to-discriminate stimuli, which are perforce perceived to be the same on many trials.

Response Bias.   The participants in Example 9a display just such a preference for "same" over "different" responses, implying that the criterial value of likelihood ratio is some value greater than 1.0. Figure 9.2 shows how the decision space is divided up by an observer who is biased toward "same" so that an observation must be at least twice as likely to come from a Different trial to evoke a "different" response.

It is possible to convert the representation of Fig. 9.2 to a one-dimensional one. In this strategy, described by Irwin, Hautus, and Francis (2001), the decision axis is the likelihood ratio β_i [more precisely,

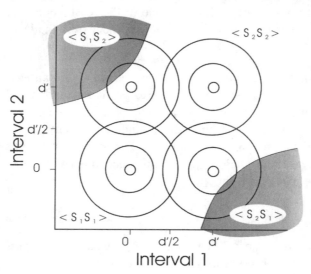

FIG. 9.2. Decision space for the same-different experiment. The decision rule is to respond "different" in the shaded area; this observer is biased toward "same."

$\ln(\beta_i)/d'$], and a Same and a Different distribution are constructed on that axis. (Neither distribution is Gaussian in shape.) The height of the Same distribution for a specific value of β_i is the sum of heights of the $<S_1 S_1>$ and $<S_2 S_2>$ distributions in Fig. 9.2 for which β_i has that value, and the height of the Different distribution is the sum of the heights of the $<S_1 S_2>$ and $<S_2 S_1>$ distributions over points for which β_i has that value. One measure of response bias is simply β_i, and another is the criterion location on the decision axis, denoted c_i and equal to $\ln(\beta_i)/d'$.

Figure 9.3 shows isobias curves for both of these measures, and it is immediately clear that they bear family resemblances to c and β, the corresponding statistics for the yes-no design introduced in chapter 2. The criterion location measure again behaves more regularly than likelihood

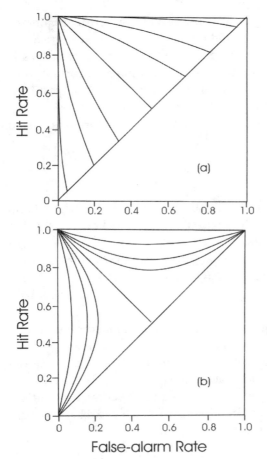

False-alarm Rate

FIG. 9.3. Isobias curves for criterion location c_i (panel a) and likelihood ratio β_i (panel b) according to the independent-observation model. (Adapted from Irwin et al., 2001, Figure 4, with permission from the author and publisher.)

ratio. Empirical isobias curves for visual (Irwin, Hautus, & Francis, 2001) and auditory discrimination (Hautus & Collins, 2003) favor c_i over β_i.

Calculating these measures is somewhat onerous, and for purposes of comparing experimental conditions it is tempting to adopt the strategy of simply using the yes-no formulas. A defense for this approach is the similarity between the curves in Fig. 9.3 and those in Fig. 2.7 for the analogous indexes. For the current example, the statistic c (Eq. 2.1), equals $-0.5[z(H) + z(F)] = -0.5(0.253 - 0.842) = 0.294$. The likelihood ratio $\beta = \phi(.6)/\phi(.2) = 1.380$. For comparison with biases observed when d' is higher or lower, the criterion measure can, as before, be normalized by dividing by $z(H) - z(F)$: $c' = 0.294/1.095 = 0.268$. All show that there is some bias toward saying "same" in these data.

ROC Curves.   By systematically varying the critical value of likelihood ratio and calculating H and F for each value, we can trace out a same-different ROC. The important characteristic of such curves is that they are approximately straight lines with unit slope on normal coordinates, so that $z(H) - z(F)$ does not change with criterion. This result allows a simple strategy for finding d' in a same-different task: First, convert $z(H) - z(F)$ to the equivalent proportion correct for an unbiased observer (Eq. 7.4):

$$p(c)_{max} = \Phi\{[z(H) - z(F)]/2\} . \tag{9.7}$$

Then insert $p(c)_{max}$ into Equation 9.3 to find d'. We have followed this logic in constructing Table A5.3, which provides d' corresponding to any value of $z(H) - z(F)$ observed in a same-different task.

We can now, finally, analyze the data matrix from the beginning of the chapter. The transformed difference $z(H) - z(F)$ equals $z(.60) - z(.20) = 0.253 + 0.842 = 1.095$, and $p(c)_{max} = .71$. The underlying d' is found from Table A5.3 (or Eq. 9.3) to be 1.86.

Our model abandons the requirement of unbiased responding, but retains another simplifying assumption: The critical value of likelihood ratio for responding "different" is the same whether the observed difference is positive or negative. Although to our knowledge this assumption is shared by all models for the same-different paradigm, it need not be correct: $P(\text{"different"}|<S_2S_1>)$ may not equal $P(\text{"different"}|<S_1S_2>)$, and the two halves of the decision contour in Fig. 9.2 may not be symmetric.

Differencing Rule

In chapter 7, we distinguished fixed and roving versions of the 2AFC experiment, according to whether two fixed stimuli recurred throughout a block of trials or the stimulus pair roved along a continuum. The roving feature is also often incorporated into same-different tasks. Suppose in our categorization experiment (Example 9a) there are four stimulus classes—S_1, S_2, S_3, and S_4—and we wish to measure sensitivity for each adjacent pair. A fixed experiment requires three separate blocks of trials, whereas the roving procedure can employ just one. An appealing feature of roving experiments is that they more closely resemble real-life situations, in which repeated presentation of the same pair of stimuli is unusual.

Sample data for a roving same-different experiment are given in Table 9.2. The participants' responses have been classified into those relevant to measuring sensitivity between S_1 and S_2, S_2 and S_3, and S_3 and S_4. Notice that some Same trials are used twice in this table: $\langle S_2 S_2 \rangle$ trials, for example, enter into both the S_1/S_2 and S_2/S_3 comparisons. This fact produces a correlation between adjacent sensitivities that would not be present in a fixed design.

TABLE 9.2   *Sample Roving Same-Different Data*

	Response	
Stimulus Pair	*"Different"*	*"Same"*
$\langle S_1 S_2 \rangle$ or $\langle S_2 S_1 \rangle$	30	20
$\langle S_1 S_1 \rangle$ or $\langle S_2 S_2 \rangle$	10	40
$\langle S_2 S_3 \rangle$ or $\langle S_3 S_2 \rangle$	35	15
$\langle S_2 S_2 \rangle$ or $\langle S_3 S_3 \rangle$	5	45
$\langle S_3 S_4 \rangle$ or $\langle S_4 S_3 \rangle$	25	25
$\langle S_3 S_3 \rangle$ or $\langle S_4 S_4 \rangle$	5	45

In 2AFC the observer's ideal response strategy is the same for both roving and fixed designs, but in same-different the response rules for the two cases differ. To see why, consider a possible sequence of stimulus pairs in roving same-different discrimination: $\langle S_1 S_2 \rangle$ on Trial 1, then $\langle S_3 S_3 \rangle$, $\langle S_2 S_3 \rangle$, $\langle S_4 S_3 \rangle$, and so on. The independent-observation decision rule portrayed in Figs. 9.1 and 9.2 requires the observer to independently assess the relative likelihood that each sound arose from both of the two stimuli in the sequence. For a set of four possible stimuli, this rule is very complex: The participant must estimate a likelihood ratio based on the 10 possible stimulus pairs listed in Table 9.2. If, as is often true, the observer does not know exactly how large the stimulus set is, the information needed for the calculation is not even available. Another strategy is needed.

Statement of Rule. The appropriate procedure is a differencing strategy like that used in comparison designs: The two observations on a trial are subtracted, and the result is compared to a criterion. If the difference exceeds the criterion, the stimuli are called "different," otherwise "same." The differencing strategy was first described by Sorkin (1962) and has been found to describe data from experiments in pitch perception (Wickelgren, 1969), speech perception (Macmillan et al., 1977), and some visual discrimination situations that we discuss presently.

Figure 9.4 illustrates the differencing decision rule; for simplicity, only stimuli S_1 and S_2 are considered. The criterion lines for a constant difference resemble the line for 2AFC (Fig. 7.1), but the decision space is more complicated. The shaded areas in the figure mark observations that lead to a "different" response under the differencing rule, which is at odds with the independent-observation rule in certain regions of the space.

An example cited by Noreen (1981) can be extended to contrast the rules. Suppose the two stimulus classes are fifth- and sixth-grade boys, and the only information available for discriminating the classes is height, which averages 54 inches in Grade 5 and 56 inches in Grade 6. Then two boys

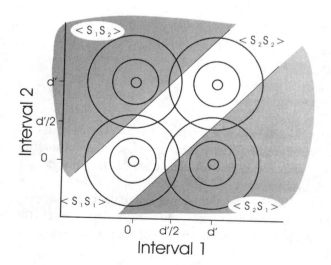

FIG. 9.4. Decision space for the same-different experiment. The effects of the two intervals are subtracted, and the absolute value of the result is compared to a criterion (differencing model). The decision rule is to respond "different" in the shaded area.

whose heights are 54 and 56 inches should probably be judged "different" (i.e., from different grades), but two boys whose heights are 58 and 60 inches should be judged "same" because both are more likely sixth than fifth graders. This example mimics a fixed design and adopts an independent-observation strategy. In the corresponding roving paradigm, boys are drawn from Grades 5 to 8, and the average heights are 54, 56, 58, and 60 inches. Again the heights of two boys are announced; now one must decide whether the two are from the same grade or 1 year apart. Using the differencing strategy, any difference of two inches or more leads to a "different" response. The strategy is not optimal—heights of 65 and 67 inches are more likely Same than Different—but it is reasonable and simple to apply. And it is the only sensible approach for a decision maker without knowledge of the stimulus range.

Because the differencing rule depends on a single variable—the difference between two observations—we can simplify the decision space by projecting the distributions onto one dimension (as we did for comparison designs in chap. 7). Let us consider the probability distributions of the difference for each type of trial. When both trials contain the same stimulus, so that the pair is either $<S_1S_1>$ or $<S_2S_2>$, the mean difference is 0. However, there are two types of Different pairs: those that, when subtracted, yield a mean difference of d', and those yielding a mean of $-d'$. The decision problem in one dimension thus involves three difference distributions on one axis, as shown in Fig. 9.5. The representation resembles that for uncertain increment-decrement detection (Example 8b), and the decision rule is the same: Respond "different" whenever the observed difference is more extreme than k, either in a positive or negative direction. As in comparison designs, however, these are difference distributions; because two independent variables with (by definition) variance 1 are being subtracted, they have variance 2.

Sensitivity and ROCs.   The hit and false-alarm rates result from combining areas under these distributions:

$$H = P(\text{"different"}|\text{Different}) = \Phi[(-k + d')/\sqrt{2}] + \Phi[(-k - d')/\sqrt{2}] \quad (9.8)$$

$$F = P(\text{"different"}|\text{Same}) = 2\Phi(-k/\sqrt{2}).$$

If k is varied, Equations 9.8 can be used to trace out an ROC; some examples are shown in Fig. 9.6. Unlike the ROCs for the independent-observation rule, these do not have unit slope, so two points with equal values of

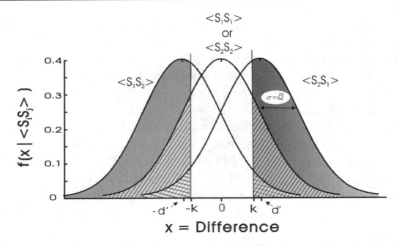

FIG. 9.5   One-dimensional decision space for the same-different experiment according to the differencing model. The representation is equivalent to that in Fig. 9.4: The decision rule is to respond "different" in the shaded area. The hit rate is the sum of all shaded areas under the right-hand (or left-hand) distribution; the false-alarm rate is the sum of the diagonally shaded and cross-hatched areas under the center distribution.

$z(H) - z(F)$ do not necessarily have the same d'. Therefore, we cannot expect to find d' via $z(H) - z(F)$, as we did for the independent-observation model. Table A5.4, modified from the tables of Kaplan, Macmillan, and Creelman (1978), gives d' for any (F, H) pair, assuming the differencing model to be correct.

Applying the differencing model to our categorization data yields the following sensitivity values: $d'_{12} = 2.16$, $d'_{23} = 3.07$, and $d'_{34} = 2.32$. What would happen if we had mistakenly applied the independent-observation model to these data? Table A5.3 yields $d'_{12} = 1.85$, $d'_{23} = 2.56$, and $d'_{34} = 2.04$. The independent-observation model implies smaller values of d', as it must, but the two models are not dramatically different for these data, leading to values of d' differing by an average of 13%. The ROCs suggest that the greatest discrepancy will occur when the probability of responding "different" is small. Indeed if $H = .10$ and $F = .01$, d' is 3.04 according to the differencing model and only 1.83 under the independent-observation model.

As Table 9.1 shows, $p(c)$ by an unbiased differencing observer is poorer than for the independent-observation rule (unsurprising because the latter is optimal). One implication is that quite high d' values correspond to less-than-perfect accuracy. This can sometimes be convenient: If participants

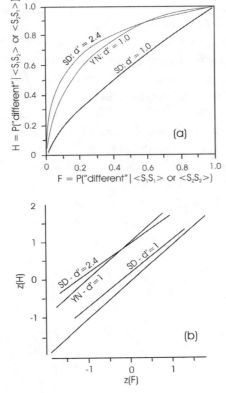

FIG. 9.6. ROCs for the same-different (SD) and yes-no (YN) experiments, according to the differencing model, on (a) linear coordinates, and (b) z coordinates.

are "too good" in yes-no or 2AFC and the stimuli cannot be adjusted, a shift to same-different can avoid a ceiling effect.

Response Bias in the Differencing Model. The likelihood ratio β_d of Different vis-à-vis Same pairs is the easiest bias measure to formulate. Its value at the point k is the average height of the two Different distributions divided by the height of the Same distribution. Assuming equal presentation probabilities for the subtypes of each stimulus class, we have

$$\beta_d = \frac{1}{2} e^{-\frac{1}{4}d'^2} \left(e^{\frac{1}{2}d'k} + e^{-\frac{1}{2}d'k} \right). \tag{9.9}$$

To develop a criterion-location measure, it is helpful to consider an alternative version of the representation in Fig. 9.5 in which the decision axis is the *absolute value* of the difference between the intervals. Only positive values can occur, of course; the Same distribution looks like the right half of a normal distribution; the Different distribution looks roughly like a normal distribution whose left tail has been cut off. Equation 9.8 still applies, and k is still the criterion location expressed as a distance from 0. A better reference point is the location at which $\beta_d = 1$; the distance from this point, which is denoted c_d, can be calculated by a method given in the Computational Appendix.

Figure 9.7 shows the isobias curves for c_d and β_d. The family resemblance between these measures and the corresponding indexes for the independent-observation model (Fig. 9.3) and the yes-no experiment (Fig. 2.7) is clear and encourages a preference for the criterion statistic.[3] Data from an auditory experiment (Hautus & Collins, 2003) also support c_d over β_d.

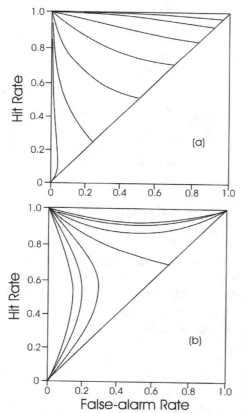

FIG. 9.7. Isobias curves for (a) criterion location c_d, and (b) likelihood ratio β_d according to the differencing model. (Adapted from Irwin et al., 2001, Figure 2, with permission from the author and publisher.)

[3]Two alternative reference points have been explored: c_{sd} compares the criterion location to the point $d'/2$, and c^*_{sd} compares it to the point at which $H = F$. By the criteria of reasonable isobias curves and fit to data, the second of these measures is a good one and the first is not (see Hautus & Collins, 2003).

As with the independent-observation model, calculation of either of these statistics is a somewhat tedious process, and the resemblance of the isobias curves to those for c and β argues for the heuristic use of those statistics for the purpose of comparing experimental conditions. For our example, the values of $\ln(\beta)$ are 0.32, 0.68, and 2.27, and the values of c are 0.29, 0.38, and 0.64. That all values are positive reflects the preponderance of "same" responses in the data.

Relation Between the Two Strategies

The independent-observation and differencing strategies are both special cases of a general situation (Dai, Versfeld, & Green, 1996). Consider what would happen if the correlation between x and y (which is 0 in the diagrams so far) were substantial—that is, if we had perceptual dependence. We assume that the amount of dependence, represented by the correlation ρ between x and y, is the same for all four stimulus sequences.

The upper panel of Fig. 9.8 shows ellipses with correlation ρ. Because the correlations (and variances) are the same in all distributions, this representation is equivalent to a mean-integral one in which the distributions are perceptually independent, but the axes intersect at an angle of $\cos^{-1}(-\rho)$ (Ashby & Townsend, 1986). The lower panel shows that when ρ is not 0 (and the angle between the axes not 90°), the spacing between the distributions is wider along the negative diagonal than along the positive one, an effect that results from the smaller standard deviation in that direction. The optimal rule for this case is not straight lines intersecting at a right angle. In fact the larger ρ is, the closer the rule is to two parallel lines perpendicular to the negative diagonal, as in the differencing model.

Some Relevant Results

Our models for same-different performance could be tested in two ways. One test concerns the shape of the ROC. The observers in the Irwin and Francis (1995a) categorization experiment on which Example 9a is based produced ROCs supporting the independent-observation model, but these researchers have also shown that observers spontaneously adopt either strategy depending on the stimulus set (Francis & Irwin, 1995; Irwin & Francis, 1995a, 1995b). The independent-observation model applied when observers compared letters varying in orientation (correct vs. reversed), whereas the differencing model was supported by data using color patches that could vary in any direction in color space (a type of roving design).

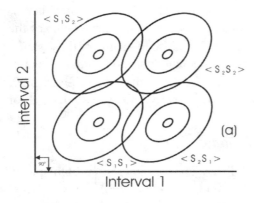

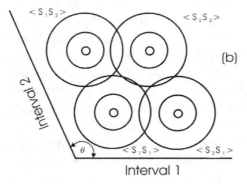

FIG. 9.8. Two equivalent representations for perceptual dependence. In (a) each of the bivariate distributions has a correlation ρ between x and y. In (b) the x and y axes meet at an angle such that $\cos(\theta) = -\rho$.

A second approach asks whether either model correctly describes the relation between same-different and other discrimination designs, and a few studies have compared same-different with performance in either 2AFC or yes-no using the differencing model. In a taste study, Hautus and Irwin (1995) found same-different d' to be just 3% higher than yes-no d'. Macmillan et al. (1988), investigating a synthetic vowel continuum, found estimated d' values to be almost exactly equal in same-different and 2AFC for both fixed and roving procedures. Chen and Macmillan (1990) found same-different d' to be 6% lower than 2AFC d' in line-length discrimination. In frequency discrimination, Creelman and Macmillan (1979) found same-different d' to be 14% lower. Creelman and Macmillan also studied a continuum of pure-tone octaves differing in relative phase, and for these stimuli the model failed: d' was 50% higher in same-different. Taylor, Forbes, and Creelman (1983) speculated about characteristics of these stimuli that might account for the discrepancy.

ABX (Matching-to-Sample)

Of the three stimuli presented on an ABX trial, the third is the focus. The first two stimuli (*A* and *B*) are standards, S_1 and S_2 in a randomly chosen order, and the observer's task is to choose which of the two is matched by the final stimulus (*X*). (A parallel notation, AX, is sometimes used for same-different. Again the first interval is not fixed as it would be in a reminder experiment; rather, *A* is a place holder for either possible stimulus.) Altogether there are four legal stimulus sequences in ABX, and they are evenly partitioned by the two responses: "A" is the correct answer for $<S_1 S_2 S_1>$ and $<S_2 S_1 S_2>$, and "B" is correct for $<S_1 S_2 S_2>$ and $<S_2 S_1 S_1>$.

Example 9b: Matching-to-Sample by Chimpanzees

In animal research, the ABX design is called matching-to-sample. Suppose we want to know whether a chimpanzee can distinguish a circle (S_2) from an ellipse (S_1). On each trial, we present the animal with an object (*X*) and two keys to press (*A* and *B*), each of which is labeled with a "sample" for the chimp to "match." The samples <*AB*> are randomly assigned to the two key positions, and thus are always in one of two orders: <ellipse, circle> or <circle, ellipse>. The third object (*X*), which may be either shape, is presented below the two keys. The chimp's task is to press whichever sample key matches object *X*. On succeeding trials, both the labels for the sample keys and the identity of the *X* stimulus are chosen anew (e.g., Spence, 1937).

Because the samples are presented unpredictably, the chimp must compare object *X* with the standards to respond correctly—this is the "matching" in "matching-to-sample." If the order were always, say, <circle, ellipse>, then comparison would not be necessary: The animal might learn to respond "A" when *X* was a circle and "B" when it was an ellipse, comparing *X* with a remembered criterion rather than the sample. Although this is a perfectly respectable discrimination design—a kind of "reminder" experiment—it is rarely performed.

Two aspects of matching-to-sample experiments are neglected in this description. First, the samples may be presented last (an XAB design), or flanking the test stimulus (AXB), as in Example 9b. Second, a delay may be imposed between the samples and the test stimulus. We describe optimal strategies that are unaffected by ordering and assume perfect memory. To capture the nature of nonoptimal processing, substantive theories must be added to psychophysical models.

The following table offers some possible data for the chimp experiment.

	Response	
Stimulus Sequence	*"A"*	*"B"*
X matches A: $<S_1S_2S_1>$ or $<S_2S_1S_2>$	30	20
X matches B: $<S_1S_2S_2>$ or $<S_2S_1S_1>$	10	40

Our analysis requires that all four sequences be equally likely, so that we can lump together the two sequences for which "A" is the correct response and the two for which "B" is correct to form the familiar 2 × 2 table. The observant reader will notice that the numerical data are the same as Example 9a, a same-different experiment. Hits are now defined as correct matches of X to the A sample, false alarms as incorrect matches of X to the A sample, so that:

$$H = P(\text{"A"}|<S_1S_2S_1> \text{ or } <S_2S_1S_2>) \tag{9.10}$$

$$F = P(\text{"A"}|<S_1S_2S_2> \text{ or } <S_2S_1S_1>) \ .$$

In this example, $H = .6$ and $F = .2$.

Summarizing discrimination data as a (false-alarm, hit) pair is, as for all designs, only a start toward finding underlying detectability. Our goal is to extract from these statistics the difference between underlying single-stimulus probability distributions. As for same-different, there are two contrasting decision rules, one using independent observations and one using differencing.

The Independent-Observation Decision Rule

In the independent-observation model, the observer has to decide two things: the order of the first two stimuli ($A - B$) and the value of the third (X). Each of these decisions corresponds to a familiar design. The subsequence $<AB>$ can be either $<S_1S_2>$ or $<S_2S_1>$ so the first two intervals, considered by themselves, compose a 2AFC task. The third stimulus X can be either S_1 or S_2 so the third interval, considered in isolation, is a yes-no experiment. Although each ABX trial contains three stimuli, there are only two independent pieces of information: the order of the samples, or standards (A and B), and the value of the third stimulus, X. If the internal variable on which A and B differ is eccentricity, then the intelligent chimp is interested in two statistics. One is the difference in eccentricity between A and B (the information needed in 2AFC), and the other is the eccentricity of X (needed in yes-no).

These two variables combine independently, producing a space similar to that for same-different, as shown in Fig. 9.9. (The model described here is that of Macmillan et al., 1977.) As usual, the figure portrays the likelihood of distributions of the internal representations. The result of comparing (i.e., subtracting) the two standards is plotted on the horizontal axis, the two possible orderings each generating a distribution, as in the 2AFC model of chapter 7. The difference distributions have means of $-d'$ and $+d'$, and variance twice as large as any single-stimulus distribution. The horizontal axis in the figure has been rescaled by dividing by $\sqrt{2}$, so that the means are at $-d'/\sqrt{2}$ and $+d'/\sqrt{2}$. The vertical axis of the figure represents the X part of a trial, on which a single stimulus drawn from one of the two distributions is presented.

A full ABX trial yields a value on each axis, and thus a point in the plane of the figure. Each of the four distributions in the figure arises from one of the four possible stimulus sequences. If the stimulus sequence is $<S_2 S_1 S_2>$, for example, then $(A-B)/\sqrt{2}$ averages $d'/\sqrt{2}$, X averages d', and the chimp's observation is drawn from the distribution at the upper right. To determine a response, the unbiased observer partitions the decision space, using vertical and horizontal criterion lines, into regions in which each response is more likely to be correct. Observations in the shaded area of Fig. 9.9, the upper right and lower left quadrants of the space, lead to an "A" response, other regions to a "B."

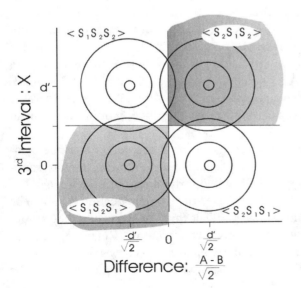

FIG. 9.9. Decision space for fixed ABX (independent-observation model). Probability distributions of joint occurrence of $A - B$ differences and X observations are shown for the four possible presentation sequences. Abscissa values are scaled by $1/\sqrt{2}$ to equate standard deviations on the two axes. The shaded region leads to the response "A" and the unshaded region to the response "B."

As always, the d' we seek to estimate is the distance between the means of the S_1 and S_2 distributions. In the ABX decision space, this is the distance between the means of the $<S_2S_1S_2>$ and $<S_2S_1S_1>$ distributions along the vertical axis. Because the decision strategy shown in Fig. 9.9 is unbiased, it is sufficient to calculate $p(c)$, which equals both H and $1 - F$, and to consider just one possible sequence. Proportion correct on an $<S_2S_1S_2>$ trial has two components, the probabilities of observations in the upper right and lower left quadrants. Each component probability is the volume over an infinite rectangular area. The analysis parallels that for the same-different design earlier in the chapter, and proportion correct can be expressed as follows:

$$p(c)_{\text{ABX, IO}} = \Phi(d'/\sqrt{2})\Phi(d'/2) + \Phi(-d'/\sqrt{2})\Phi(-d'/2) \ . \tag{9.11}$$

Equation 9.11 can be used to find proportion correct from d'. What the investigator usually wants is the inverse function, which calculates d' from proportion correct. Table A5.3, based on a table in Kaplan et al. (1978), provides a solution to this problem.

The chimp in our matching-to-sample example neglected to adopt the unbiased decision rule shown in Fig. 9.9; that is, the animal's likelihood-ratio criterion is some value other than 1.0. Boundaries in the decision space for which likelihood ratio is constant but not equal to 1.0 resemble those calculated for the same-different design (see Fig. 9.2). For each possible value of likelihood ratio, the hit and false-alarm rates can be computed by numerical integration. When this is done for many values of likelihood ratio, an ROC curve results. It turns out that the ROC has unit slope, so sensitivity depends only on $z(H) - z(F)$, and (as was true for same-different) d' can be determined by a two-step procedure. First, find $z(H) - z(F)$ using Table A5.1 and then convert to d' by using Table A5.3. For our chimps, $z(H) - z(F) = z(.60) - z(.20) = 1.095$. According to Table A5.3, $d' = 1.57$. A given performance level, notice, is more difficult to reach in ABX than in yes-no.

The three types of bias measures discussed in chapter 2 can all be computed from ABX data. Absolute and relative criterion location are mimicked by $0.5[z(H) + z(F)]$ with or without dividing by $z(H) - z(F)$. The association is not precise because the idea of criterion location is, as can be seen in Fig. 9.9, a two-dimensional one. The third measure, likelihood ratio, is conceptually simple but computationally unpleasant. Remember that, as in the one-interval experiment, all these measures convey exactly the same information when sensitivity is constant; when it is not, a choice must be made along the lines sketched in chapter 2. Bias measures derived explicitly for the ABX task have not been developed.

Roving ABX: Another Differencing Model

The fixed versus roving distinction applies to ABX. For our chimpanzee, the issue is whether every trial contains only a particular pair of circle and ellipse or whether trials with ellipses of different degrees of eccentricity might be intermixed.

In the roving design, the decision rule illustrated in Fig. 9.9 will not work. Suppose there are three possible stimuli, a circle, a broad ellipse, and a narrow ellipse, and a particular trial happens to contain the triplet <circle, broad ellipse, broad ellipse>. According to our model, the first two intervals are subtracted, giving, on the average, a negative value—we are in the left half of Fig. 9.9—but which quadrant are we in? The observer does not know whether an obtained value of X for the third interval should be treated as a large value of eccentricity (relative to a circle) or a small one (relative to a narrow ellipse), and thus does not know whether to respond "A" or "B."

A better decision procedure for the roving design (first described by Pierce and Gilbert, 1958) is to compare each sample A and B directly with X. As for same-different, we contrast this differencing rule with the independent-observation strategy. The differencing participant calculates two differences, $A - X$ and $B - X$, and then faces the decision problem shown in Fig. 9.10. Because both axes depend on the third (X) interval, they are correlated, and the distributions are elliptical. For the unbiased observer,

$$p(c)_{\text{ABX, diff}} = \Phi(d'/\sqrt{2})\Phi(d'/\sqrt{6}) + \Phi(-d'/\sqrt{2})\Phi(-d'/\sqrt{6}) \ . \quad (9.12)$$

As with the independent-observation strategy, varying the critical value of likelihood ratio generates an ROC with approximately unit slope. This means that $z(H) - z(F)$ has the same value no matter what the criterion is. Table A5.3 contains a column for finding d' in the ABX task, assuming the differencing rule. If the false-alarm/hit pair (.2, .6) occurs in a roving ABX experiment, the corresponding true d' is 1.76. For all values of $z(H) - z(F)$, the differencing rule gives a higher value of d' than the independent-observation rule (Eq. 9.11). If the stimuli (and therefore d') are held constant, proportion correct is lower according to the differencing model because of the additional variance contributed by doing two subtractions rather than one.

The alternatives for measuring bias are the same as for the independent-observation model.

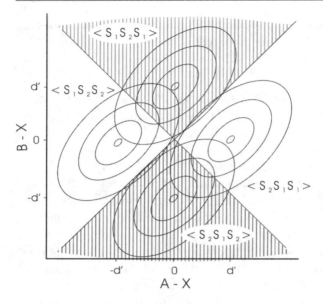

FIG. 9.10. Decision space for roving ABX (differencing model). Probability distributions of joint occurrence of $A - X$ and $B - X$ values are shown for the four possible presentation sequences. Elliptical equal-probability contours result because the axes are correlated positively: The observed value of the third presentation, X, contributes to each value. The shaded region leads to the response "A" and the unshaded region to the response "B."

Some Relevant Results

Do these models successfully account for performance in the ABX design? Hautus and Meng (2002) conducted a series of ABX experiments in which observers discriminated Gaussian *distributions* of circle size, number value, and tone amplitude. The use of distributions allowed the observers' decision bounds to be estimated (see chap. 12 for more detail on this experimental approach). In all experiments, the differencing strategy provided a good description of these bounds, although extensive experience and feedback should have permitted use of the optimal independent-observation strategy. Hautus and Meng speculated that differencing was preferred because of its minimal cognitive demands.

Two experiments that have compared ABX and 2AFC performance are modestly encouraging. Creelman and Macmillan (1979) found the ratio of d' values to be 0.85 for auditory frequency and 1.11 for auditory phase (for both dimensions, the ABX/yes-no ratio was substantially higher). Macmillan (1987) found a ratio of 0.98 for a pluck-bow continuum.

Threshold Analysis

"Nonparametric" (threshold) analysis of ABX grows from the assumption of discrete internal states, as delineated in chapter 4. In this view (Pollack &

Pisoni, 1971), each of the three presentations on a trial is covertly classified as either S_1 or S_2, and a decision to respond "A" or "B" is based on these classifications. Proportion correct is related to p_1 and p_2, the probabilities that stimuli S_1 and S_2 are covertly classified as S_2, by

$$p(c)_{ABX} = \tfrac{1}{2}[1 + (p_2 - p_1)^2] . \qquad (9.13)$$

The flaws in proportion correct as a performance measure, outlined earlier for other designs, also undercut its use to summarize ABX data. Specifically, the ROC implied by this measure is strongly nonlinear on z coordinates. Because bias (toward the "B" response) is common in temporal ABX, this is a significant shortcoming. For a constant sensitivity, $p(c)$ is smaller for biased than for unbiased responding, so it is easy to mistake criterion changes for sensitivity effects.

A comparison of Equations 9.5 and 9.13 reveals that, according to threshold theory, the value predicted for proportion correct in same-different is exactly the same as in ABX. Experiments that have compared the two paradigms (Creelman & Macmillan, 1979; Pastore, Friedman, & Baffuto, 1976; Rosner, 1984) generally have not supported this prediction, but have instead found $p(c)$ to be higher in ABX, consistent with SDT analysis.

Oddity (Triangular Method)

Another three-stimulus task is the oddity design. The observer is presented with a "triangle" <*ABC*> of stimuli, two of which are alike, and is asked to locate the "odd" one, which may be any of the three. The identity of the minority stimulus is not known to the observer and can be either S_1 or S_2. There are six possible stimulus sequences: <$S_1S_2S_2$> and <$S_2S_1S_1$>, for both of which the correct response is "A"; <$S_2S_1S_2$> and <$S_1S_2S_1$>, for which the answer is "B"; and <$S_2S_2S_1$> and <$S_1S_1S_2$>, for which the answer is "C."

Oddity offers a new complication: Three rather two responses are allowed. To our knowledge, no models for characterizing response bias have been developed for this design. Oddity is not restricted to three intervals, but could include any number (although "triangular method" would be a poor description of four-interval oddity). In practice, three interval is the most popular variant.

Example 9c: Taste Discrimination

Oddity is a frequently used design in "sensory evaluation" experiments conducted by food scientists to measure sensitivity to differences in taste and smell. We consider an enjoyable experiment of this sort, in which tasters attempt to distinguish between two wines: a Burgundy and a claret. Professional wine tasters, it should be said, would be unlikely to use this method because it does not require that they be able to say which wine was which. The oddity task is suitable for the enthusiastic novice, who might be learning the aspects of taste and smell that differentiate wines, but still would find identification of the dimensions of difference difficult.

On each trial, the taster receives three wine samples, two of one type and one of the other. Whether the odd glass has Burgundy or claret is randomly decided for each trial, as is the location of the odd glass among the three. Because there are six possible stimulus sequences and three responses, the data from this study are best summarized in a 6×3 matrix, but in practice the overall proportion correct is almost always reported. Let us suppose our tasters, mimicking the results of an experiment with "aqueous solutions of simple compounds" by Byers and Abrams (1953; described by Frijters, 1979a), are correct on 21 of 45 trials.

Measuring Sensitivity

A decision rule for the oddity task has been described by Frijters (1979b). The observer compares each pair of presentations in the triplet, determines the pair with the smallest difference, and chooses the response corresponding to the remaining stimulus. Thus, if glasses A and B are most similar in taste, glass C is most different from the others, and response "C" is given. Because the observer knows nothing of the dimensions of judgment, the absolute differences are used. In the language of our other models, this is a differencing, not an independent-observation rule.

The problem for the observer is portrayed in Fig. 9.11. The dimensions of the space are two of the differences computed by the observer, those between the effects of Intervals A and B and between Intervals B and C. Each triplet is composed of samples from S_1 (Burgundy) and S_2 (claret). The distance between a single S_1 stimulus and a single S_2 is, as always, d'. Hence, the six possible sequences are readily located in the space. The distributions corresponding to each stimulus are not circular because both axes depend on stimulus presentation B, so that the two dimensions covary negatively.

The taster's decision rule is this: Find the smallest of the three differences between pairs of stimuli, and select the response corresponding to the stimulus that is *not* in that pair. The decision boundaries arising from this rule are shown in Fig. 9.11. At first, it may seem surprising that the areas allocated to the three responses are not equal. The area in which response "B" is

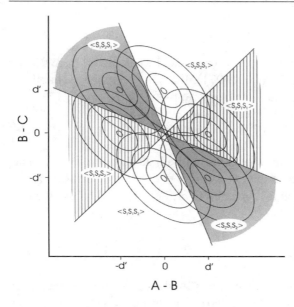

FIG. 9.11. Decision space for the oddity (ABC) task. The joint probability distributions of $A - B$ and $B - C$ observations for the six possible presentation sequences are shown. Elliptical equal-probability contours result from correlation between the axes (B contributes to both). Decision boundaries separate unequal areas of the decision space because of this asymmetry. The region with vertical shading leads to response "A," the dark region to response "B," and the unshaded region to response "C."

appropriate is smaller only because of the covariance noted earlier; the model does not predict any asymmetry in response rates.

Craven (1992) calculated sensitivity as a function of proportion correct for this decision rule and its extensions to m-alternative oddity, m ranging from 3 up to 32. The results are given in Table A5.5. An observer who correctly chooses the odd wine on 21 of 45 trials [$p(c) = .47$] has a d' of 1.31.

This rule is clearly in the differencing family; is there an independent-observation model for this paradigm? Versfeld, Dai, and Green (1996) derived predictions for such a model. As in the independent-observation rule for the same-different paradigm, the observer does not subtract values, but instead computes the likelihood of the multi-interval observation under each of the two hypotheses and bases a decision on these likelihoods. For $m = 3$, the representation is thus four-dimensional (the three intervals plus likelihood), and we do not attempt to illustrate it here. Table A5.6 gives values of $p(c)$ for both the differencing and independent-observation models, for $m = 3$, 4, and 5. For our wine taster, a $p(c)$ of .47 has a d' of 1.10; this is lower than under differencing assumptions because an optimal model requires a lower d' than a nonoptimal one to reach any given level of performance.

Threshold Analysis

Proportion correct by an unbiased observer can be calculated from threshold assumptions (Pollack & Pisoni, 1971). If p_1 and p_2 are the probabilities of covertly identifying S_1 and S_2 as being S_2, then

$$p(c) = \tfrac{1}{3}[1 + 2(p_2 - p_1)^2] \ . \tag{9.14}$$

Performance according to this model is intermediate between the differencing and independent-observation rules. If $p_2 = 1 - p_1$, then Equation 9.14 can be solved for p_2 and this value converted to d'. For $p(c) = .47$, as in the running example, $d' = 1.20$.

Summary

In a same-different experiment, a pair of stimuli is presented on each trial, and the observer decides whether its two elements are the same or different. Two stimuli generate four possible stimulus pairs in such an experiment. The optimal strategy for the observer is to treat the two observations independently. Even with this approach, the task is more difficult than yes-no. When the stimulus level is roving, the optimal strategy may not be available, and a differencing rule may be used instead. In this approach, only the difference between the two observations on a trial is used in making a decision. Performance is poorer than with the independent-observation strategy, especially in remote regions of ROC space.

In an ABX experiment, three stimuli are presented on each trial; the third presentation matches one of the first two, and the observer's task is to decide which. With two stimuli, four stimulus triplets are possible in this experiment. The optimal independent-observation strategy is to independently assess (a) the difference between *A* and *B* and (b) *X*. When stimulus level is roving, the optimal strategy may not be available, and a differencing rule may be used instead. In this approach, two differences contribute to the decision: $A - X$ and $B - X$. Performance is poorer than with the independent-observation model.

In the three-alternative oddity task, two of the three presentations are the same, and the observer must select the different one. Six stimulus triplets are possible. The differencing model proposes that the observer finds the smallest of the three differences and chooses the response corresponding to the stimulus that does not contribute to it. The independent-observation model requires covert identification of all presentations. Both models assume unbiased responding.

Nonparametric (threshold) analysis—use of proportion correct to summarize data—is identical with the independent-observation approach for an unbiased observer in both same-different and ABX. In the oddity task, it predicts performance intermediate between the models. For biased observers, threshold assumptions militate against the use of the model in all tasks.

Methods appropriate for finding sensitivity in these paradigms are given in Chart 6 of Appendix 3, those for finding response bias in Chart 7.

Computational Appendix

*Finding d' From Unbiased p(c) in Same-Different
Independent-Observations Model*

From Equation 9.2,

$$p(c)_{\text{SD,IO}} = [\Phi(d'/2)]^2 + [\Phi(-d'/2)]^2 .$$

This equals

$$p(c)_{\text{SD,IO}} = 2[\Phi(d'/2)]^2 - 2\Phi(d'/2) + 1 .$$

Solving for $\Phi(d'/2)$ by using the quadratic formula leads to:

$$\Phi(d'/2) = \tfrac{1}{2}[1 + [2\,p(c)_{\text{SD,IO}} - 1]^{1/2}] .$$

Then apply the z transformation to both sides and multiply by 2 to get

$$d' = 2z[\tfrac{1}{2}(1 + [2\,p(c)_{\text{SD,IO}} - 1]^{1/2})] ,$$

which is Equation 9.3.

*Relation Between Yes-No and Same-Different
in Threshold Theory*

From Equations 9.5 and 9.6,

$$p(c)_{\text{same-different}} = \tfrac{1}{2}[1 + (p_2 - p_1)^2]$$

and

$$p(c)_{\text{yes-no}} = \tfrac{1}{2}[1 + (p_2 - p_1)] .$$

Solving the second equation for $p_2 - p_1$ gives

$$p_2 - p_1 = 2\,p(c)_{\text{yes-no}} - 1 .$$

Substituting into the expression for $p(c)_{\text{same-different}}$,

$$p(c)_{\text{same-different}} = \tfrac{1}{2}\{1 + [2\, p(c)_{\text{yes-no}} - 1]^2\}\ .$$

Expanding this yields

$$p(c)_{\text{same-different}} = p(c)^2_{\text{yes-no}} + [1 - p(c)_{\text{yes-no}}]^2\ ,$$

which is Equation 9.4.

Criterion Location Measure c_d in Same-Different Differencing Model

The likelihood ratio in the differencing model, as given in Equation 9.9, can be rewritten:

$$\beta_d = \tfrac{1}{2} e^{-\frac{1}{4} d'^2} \left(e^{\frac{1}{2} d'k} + e^{-\frac{1}{2} d'k} \right) \tag{9.15}$$

$$= e^{-\frac{1}{4} d'^2} \cosh(d'k / 2)\ ,$$

where cosh is the hyperbolic cosine. The value of the decision axis at which $\beta_d = 1$ can be found be solving this equation for k:

$$k = \left[\frac{2}{d'} \cosh^{-1}\left(e^{\frac{1}{4} d'^2} \right) \right]^{-1}\ , \tag{9.16}$$

where $\cosh^{-1}$ is the inverse hyperbolic cosine. To find c_d, first estimate k for the data at hand from Equation 9.8 (which requires iteration), then subtract from it the value in Equation 9.16.

Problems

9.1. Interpret the following stimulus–response matrixes (repeated from Problem 7.1) as arising from a same-different experiment. (Different pairs correspond to the top row, "different" responses to the left-hand columns.) Calculate d' assuming (a) the independent-observation model to be correct, and (b) the differencing model to be correct.

Matrix A		Matrix B		Matrix C		Matrix D	
12	8	18	2	4	16	9	6
8	12	14	6	1	19	2	1

9.2. You observe $H = 1 - F = .9$ in a yes-no experiment, and you construct a same-different experiment with the same set of stimuli. If the observer continues to be unbiased, what do you expect $p(c)$ to be according to the threshold model? The independent-observation model? The differencing model?

9.3. Repeat Problem 9.2, except now assume that the observer is *not* unbiased in same-different, but has $F = .05$.

9.4. The following table shows data from a roving same-different experiment. Assuming the differencing model, find d' and c_d for both the $<S_1S_2>$ and $<S_2S_3>$ pairs. Which is weighted more heavily in these estimates, Same trials or Different trials? Looking at the entire data matrix, find $p(c)*$ and compare it to the average of the hit and correct-rejection rates. Why are these not the same?

	Response	
Stimuli	*"Different"*	*"Same"*
$<S_1S_1>$	6	14
$<S_1S_2>$	12	8
$<S_2S_1>$	10	10
$<S_2S_2>$	4	16
$<S_2S_3>$	14	6
$<S_3S_2>$	12	8
$<S_3S_3>$	2	18

9.5. You observe $p(c) = .95$ (symmetric bias) in a fixed same-different task. What is d'? What would you predict $p(c)$ to be in yes-no? 2AFC?

9.6. In the differencing model for same-different, the observer's decision is assumed to be based on the difference between the observations from the two intervals. Can you devise a decision rule that uses the *sum* of the observations? How well will someone using this rule perform compared with someone using the differencing rule?

9.7. Interpret the matrixes of Problem 9.1 as arising from an ABX experiment. ($A = X$ corresponds to the top row, "A" responses to the left-hand columns.) Calculate d' assuming (a) the independent-observation model to be correct, and (b) the differencing model to be correct.

9.8. Suppose $d' = 1$ in a fixed-level experiment and the participant is unbiased. What is $p(c)$ in 2AFC, ABX, same-different, and oddity? Repeat for $d' = 2$. Is the ordering of conditions the same at both levels?

9.9 Suppose $d' = 1$ in a roving-level experiment and the participant is unbiased. What is $p(c)$ in 2AFC, ABX, same-different, and oddity? Repeat for $d' = 2$. Is the ordering of conditions the same at both levels? The same for roving as for fixed paradigms?

9.10. Repeat Problems 9.8 and 9.9, but assume that $F = .1$ and find H for each paradigm.

10

Identification of Multidimensional Objects and Multiple Observation Intervals

In an *identification* experiment,[1] a single stimulus from a known set is presented on each trial, and it is the observer's job to say which it was—to identify it. The purposes of such experiments vary, but usually include obtaining an overall index of performance, as well as a measure of sensitivity for each stimulus pair and bias for each response.

If there are only two stimuli, identification is simply the yes-no task of chapters 1 and 2, and performance can be summarized by one sensitivity and one bias parameter. The nature of the stimuli is unimportant—it does not even matter if they differ along one physical dimension (lights of different luminance) or many (X-rays of normal and diseased tissue). With more than two stimuli, the task is easily described: One stimulus from a set of m is presented on each trial, and the observer must say which it was. From the participant's point of view, there is nothing more to say, but to extend the analysis to more than two stimuli the dimensionality of the representation must be known. If all stimuli differ perceptually on a single dimension, then $m - 1$ sensitivity distances between adjacent stimuli and $m - 1$ criterion locations can be found along it, as we saw in chapter 5. Perceptual distances for all other pairs of stimuli are easily calculated as the sum of the stepwise distances between them. To characterize overall performance, it is natural to add sensitivity distances across the range.

---

[1]The term *identification* has another meaning in speech perception, where it describes what we have termed a two-category classification experiment, in which psychometric functions are collected. Paradigms like those in this chapter are sometimes distinguished by being termed *absolute identification*.

The assumption of unidimensionality is a restrictive one, and in this chapter we consider two multidimensional cases. In the first, all members of the stimulus set are independent of each other and may be thought of as being processed by different channels. In our perceptual-space models, each stimulus produces a mean shift along a different dimension. An important application is to the special case in which identification is of intervals in *m-alternative forced-choice* experiments. In a second experimental situation, the *feature-complete factorial design*, values on each of two or more dimensions are manipulated independently. This design is useful in assessing interactions between dimensions, a topic we introduced (using simple discrimination experiments) in chapter 8.

Object Identification

Example 10a: Letter Recognition

Consider a letter recognition task: The observer fixates the center of a video terminal on which is displayed, briefly, a single letter followed by a "mask" that serves to disrupt retinal storage of the stimulus. One of just four letters can occur: N, O, P, or S. The task is to press a computer key corresponding to the letter shown. Let us suppose that an observer obtains $p(c) = .5$ in this task.

We adopt the simplifying assumption that these four letters are processed by independent channels. (Although this is too strong a requirement, it is certainly better than assuming that the letters differ along a single dimension.) The decision space contains $m = 4$ distributions, each removed from a common origin in a different dimension. Using the notation of previous chapters, the m sequences to be distinguished can be written $<S_1N_2N_3N_4>$, $<N_1S_2N_3N_4>$, $<N_1N_2S_3N_4>$, and $<N_1N_2N_3S_4>$.

Sensitivity (Assuming No Bias)

The simplest (and most optimistic) calculations assume not only that each stimulus activates a separate, orthogonal channel, and that each is equally far from the Null stimulus N, but also that there is no bias. In this case, $p(c)$ can be used to summarize accuracy. An SDT analysis that relates the proportion correct to d' was developed by Elliott (1964) and improved by Hacker and Ratcliff (1979). Table A5.7 makes the latter calculations available. For each value of proportion correct, the columns show the associated value of d' for different numbers of alternatives. For our observer, $p(c) = .5$ and $m = 4$ implies a d' of 0.84. The $m = 2$ column gives values for the two-choice experiment. For example, if $p(c) = .75$ when $m = 2$, then $d' = 0.95$. The table shows negative values of d' for $p(c)$ less than $1/m$, because to score reliably below chance the observer must know enough systematically to avoid the correct alternative.

For Choice Theory, again assuming no bias, $\ln(\alpha)$ can be calculated using an equation given by Luce (1963a, p. 140):

$$\ln(\alpha) = \frac{1}{\sqrt{2}} \ln\left[\frac{(m-1)p(c)}{1-p(c)}\right] . \tag{10.1}$$

Notice that $\ln(\alpha) = 0$ when $p(c) = 1/m$ (the chance level) and reduces to the unbiased case of Equation 7.3 when $m = 2$. According to Equation 10.1, a score of .5 in 4AFC and a score of .75 in 2AFC each indicate that $\ln(\alpha)$ equals 0.78.

In the general Choice Theory model, sensitivity is related to the product of the odds ratios that, for each cell, compare the probability of the response actually given to the probability of the correct response:

$$\ln(\alpha) = \frac{1}{\sqrt{2m(m-1)}} \ln\left[\prod \frac{P(R_i|S_i)}{P(R_j|S_i)}\right] . \tag{10.2}$$

The product $\prod$ is over all cells in the stimulus–response matrix. When $i = j$, the ratio $P(R_i|S_i)/P(R_j|S_i)$ equals 1 and may be ignored, so the product is effectively over all the nondiagonal cells (those in which $i \neq j$). Equation 10.1 is the special case in which all the proportions of correct responding $P(R_i|S_i)$ are equal to p and all proportions of incorrect responding $P(R_j|S_i)$ (for $i \neq j$) are equal to $(1 - p)/(m - 1)$.

A Model Assuming "Bias Constancy": The Constant-Ratio Rule

Luce (1959) developed Choice Theory from an axiom about the relation among recognition tasks using different-sized subsets from a common universe (see chap. 4). According to the choice axiom, the ratios of response frequencies in a confusion matrix do not depend on the number of stimuli in the experiment. This *constant ratio rule* (Clarke, 1957) can be applied, for example, to the data from Example 5f, in which the four stimuli were tones differing in intensity. Table 10.1 gives the number of responses out of 50 presentations for each stimulus in a set of four.

If we were to eliminate Stimulus 4, according to the constant ratio rule we should find the results in the lower part of Table 10.1. The proportions in this table are calculated by dividing original frequencies by the total frequency in the first three columns of each row. The response proportions for Stimulus 3, for instance, are equal to the frequencies 11, 10, and 12, each divided by their total, 33.

TABLE 10.1   *Results of an Identification Experiment With Four Stimuli and Four Responses*

	Response			
(a) Original Data				
Stimulus	1	2	3	4
1	39	7	3	1
2	17	12	10	11
3	11	10	12	17
4	3	5	9	33
(b) Proportions in Three-Stimulus Identification Predicted by the Constant Ratio Rule				
	1	2	3	
1	.80	.14	.06	
2	.44	.31	.26	
3	.33	.30	.36	

Because the constant ratio rule can extract a 2×2 matrix from a larger one, it can be used to calculate sensitivity for discriminating any stimulus pair. In an experiment involving only Stimuli 2 and 3, the constant ratio rule predicts $P(R_2|S_2) = 12/(12 + 10) = .55$ and $P(R_2|S_3) = .45$. Either SDT or Choice Theory can be used to calculate sensitivity. The predicted value of $\ln(\alpha)$ is $\ln[(.55 \times .55)/(.45 \times .45)]^{\frac{1}{2}} = 0.2$. Predicted $d'_{2,3} = z(.55) - z(.45) = 0.25$—lower than that predicted under unidimensional assumptions in chapter 5, where we found $d'_{2,3} = 0.4$ for this example. Hodge (1967; Hodge & Pollack, 1962) concluded that the constant ratio rule was more successful when applied to multidimensional than one-dimensional stimulus domains.

The constant ratio rule is a variant of Choice Theory, in which bias is presumed not to depend on the stimulus subset being studied (Luce, 1963a). This is a strong and (Luce suggested) uncongenial assumption. One way in which the assumption could be correct is for participants to be unbiased in all conditions.

A Complete Model

Bias assumptions can be avoided by using a more systematic approach. A complete Choice Theory solution, which calculates bias parameters for each response and discriminability measures for each pair of stimuli, is provided by Smith (1982b, Appendix B).

Interval Identification: m-Alternative Forced Choice (mAFC)

Now that we know how to model the identification of the correct object in a set of any size, it is easy to translate to the identification of one *interval* in which a stimulus might be presented. The analogous task is one in which there are m spatial or temporal intervals, one containing S_2 and the others S_1. The analytic problem is formally the same as for identification of objects, just as the same-different discrimination task was formally the same as divided attention.

Example 10b: Multiple-Choice Exams

An obvious educational application of mAFC is the venerable multiple-choice exam, in which one correct and $m - 1$ incorrect choices are provided for each question. We wish to estimate true sensitivity for a student for whom $p(c)$ in a four-alternative exam is .5, perhaps for comparison with another student who scores .75 on a two-alternative version. To make such comparisons possible, our models must apply to any number of alternatives.

Representation and Sensitivity

In the initial statement of the 2AFC problem (Fig. 7.1), each interval corresponded to a separate dimension in the decision space. This representation is also appropriate for $m > 2$ intervals, so there are as many dimensions in the representation as there are intervals in the task. The optimal unbiased strategy is to choose the interval with the largest observation. In 2AFC, this rule is equivalent to basing the decision on $A - B$ and using a criterion of 0, so the task can be analyzed as a comparison design. No such shortcut is possible for $m > 2$.

The models of the previous section apply directly to the *m*AFC problem, and the assumptions of equal sensitivity and independent effects for all alternatives are apparently quite reasonable. If we are still willing to assume unbiased responding (a less compelling assumption), we can use Equation 10.1 to convert $p(c)$ to a distance measure. The calculations of the previous section allow estimation of SDT and Choice Theory sensitivity parameters. Thus, $p(c) = .5$ in 4AFC corresponds to a d' of 0.84 and a $\ln(\alpha)$ of 0.78. According to SDT, the comparison student for whom $p(c) = .75$ in 2AFC is superior: $d' = 0.95$ (Eq. 7.1). According to Choice Theory, $\ln(\alpha) = 0.78$ for both students.

For forced-choice experiments with unbiased decision rules, distributions consistent with Choice Theory are not logistic, but double-exponential (Yellott, 1977). Tables of *m*AFC performance for logistic distributions have been published as well (Frijters, Kooistra, & Vereijken, 1980).

Response Bias in mAFC

The Choice Theory model for sensitivity given by Equation 10.2 also specifies $m - 1$ bias parameters. Each response R_j has a corresponding bias b_j, but only the ratios between biases can be estimated:

$$\ln\left(\frac{b_i}{b_j}\right) = \frac{1}{m}\ln\left[\prod \frac{P(R_i|S_k)}{P(R_j|S_k)}\right] . \tag{10.3}$$

The product is over all possible stimuli S_k.

As we have seen, bias is customarily ignored in analyzing *m*AFC data. That it does not therefore go away is shown in some 4AFC experiments of Nisbett and Wilson (1977), whom we quote:

> In both studies, conducted in commercial establishments under the guise of a consumer survey, passersby were invited to evaluate articles of clothing—four different nightgowns in one study ... and four identical pairs of nylon stockings in the other .... [T]he right-most object in the array was heavily overchosen. For the stockings, the effect was quite large, with the right-most stockings being preferred over the left-most by a factor of almost four to one.(p. 243)

This study provides an interesting insight into the unconscious nature of response bias:

> When asked the reasons for their choices, no subject ever mentioned spontaneously the position of the article in the array. And, when asked directly about a possible effect of the position of the article, virtually all

subjects denied it, usually with a worried glance at the interviewer suggesting that they felt either that they had misunderstood the question or were dealing with a madman. (pp. 243–244)

Nisbett and Wilson's experiments were unusual in that d' sometimes equaled 0, so that there was no basis for choice in the task they put to their participants. In other conditions, however, the stimuli really did differ. As in 2AFC, proportion correct is highest when bias is least, so the effect of asymmetrical responding is to depress measures of sensitivity that ignore the possibility of bias.

Interval effects were found in psychophysical tasks by Johnson, Watson, and Kelly (1984), who observed that $p(c)$ was higher for the third interval of a 3AFC design than for the first. Such a result could arise from either bias or sensitivity changes across intervals. Bias effects can be diagnosed by application of Equation 10.3; to uncover sensitivity effects requires a more complex model.

mAFC Compared With 2AFC and Yes-No

Do our equations and tables for *m*AFC accurately describe the relations among two-, three-, and higher-choice paradigms? We know of few data that address this question, but in an early experiment on tone detection Swets (1959; cited in Green & Swets, 1966, pp. 111–113) found performance to be well predicted by SDT for up to eight choices. Equation 10.1 makes similar, and thus equally compelling, predictions for Choice Theory.

The Boundary Theorem.   Detection theory models make, as always, distributional assumptions. Shaw (1980) showed that if the decision rule is unbiased, a lower limit can be placed on *m*AFC performance regardless of the shape of the underlying distribution. Her *boundary theorem* relates this lower limit, in *m*AFC, to 2AFC performance by a generalization of the area theorem:

$$p(c)_{mAFC} \geq [p(c)_{2AFC}]^{m-1} \ . \tag{10.4}$$

Table 10.2 compares this lower bound with the level of performance predicted from SDT. For moderate to high sensitivities, the two values are quite close.

Threshold Measures (Correction for Guessing) in mAFC

We saw earlier that 2AFC data could be "corrected for guessing" (Eq. 7.8). An equivalent correction has also been applied to *m*AFC; because the guessing rate is $1/m$,

TABLE 10.2   *Gaussian Predictions (Table A5.7) and Boundary-Theorem Limits (Eq. 10.4) for Proportion Correct in mAFC by an Unbiased Observer*

d'	$p(c)_{2AFC}$	$p(c)_{3AFC}$		$p(c)_{4AFC}$	
		Gaussian	Boundary	Gaussian	Boundary
0.5	.64	.48	.41	.39	.26
1.0	.76	.63	.58	.55	.44
2.0	.92	.865	.85	.82	.78
3.0	.983	.969	.966	.956	.950

$$q_{mAFC} = \frac{p(c) - \frac{1}{m}}{1 - \frac{1}{m}} = \frac{mp(c) - 1}{m - 1} . \tag{10.5}$$

This correction has been applied to standardized multiple-choice examinations of the sort used for college admission. The idea, apparently, is that students who answer more questions are probably guessing more often on problems that they cannot solve, and that the "correction" will rectify an otherwise unfair advantage of this strategy.

As we saw in chapter 4, the correction for guessing implies a threshold model. In the context of multiple-choice exams, this means that students are assumed either to know the answer or guess at random. This denial of the possibility of partial information will strike the experienced test taker as unrealistic. Often several alternatives in a four-alternative test item can be rejected, and even a completely random choice among the others will lead to a score above chance. Indeed Equation 10.5 expresses a quite different relation among *m*AFC paradigms from our detection theory models. For example, q is .50 for $p(c) = .75$ in 2AFC, but equals only .33 for $p(c) = .50$ in 4AFC; recall that both SDT and Choice Theory assess these performances as quite similar. Because the sparse existing data (Swets, 1959) support the detection theory models, they reject the correction for guessing, which should probably not be used.

Comparisons Among Discrimination Paradigms

With the treatment of *m*AFC just completed, we have covered the last discrimination design in the book, and the time has come to compare them all. As measured by d', all tasks yield the same outcome—or so detection theory asserts. What differences should we expect among paradigms if performance is measured by proportion correct?

The "Best" Paradigm?

We do not need to compare all paradigms and models with each other. The rating design can be ignored because performance in that experiment is the same as in yes-no. We arbitrarily omit mAFC and oddity paradigms with more than three intervals. The remaining designs are: yes-no, 2AFC, 3AFC, same-different, ABX, and oddity. Except for yes-no, any design can be used in either a fixed or roving experiment, and some designs have distinct models (independent-observation versus differencing) for these two applications. We consider fixed and roving experiments separately.

To determine $p(c)$ for these paradigms, we assume that observers are unbiased—we must make some assumption about bias, and we have already adopted this one in our analysis of 3AFC and oddity. Figures 10.1 (for fixed experiments) and 10.2 (for roving) show the results for d' between 0 and about 5. The designs vary greatly in performance level as measured by $p(c)$, and the magnitudes of the differences depend on d'. Performance in ABX is well below that for yes-no at low sensitivity levels, for example, but very close when d' is large. Three-interval forced-choice, in which $p(c)$ is perforce lower than in yes-no when d' is small, outperforms that task easily at high accuracy levels.

Experimenters frequently measure discrimination $p(c)$ for stimulus differences of varying size, generating a "psychometric function," as discussed in chapter 5. That the curves in Figs. 10.1 and 10.2 are not linearly related to each other implies that the shapes of these psychometric functions differ across paradigms. Our models offer a solution: Plots of d' versus stimulus difference should be invariant with paradigm.

The Importance of Knowing What the Design Is

It may take a little thought to be sure what design has actually been used in someone else's discrimination experiment or even one's own. One confusing factor is the lack of agreement on terminology. For example, "same-different" is used by some writers to refer to what we have called "reminder" experiments, and "forced-choice" has been used to describe experiments other than mAFC simply because the observer is obliged to respond on every trial. A second difficulty is in deciding among competing models for the same paradigm—for example, independent-observation versus differencing strategies—when either is possible.

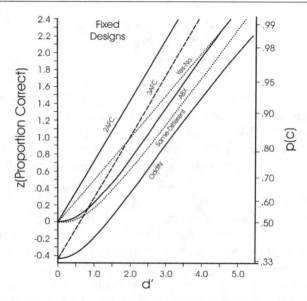

FIG. 10.1.   Proportion correct as a function of d' for six different experimental paradigms with fixed stimuli. Independent-observation models are assumed for ABX, same-different, and oddity. Unbiased responding is assumed for all paradigms.

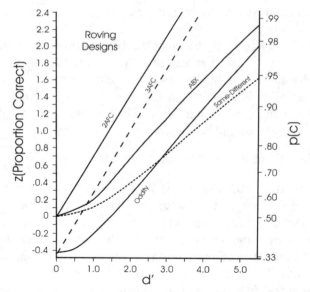

FIG. 10.2.   Proportion correct as a function of d' for five different experimental paradigms with roving stimuli. Differencing models are assumed for ABX, same-different, and oddity. Unbiased responding is assumed for all paradigms.

A third problem arises when investigators invent paradigms that appear new, but (at least as far as detection theory is concerned) are not. For example, consider the design (discussed in chap. 7) in which the possible presentations are <*AABA*> and <*ABAA*>. The observer is instructed to decide whether the second or third interval is unlike the others. Analytically, the end intervals are reminders and are ignored by an optimal participant, so this design is simply 2AFC. Of course it is an empirical question whether performance will be the same: Perhaps the presence of the end intervals and the instructions will serve to encourage inappropriate difference judgments and lower performance, or perhaps the reminders will reduce memory variance and actually improve accuracy.

Occasionally an incorrect characterization of the paradigm leads to substantive confusion. In chapter 9, we mentioned an experiment by Byers and Abrams (1953) in which proportion correct was .47 in a three-alternative oddity task. These investigators reported a "paradox": When the tasters were presented a second time with the 24 triplets to which they had not responded correctly and were asked to choose the weakest or strongest stimulus (whichever was appropriate), they were successful in 17 (71%) of these cases. The paradox lay in the ability to give relatively accurate reports to triplets that had previously not been discriminable.

The paradox depends on the assumption that incorrect responses are based on a total lack of knowledge (see Frijters, 1979b). The use of $p(c)$ to compare the original oddity task with the later 3AFC introduces a threshold model. According to our continuous differencing model (Table A5.5), the sensitivity corresponding to $p(c) = .71$ is $d' = 1.28$, in good agreement with the d' of 1.31 found earlier for oddity. Our analysis leads to a prediction: The same level of performance, 71%, should be found for the 21 stimulus triplets that were correctly responded to in the initial oddity task.

Some final thoughts on comparing discrimination paradigms are contained in this chapter's Essay.

Simultaneous Detection and Identification

In some situations, detection and identification are both interesting. (Obviously, the detection must be under uncertainty or else there is nothing to identify.) In the laboratory, participants may try to detect a grating that has one of several frequencies and also to identify which grating was seen. In eyewitness testimony, the witness must both "detect" whether a perpetrator is present (in the lineup, or in court) and also identify which person that is.

*Example 10c: Measuring Detection and Identification
Performance*

The next example mimics the X-ray detection/spatial identification task of
Starr, Metz, Lusted, and Goodenough (1975). Possible data are shown in Table 10.3a. Two of the table's three rows are familiar from chapter 3: The top
row gives the proportion of responses (R) in each rating category when a
shadow was present in one quadrant of the X-ray stimulus, and the bottom
row gives false-alarm data from trials on which no signal was presented. The
second line of the table, which is new to this design, gives the proportion of
Signal trials that were assigned a particular rating *and* whose location was
correctly identified. The notation $P(R\&C|S)$ means "the probability of the rat-
ing and a correct recognition, given a Signal presentation."

Cumulating these proportions to give the coordinates of an ROC, as in
chapter 3, leads to Table 10.3b, and to two curves, for detection and combined detection/identification. There is only one set of false-alarm probabilities—it makes no sense to ask the likelihood of being right in identification
when no signal is present. Figure 10.3 shows the two performance curves:
the familiar ROC and (below it) the new *identification operating characteristic (IOC)*.

TABLE 10.3   *Detection and Detection/Identification Responses*

(a) Proportions for Five Rating Categories					
Rating	5	4	3	2	1
$P(R|S)$	.10	.25	.26	.25	.14
$P(R\&C|S)$	.08	.24	.21	.10	.06
$P(R|N)$	.01	.07	.15	.42	.35
(b) ROC and IOC Curve Coordinates Accumulated Across Rating Categories					
$P(R|S)$	.10	.35	.61	.86	1.00
$P(R\&C|S)$	.08	.32	.53	.63	.69
$P(R|N)$	.01	.08	.23	.65	1.00

Relation Between Identification and Uncertain Detection

An independent-observation model can be used to predict the identification
operating curve of Fig. 10.3 from the uncertain-detection ROC
(Benzschawel & Cohn, 1985; Green, Weber, & Duncan, 1977; Starr et al.,

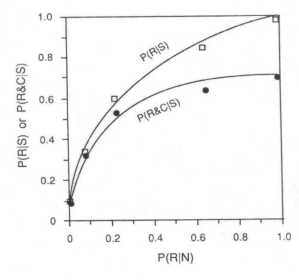

FIG. 10.3. ROC [$P(R|S)$] and IOC [$P(R\&C|S)$] for the data of Table 10.3. The IOC (identification operating characteristic) plots the proportion of trials on which identification and detection responses were both correct.

1975). Within this model, there is a natural decision rule: The channel with the maximum output determines the identification response and is compared to a criterion to determine the detection response. Integration models are not so easily adapted to identification.

To understand the relation between the two operating characteristics, consider the decision space. Figure 10.4 shows a single detection boundary of the independent-observation type used in uncertain detection (as in Fig. 8.6a). The identification boundary is symmetric because the observer is simply choosing the dimension (channel) with the larger output. The two criteria divide the space into four regions, those in which the observer responds "yes-1" (there was a signal, and it was S_1), "yes-2," "no-1," and "no-2." All four regions are labeled in the figure.

Now we compare detection and identification for S_1. (Because the stimuli are equally detectable and the identification decision rule is symmetric, we need to think about only one of the two signals.) The probability of both detecting and correctly identifying S_1—the height of the IOC—is that part of the S_1 distribution in the "yes-1" area. The probability of just detecting it—the height of the ROC—includes both the "yes-1" and "yes-2" areas and must therefore be larger.

To trace out the IOC and ROC by increasing the false-alarm rate, the detection criterion curve is moved down and to the left. When the curve has been moved as far as possible in this direction, both the false-alarm rate and the detection (ROC) hit rate will equal 1. The identification (IOC) success

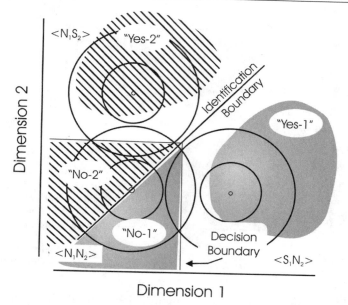

FIG. 10.4.   Decision space showing criteria for the simultaneous detection and identification task. Observer gives both a detection response ("yes" or "no") and an identification response ("1" or "2" was presented). The space is therefore divided into four regions, one for each response.

rate will equal the proportion correct by an unbiased observer in mAFC, as can be seen by comparing Fig. 10.4 with Fig. 7.1. For $m = 2$, the area theorem (see chap. 7) implies that the asymptote of the IOC will equal the area under the ROC. Green et al. (1977) generalized the 2AFC area theorem to the case of m signals.

Subliminal Perception

Of the various stimulus–response events that can occur in simultaneous detection and identification, one has attracted special attention. Is it possible for an observer to identify an undetected stimulus? If so, how is the combination to be understood? Such a result has generally been considered paradoxical at best; early attempts to demonstrate "subliminal" (literally, below the threshold) perception were driven by a search for "motivational" determinants of perception.

A reexamination of Fig. 10.4 shows that identification without detection, on some trials, is to be expected when orthogonal signals are used

(Macmillan, 1986). All observations to the left of the detection boundary lead to "no" responses; if they arise from either $<S_1 N_2>$ or $<N_1 S_2>$, they correspond to failures to detect. Yet such observations—all those in the "no-1" and "no-2" regions—fall more often than not on the correct side of the identification boundary (i.e., nearer to the Signal distribution responsible for the observation than to the other), so identification performance will clearly be above chance.

At least some psychophysically oriented tests of identification without detection have supported this interpretation. In one that did not, Shipley (1960) presented tones having one of two frequencies in a 2AFC uncertain detection experiment and also asked her observers to state on each trial which signal had been presented. She found chance recognition performance on trials for which the detection response was incorrect. But Lindner (1968) was able to reverse Shipley's results by explaining to his subjects the nonreality of thresholds. He found that the proportion of correct identifications increased with criterion (as the IOC suggests), and that identification was above chance on incorrect detection trials.

Subliminal perception results seem surprising to the degree that an inappropriate, threshold model guides theorizing and experimentation: Failure to detect is understood as a drop below threshold, rather than below criterion. This interpretation meshes well with the idea that the threshold is the dividing line between consciousness and its absence. As noted in chapter 4, however, detection theory has no construct corresponding to consciousness. It is true that instructions of which participants presumably are conscious can lead to criterion changes, but the converse implication need not hold.

Using Identification to Test for Perceptual Interaction

GRT Analysis of Identification

Identification experiments are a valuable tool for testing whether perceptual dimensions interact or are perceived independently. General Recognition Theory has clarified various types of independence (Ashby & Townsend, 1986) and provides two general approaches to testing it with identification designs. We consider one such method here.[2]

---

[2]The method we do not discuss, hierarchical model fitting (Ashby & Lee, 1991), is more computationally intensive. A set of models is constructed in which more complex models are "nested" within and tested against simpler ones. For example, a model that includes decisional separability might be compared with one that does not; failure to find a statistically significant improvement in fit for the latter model is considered evidence for decisional separability.

In the basic stimulus set for testing independence, each value of one dimension is factorially combined with each value of the others. In two dimensions (the only case we consider), choosing two values on each dimension leads to four stimuli, two on one and three on the other to six, and so forth. As in all identification experiments, one stimulus is presented on each trial and the task is to assign a unique label to each stimulus; this may be done by reporting the value on each dimension separately. Such an experiment implements the *feature-complete identification design.*

In chapters 6 and 8, we distinguished three meanings of independence. *Perceptual independence (PI)* is the independence of two variables and applies to a single stimulus. If *X* and *Y* are perceptually independent, their joint distribution is the product of the marginal distributions,

$$f(x, y) = g(x)g(y) , \tag{10.6}$$

and has circular or elliptical equal-likelihood contours that display no correlation. *Perceptual separability (PS)* refers to sets of stimuli, and it is present if the marginal distributions on one dimension, say *X*, are the same for different values of *Y*, that is,

$$g(x)_{y=1} = g(x)_{y=2} \tag{10.7}$$

and so forth for other values of *Y*. *Decisional separability (DS)* also refers to sets of stimuli and means that the decision criterion on one variable does not depend on the value of the other. When decisional separability occurs, decision bounds are straight lines perpendicular to a perceptual axis.

These independence qualities, or their opposites, are theoretical characteristics of the perceptual representation. Certain empirical features of the identification data provide information about each type of independence. We introduce a GRT method, *Multidimensional signal detection analysis (MSDA)* (Kadlec & Townsend, 1992a, 1992b) that can be implemented using a straightforward computer program (Kadlec, 1995, 1999).[3] It is helpful to refer to an example.

Example 10d: Perception of Curvature and Orientation

Kadlec (1995) asked her observers to identify stimuli that varied in curvature and orientation (and also location, which we ignore here). There were two

---

[3]Kadlec's program *msda_2a* is available at http://castle.uvic.ca/psyc/kadlec/research.htm.

levels of each variable, and thus four possible stimuli. The data from 200 trials per stimulus filled a 4 × 4 contingency table as shown in Table 10.4.

TABLE 10.4   *Stimulus–Response Matrix for Identification*
of Curve/Orientation Stimuli

	Response Pair			
Stimulus	*"Curvature 1; 50°"*	*"Curvature 1; 55°"*	*"Curvature 2; 50°"*	*"Curvature 2; 55°"*
Curvature 1, 50°	172	13	11	4
Curvature 1, 55°	82	98	12	8
Curvature 2, 50°	2	2	156	40
Curvature 2, 55°	1	15	54	129

The MSDA technique includes several distinct analyses; we illustrate the
approach by examining a "macroanalysis" of perceptual and decisional
separability. The question to be asked is whether judgments of curvature are
perceptually or decisionally independent of orientation. Three aspects of
the data are relevant:

1. Marginal response rates. Does the probability of a particular curvature response depend on the orientation? In the table, first look just at
the cases for which orientation was 50° (Rows 1 and 3). The hit rate for
curvature (probability of using response "1" for curvature 1) is (172 +
13)/200 = .925, and the false-alarm rate is (2 + 2)/200 = .02. Compare
these with the same proportions for cases in which orientation was 55°,
which are (82 + 98)/200 = .90 and (1 + 15)/200 = .08. Are the hit and
false-alarm rates invariant? Use of the MSDA program reveals that the
false-alarm rates are reliably different, but the hit rates are not.

2. Marginal d' values. The hit and false-alarm rates can be used to
find curvature d' for both values of orientation; the values are 3.49 and
2.69, which are significantly different.

3. Marginal criterion values. The hit and false-alarm rates can be
used to find curvature criterion values (relative to the means of the cur-
vature-1 distributions) for both values of orientation; the values are
1.44 and 1.28, which are not significantly different.

What can we conclude from these calculations about independence of curvature and orientation? Table 10.5 (from Kadlec, 1995; Kadlec & Townsend,
1992b) summarizes the implications of the data. The left-hand columns give

TABLE 10.5   *Inferences About Perceptual and Decisional Separability From Identification Data*

Observed Results			Conclusions	
Marginal Response Invariance?	Marginal d' Equal?	Marginal Criteria Equal?	Perceptual Separability	Decisional Separability
T	T	T	yes	yes
T	T	F	yes	no
T	F	T	no	yes
T	F	F	no	no
F	T	T	yes	possibly no
F	T	F	yes	no
F	F	T	no	unknown
F	F	F	no	unknown

possible outcomes of the three statistical comparisons, in which the marginal statistics can be equal (T, or true, in the table) or not (F, or false). Conclusions about separability are in right columns. Notice that if the marginal responses are invariant, then perceptual separability is associated with equal marginal d' and decisional separability with equal criteria. In the absence of marginal response invariance, as in the example, conclusions are less firm. The next-to-last row of the table tells us that Kadlec's data do not display perceptual separability and are inconclusive about decisional separability.

This example portrays only part of the MSDA method. For example, we have considered the macroanalysis of curvature across orientation, but not orientation across curvature (see Problem 10.3). Completely different calculations are needed to evaluate perceptual independence. Identification tasks build on a detailed theoretical analysis (Ashby & Townsend, 1986; Kadlec & Townsend, 1992b) and are a powerful tool for analyzing interaction and independence.

Essay: How to Choose an Experimental Design

In this section, we offer some final comments on discrimination paradigms. Are all the designs we have described—yes-no, identification, and several examples of comparison and classification—equally useful? Although all should, in principle, yield the same d' values, many factors influence the choice of a paradigm. We begin with considerations that derive from our detection theory models and then discuss the possibility that tasks differ in the cognitive processes they require, and thus are not related as detection theory says they should be after all.

Detection Theory Factors

Figures 10.1 and 10.2 suggest one important consideration in choosing a design: level of performance. An observer with a particular value of sensitivity will do best, in proportion correct terms, in a 2AFC task, and worst in oddity. Knowing this, which (if either) should the experimenter select?

Many sensory psychologists would opt for 2AFC, arguing that this produces the "best" performance of which the observer is capable. But considering only the detection theory models (as we are doing in this section), all paradigms yield the *same* performance, that is, the same d'. Showing that a participant can obtain $d' = 1.5$ in same-different is just as impressive as the same demonstration in 2AFC, even though $p(c) = .86$ in 2AFC and only .55 in same-different (differencing model).

A more important consideration is the possibility of floor and ceiling effects. Most experiments aim to compare the discriminability of several different stimulus pairs, a goal that is hard to realize if $p(c)$ is near chance or near perfect. When $p(c)$ is near chance, making the task more difficult cannot produce a corresponding drop in performance; when $p(c)$ is near perfect, not only can improvements not be seen, but detection theory measures cannot even be calculated. Thus, 2AFC (and other high-performance paradigms) are desirable when sensitivity is low, but oddity and its low-performance relatives are desirable when sensitivity is high.

Processing Differences

In assuming optimal processing, detection theory models make a significant simplification. Although people surely fall short of the ideal, inefficiency itself is not usually a serious problem in application. (In chap. 12, we discuss methods for investigating such inefficiencies in their own right.) What *is* worth worrying about is the possibility that inefficiency characterizes some designs more than others, so that the relation among paradigms is not as expected.

Three predictions of the models we have described have proved overly simple in this way: (a) One-interval experiments often yield poorer performance than do other designs, (b) discrimination deteriorates with interstimulus interval, and (c) sensitivities measured in fixed and roving experiments differ by more than the models predict. Models have been proposed, and successfully tested, that account for each of these effects in terms of memory limitation (see chap. 7).

Many experimenters entertain untested beliefs about relative performance across designs. For example, same-different, ABX, and oddity designs are generally thought to be easier for participants to understand than 2AFC or 3AFC. Also, people who serve as participants in psychophysical experiments sometimes develop strong opinions about the decision rule that best describes behavior. For example, observers in ABX (and even in 2AFC) sometimes report ignoring the first interval. Participants in multi-interval designs may report covert classification of the sort postulated by threshold theory. Models quantifying these and other types of nonoptimal processing have been developed and may seem more "psychological" than the normative decision rules of detection theory.

Introspection is a useful source of ideas, but not of experimental truths. Because the mental processing of which we are aware may not be significant in determining our performance, quantitative tests are necessary for intuitively appealing and unappealing theories alike. Furthermore, substantive theorizing is most likely to succeed when it starts from a solid methodological base. With our present understanding of the in-principle relation between, say, 2AFC and same-different, the folly of building a memory model to explain differences in $p(c)$ between them is evident.

Finally, a corollary caution for the innovator: New designs need detection analysis just as much as old ones. Driven by the demands of new content areas, investigators continue to invent new ways to measure sensitivity. Before results obtained with the new technique can be compared with older data, a model of the new design is essential.

Summary

In identification tasks, observers provide distinct labels for each of $m > 2$ possible stimuli. If all stimuli are assumed independent, detection theory analyses can estimate sensitivity for each stimulus and (for some models) bias parameters for each response. Data from one identification task can be used to predict the results of an experiment using a subset of the same stimuli using the constant ratio rule.

Multi-interval forced-choice, in which sequences of length m longer than 2 are constructed that contain one sample of S_2 and $m - 1$ samples of S_1, is a special case of identification in which intervals rather than objects are identified. Such experiments can be analyzed with or without a no-bias assumption using Choice Theory. An unbiased SDT model has also been developed.

Experiments in which stimuli are both detected and identified can be analyzed using identification operating characteristics, which are theoretically related to receiver operating characteristics (ROCs) for the same data.

Identification of stimuli constructed factorially from values on two or more dimensions provide data from which various types of perceptual interaction can be evaluated.

Methods appropriate for finding sensitivity in these paradigms are given in Chart 6 of Appendix 3, those for finding response bias in Chart 7.

Problems

10.1. Predict identification performance for a set of 25 orthogonal stimuli if detection d' for each is 1.0. What is the identification performance for a subset of five of these? two of these?

10.2. Suppose $p(c) = .75$ in a 2AFC experiment. What should $p(c)$ be in a 3-, 4-, 8-, 32-, and 1,000-alternative forced-choice experiment according to SDT? according to Choice Theory? What are the *minimum* values according to Shaw's boundary theorem? (Assume unbiased responding.)

10.3. Here is a confusion matrix for three speech sounds:

	R_1	R_2	R_3
S_1	60	10	10
S_2	10	60	10
S_3	10	10	60

(a) Using the constant ratio rule, predict d' for an experiment with only two stimuli from the set. Do this for each possible stimulus pair. (b) Use the d' values from part (a) to infer the perceptual representation. How many dimensions are required? (c) What would happen if the methods of chapter 5 were applied to these data?

10.4. A three-alternative experiment yields the following response frequencies:

	R_1	R_2	R_3
S_1	4	3	2
S_2	2	6	2
S_3	1	1	8

(a) Estimate d' and $\ln(\alpha)$, ignoring bias. (b) Using the general Choice Theory model (Equations 10.2 and 10.3), estimate $\ln(\alpha)$, $\ln(b_1/b_2)$, $\ln(b_1/b_3)$, and $\ln(b_2/b_3)$.

10.5. In fixed experiments, $p(c)$ is lower in 3AFC than in yes-no for small d', but greater for large d'. At what value are the two equal? Answer the same question for 4AFC, 8AFC, 32AFC, and 1,000AFC.

10.6. In roving paradigms, at what d' does $p(c)$ in 3AFC equal d' in ABX? in oddity? Answer both questions for 4AFC, 8AFC, 32AFC, and 1,000AFC.

10.7. Use MSDA to evaluate the perceptual and decisional independence of orientation over curvature. (Mimic the analysis of curvature over orientation, and summarize the results qualitatively rather than conducting actual hypothesis tests.)

III

Stimulus Factors

One way to characterize the shift in the attitude of psychologists toward their work that came with the cognitive revolution is as a decline in interest in "the stimulus." In the behaviorist period, understanding the effect of presenting a conditioned or unconditioned stimulus, or a reward, was central, and that effect was usually a more or less overt "response." In the cognitive era, the focus has shifted to representations and processing, both nonobservable, and in this respect detection theory is a prototypical cognitive enterprise. In this book, we have repeatedly asked how experimental situations are represented internally, and what sorts of decision processes are applied to them. Details of the stimuli being used have been missing, and in our treatment of data they have not been missed.

This story line is too simple, however, and in the next two chapters we look at two important detection theory scripts that offer the stimulus a lead role. Chapter 11, "Adaptive Methods for Estimating Empirical Thresholds," summarizes strategies for determining a stimulus whose detectability or discriminability is at a preset level. Finding the stimulus corresponding to a performance level is the inverse of the one-dimension problems in Part I and assumes the same kinds of representations. The stimulus sets to which adaptive methods have most often been applied are simple perceptual ones, although advancing technology is broadening the scope.

Chapter 12, "Components of Sensitivity," is an introduction to the use of detection theory in partitioning discriminability between the stimulus and its processing, and among different types of processing. One of the first applications of SDT was in comparing the performance of human listeners to *ideal observers*, hypothetical processors who make optimal use of the information in the stimulus in making their decisions. In this early work, sensory applications dominated, but more recently the approach has advanced into cognitive and even social domains.

11

Adaptive Methods for Estimating Empirical Thresholds

Detection theory provides tools for exploring the relation between stimuli and their psychological magnitudes. In the examples discussed so far, stimulus parameters have been chosen for their inherent interest, and the dependent variable has been d', $\ln(\alpha)$, or some other measure of performance.

Often it is natural to turn this experimental question around and try to find the stimulus difference that leads to a preselected level of performance. Such a stimulus difference we have called the *empirical threshold* or simply the *threshold*. For example, an experimenter may seek a physical difference just large enough so that an observer in 2AFC obtains a d' of 1.0 or 1.5, or (equivalently, for an unbiased observer) so that $p(c)$ equals .76 or .86. Empirical thresholds are unrelated to those of threshold theory (discussed in chap. 4); indeed, they can be measured in either detection theory (d') or threshold theory [$p(c)$] terms. The double meaning of *threshold*, although unfortunate, is unavoidable and need cause no confusion.[1]

Measuring a threshold requires access to a set of stimuli that range, on some physical variable, from too small to too large for the desired level of performance. A field in which threshold measurement has been widely used is audiology, which assesses sensitivity as an aid to the diagnosis of hearing problems. Audiologists estimate thresholds by straightforward manipulation of tone intensity using a Békésy Audiometer (von Békésy, 1947). The intensity of the tone being detected is either continuously increased or continuously decreased, and the observer is told to press a switch whenever the stimulus is audible. The switch is connected to an automatic attenuator in such a way that holding down the switch decreases the intensity and letting it

[1]The dual use of the term *has* caused confusion, in our view, in treatments of "subliminal perception." See chapter 10 for a discussion of this issue.

go increases the intensity. A graphic recorder marks the resulting up–down swings in stimulus intensity over time. Threshold is ordinarily determined by freehand averaging of the extremes, and clinical workers in audiometry expect accuracy of 5 dB from the result. For clinical diagnosis of disruptive hearing loss, this degree of accuracy is sufficient.

In some other areas to which detection theory has been applied, stimulus measurement and control are not simple, and not all sensitivity experiments can be converted into threshold ones. For example, memory for words is affected by similarity in meaning within a list, but list similarity is difficult to quantify (M. B. Creelman, 1966), and the prospects for measuring a "threshold" of semantic relatedness are not very bright. The examples in this chapter are from sensory experiments, and the stimulus variables are simple attributes of auditory and visual signals.

Two Examples

Our examples illustrate problems for which threshold measurement makes more sense than sensitivity measurement. In the first, two conditions leading to very different sensitivities are compared; if the same stimulus difference were used in each condition, at least one would necessarily give either perfect or chance performance. In the second, no correspondence function is defined by the experimenter. Sensitivity can therefore not be estimated by the methods in Parts I and II of this book, but threshold estimation is still possible.

Example 11a: Auditory Thresholds at Different Air Pressures

What effect does a difference in air pressure across the eardrum have on hearing? Creelman (1963a) addressed this question using tones of several different frequencies as stimuli. The detectability of pure tones changes greatly with frequency; because measurement of either very small or very large sensitivities is difficult, using the same tone intensity for all frequencies was impractical. Suppose d' values for 100- and 1,000-Hz tones of the same intensity are in the ratio 1 to 10. If the actual values are 0.5 and 5.0, then $p(c)$ by an unbiased observer in 2AFC will equal .64 and .9998. The second of these numbers means that only one error will be made in about 5,000 trials. Few experimenters wish to squander 5,000 trials on a single sensitivity estimate, and in any case a single error in that span could easily be due to a motor slip, attention lapse, or some other nonsensory factor. On the other hand, if the stimulus intensity is reduced so that d' equals 0.1 and 1.0, $p(c)$ will be .53 and .76. The first of these numbers is uncomfortably close to chance; a further halving of d' from 0.1 to 0.05 will change $p(c)$ by less than two percentage points. Clearly, the problem requires that stimulus intensity not be held constant.

Another important consideration in Creelman's study was the length of an experimental session. Even small differences in air pressure across the eardrum cannot be maintained for long, so relatively short runs and rapid threshold estimation were essential.

Creelman chose an *adaptive procedure* to estimate thresholds for all stimulus conditions; that is, the intensity of the stimulus being detected was changed every few trials in response to the listener's performance. Such a procedure can yield useful data in a short experimental run. Because calculating d' from a small number of trials is problematic, the sensitivity target was a value of proportion correct, $p(c) = .80$. We have seen that $p(c)$ is most acceptable as a sensitivity measure if performance is unbiased (chap. 4), and that 2AFC tends to produce unbiased responding (chap. 7). Creelman's experimental paradigm was 2AFC.

Example 11b: Brightness Matching by Pigeons

Blough (1958) wished to measure equal-brightness contours for visual stimuli of different wavelengths using pigeons as observers. He first trained birds to peck at a button corresponding to the brighter of two illuminated disks. When training was complete, the two disks were illuminated with lights of different wavelengths, one of 450 nm (blue), the other 600 nm (yellow). When both lights were presented at an intensity of 100 units, the pigeon pecked the yellow button, indicating that the yellow spot was brighter. The experimenter then increased the intensity of the blue light to 110 units for the next trial. Whenever the yellow button was pecked, the blue light was made 10 units more intense; whenever the blue key was pecked, the blue light was decreased in intensity by 10 units. After a block of trials, the average level at which the blue light was presented provided an estimate of the "threshold" intensity needed to match the yellow one in brightness.

Although two lights were presented in Blough's experiment, the design was not 2AFC, but yes-no with a "reminder" (chap. 7). The pigeon could, at least in theory, compare the brightness of the blue light to a remembered criterion corresponding to the (constant) intensity of the yellow light. An important difference between our examples concerns the events controlling the change in stimulus level. In the hearing sensitivity study, intensity was adjusted in response to the observer's sensitivity to the experimenter-defined correspondence. In the brightness matching experiment, intensity depended only on the response: No objective correspondence existed. In both cases, performance was vulnerable to the effects of response bias, but with different consequences. In the 2AFC task we needed to assume bias to be slight to have faith in $p(c)$ as an index, whereas in the matching task whatever bias existed was part of the phenomenon being investigated.

Psychometric Functions

Definitions and Illustrations

Adaptive procedures work because, over some range, sensitivity (or, in the matching case, sensory magnitude) increases with stimulus level. The experimenter knows, therefore, that performance will rise if the stimulus value increases and fall if it decreases. The underlying relation between sensitivity and stimulus level, the *psychometric function*, was introduced in chapter 5. Figure 11.1 presents illustrative psychometric functions for the two examples just described.

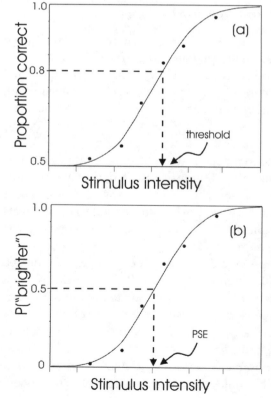

FIG. 11.1. Two psychometric functions. (a) Proportion correct versus stimulus intensity in a 2AFC tone-detection experiment, with threshold corresponding to $p(c) = .8$. (b) Proportion of "brighter" judgments versus stimulus intensity in a yes-no brightness matching experiment, with the point of subjective equality (PSE) corresponding to $P(\text{"brighter"}) = .5$.

In the first panel, proportion correct is plotted against tone intensity for one condition of our auditory detection example. The graph represents the outcome of a conventional, nonadaptive experiment: A number of 2AFC trials are presented at each of six intensities, and $p(c)$ is estimated for each. The threshold we seek is the intensity for which $p(c) = .8$. To estimate it, we draw a smooth curve (exactly *what* curve we discuss later) through the points. The threshold equals the stimulus level that corresponds to $p(c) = .8$ on this curve.

A psychometric function for the brightness matching example is shown in the second panel. Because this experiment has a yes-no decision (with a reminder), the proportion of "brighter" responses ranges from 0 to 1 rather than from .5 to 1. The "threshold" in this case is usually chosen to be the 50% point, the intensity for which judgments of "brighter" and "dimmer" are equally likely. As we learned in chapter 5, this type of threshold is called the *point of subjective equality (PSE)*. Remember that this experiment has no objective correspondence; therefore, there is no way to plot the data that takes account of response bias. The procedure for finding the PSE in Example 11b is formally the same as that for finding the threshold in Example 11a, but the result is likely to be tainted by response bias.

The Shape of the Psychometric Function

General Considerations.   Fitting a normal ogive (i.e., the Gaussian distribution function) to data of the type shown in Fig. 11.1 is traditional (Woodworth, 1938); like many traditions, the procedure still commands respect, but not obeisance. We ask here whether this is truly the appropriate form of the function and, if so, under what circumstances and for what reason.

The form of the function does not follow directly from detection theory as we have described it so far. The psychometric function is a plot of sensitivity against stimulus value, whereas the underlying distributions in a decision space take values along an internal, psychological dimension. Predictions about psychometric function shape can be made when the detection theory approach is joined to a model for stimulus transduction. To take the simplest example, if stimulus intensity is converted linearly into mean location on the decision axis and variance is constant, then the likelihood of judging a stimulus "brighter" can be obtained by moving a Gaussian distribution from low to high values relative to a fixed criterion.

If this same approach is applied to 2AFC experiments, two distributions that move with respect to each other must be considered. An unbiased observer places a criterion halfway between the two means, and proportion correct equals the area under either of these distributions on the correct side of the criterion. As the stimulus difference decreases toward zero, a normal ogive is traced out, but the curve ends at $p(c) = .5$, not $p(c) = 0$, and it looks like the upper half of the curve in Fig. 11.1b.

Some 2AFC data do take this form, but a probably greater number resemble instead the complete curve of Fig. 11.1a. There are two reasons for this variability: differences in the stimuli and/or their processing, and differences in how the stimuli are measured. An example of the first was provided by Foley and Legge (1981), who found 2AFC functions resembling full ogives in a visual detection task, but half-ogive curves in a discrimination task with the same stimuli. The second reason is that if the stimulus variable is monotonically but nonlinearly transformed, the shape of the psychometric function must be affected. The very common logarithmic transformation used in vision and hearing (the decibel is such a transformation) tends to turn steep functions like the half ogive into shallower ones.

The exact form of the psychometric function cannot be specified in the absence of a stimulus theory. Many such theories have been proposed, and we provide some illustrations in chapter 12 (e.g., quantum theory in vision, Cornsweet, 1970; ideal-observer theory in audition, Green & Swets, 1966). Except when stimulus theories are being used, it is appropriate to choose a shape for the psychometric function on the basis of experience and convenience. Such criteria have led to three prominent candidates, whose credentials we now consider.

Specific Quantitative Functions.   The functions to be discussed are mathematically different, but have similar shapes (Fig. 11.2). Each has two parameters of substantive interest: One reflects the location of the function along the x-axis and primarily determines the threshold (or PSE); the other is a measure of slope, which indicates the rate of change in response probability over the range.

The cumulative normal (Gaussian) distribution Φ has, as already mentioned, long been used to describe psychometric functions. The Gaussian distribution function cannot be written as an algebraic expression, but is the integral of the normal density (given in Eq. 2.5). The mean and standard deviation of the variable corresponding to the distribution determine the psychometric function's threshold and slope. *Probit analysis* (Finney, 1971) is a set of procedures for fitting Φ to data that take account of differences in binomial variance at different points on the curve.

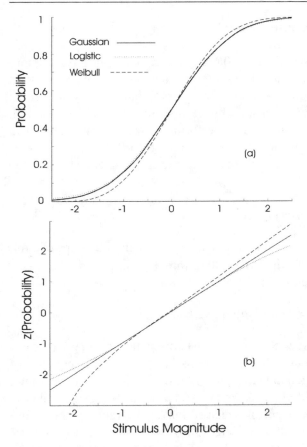

FIG. 11.2. Cumulative normal, logistic, and Weibull functions compared on (a) linear, and (b) normal coordinates. The three curves have been scaled to have similar slopes and intercepts.

The logistic distribution function is

$$p(x) = \frac{1}{1 + e^{-\frac{(x-\mu)}{\theta}}} \ , \tag{11.1}$$

where μ is the threshold parameter and θ is the slope parameter. We saw in chapter 4 that the underlying distributions implied by Choice Theory for the yes-no task are logistic in form.

The third and final candidate is the Weibull function

$$p(x) = 1 - e^{-\left(\frac{x}{\alpha}\right)^{\beta}} \ . \tag{11.2}$$

The parameter α corresponds to the threshold and β to the slope. The Weibull function has valuable theoretical properties (Green & Luce, 1975;

Quick, 1974) and is extensively used in vision research (Graham, 1989; Nachmias, 1981; Pelli, 1985).

All these functions increase from 0 to 1 as x increases. The lower asymptote of real psychometric functions is often greater than 0, however, for two reasons. First, in a yes-no experiment the observer may well produce some "yes" responses to the weakest stimulus; if this stimulus is a blank, or null, these are false alarms. Second, chance performance may be higher than 0: in 2AFC it is .5, and in mAFC it is $1/m$. In either case, the curve is often "corrected for guessing" so that the observed function $P(x)$ is related to the true function $p(x)$ by

$$P(x) = \gamma + (1 - \gamma)p(x) .$$
(11.3)

The consequence of this rescaling is that the function has the shape of a full ogive, but ranges only from γ to 1—for example, from .5 to 1 in 2AFC (McKee, Klein, & Teller, 1985).[2]

Adaptive Versus Nonadaptive Methods

To obtain a threshold, the experimenter must present the observer with stimuli at different levels, but has many options in choosing the sequence of stimuli. There are two general strategies: Decide in advance which stimuli to use and how many of each to present; or decide about the next stimulus on the basis of the observer's performance so far. The advance-planning approach has the longer history (Urban, 1908); we commented on several variants and presented appropriate data analysis techniques in chapter 5. The adaptive approaches, described in this chapter, do not require that the experimenter know beforehand what stimuli are most relevant.

Psychometric functions take on useful values (neither near chance nor near perfect) over a narrow range, and the investigator can locate the critical region only with the cooperation of the observer. Because the goal is usually to locate the threshold, adaptive methods concentrate testing on stimuli near it. An adaptive psychophysical procedure is a collaboration in which the experimenter adjusts the stimulus in a detection or discrimination task on the fly, presenting on each trial a stimulus that is likely to yield information about the location of the desired stimulus level appropriate for the observer.

---

[2]The form of the observed psychometric function is sometimes further modified to take account of "lapses," trials on which the observer fails to give the correct response to large values of x.

The Tracking Algorithm: Choices for the Adaptive Tester

To define an adaptive procedure, the experimenter must answer five separate questions: (a) Under what conditions should testing end at the present level and shift to a new one? (b) What target performance should be sought? (c) When the stimulus level changes, what new level should it change to? (d) When does an experimental run end? and (e) How should an estimate of threshold be calculated? Because these questions are largely independent of each other, an adaptive method can be constructed out of virtually any set of answers. In this section, we provide multiple-choice alternatives for each question based on answers given by past investigators.

Decision Rules: When to Change the Stimulus Level

Rules for deciding to change the stimulus level operate in one of three ways. A new stimulus may be presented on every trial, when the trial-by-trial results match a set sequence, or when the observer's performance deviates from its *target* by a specified amount. The target is the response proportion sought by the experimenter.

After Each Trial.   In the pigeon brightness-matching experiment, the blue stimulus is changed after every trial. When the blue patch is brighter than the yellow, its magnitude is decreased; when it is dimmer, it is made more intense. The proportion of "brighter" responses after one trial is either 1, which is higher than the target, or 0, which is lower. In the long run, the procedure narrows in on the neutral stimulus to which the observer is as likely to respond "brighter" as "dimmer." The proportion of "brighter" responses at that threshold stimulus equals the target proportion $p(T)$, and

$$p(T) = P(\text{"brighter"}) = P(\text{"dimmer"}) \ . \tag{11.4}$$

Because $P(\text{"dimmer"}) = 1 - P(\text{"brighter"})$—there are only the two possible responses— $p(T)$ equals .5, and the method estimates the 50% point.

Two other methods change the level on every trial, but are more flexible in the target level they track. Kaernbach's (1991) method can estimate any target percentage by systematically varying the size of the increasing and decreasing steps. Maximum likelihood procedures consider the entire run history in deciding on the new level. We discuss both of these approaches further in the next section.

When Results Match a Predetermined Pattern.   In the *up–down transformed-response (UDTR)* method (Wetherill & Levitt, 1965), the sequence of correct and incorrect trials at the current stimulus level is compared after each trial to a list of possible patterns. Some patterns require an upward change in stimulus level, others a downward change. If there is a match, the stimulus level is changed appropriately, testing is started again, and a new record of trial results is started. If there is no match, another trial is run with the same stimulus, and the pattern of results is extended using the result of that new trial.

A favorite UDTR rule, often applied to 2AFC experiments, has $p(T) = .71$: A single incorrect trial leads to a more intense stimulus, a sequence of two correct trials to a less intense one. Let us verify that this rule does indeed track the 71% point, using the logic applied earlier to the 50% target. If $p(T)$ is the probability of a correct response, the likelihood of two correct trials in a row is $[p(T)]^2$. At threshold, the patterns that yield a decision to decrease the stimulus must be as likely as the patterns that call for an increase, so $[p(T)]^2 = .5$ and $p(T) = \sqrt{.5} = .71$.

Levitt (1971) listed several other sets of sequences along with the probabilistic equations for $p(T)$ of each set. In one subset, a single incorrect response leads to an increase in level (as in the rule just described), but the number of correct responses needed for a decrease is greater than two. If three successive correct responses are required, $p(T) = .5^{1/3}$ or .79; if four are required, $p(T) = .5^{1/4}$ or .84. The experimenter can choose whichever member of this family of rules tracks the desired level of accuracy.

When Performance Deviates From the Target by a Critical Amount. The optimal rule for determining whether an observed proportion differs from a target was described by Wald (1947). In Wald's application, a factory wants to shut down an assembly line if the proportion of defective units exceeds some limit, such as one tenth (we assume we are discussing defective dolls, not defective automobile steering assemblies). The aim is to react as quickly as possible if quality slips to a lower proportion, with some acceptable likelihood of error. In adaptive psychophysics we are interested in correct responses instead of satisfactory units; $p(T)$ is the proportion correct at threshold, and we want to know if the current stimulus gives either better or worse performance.

Figure 11.3 illustrates the Wald rule for our auditory detection example. For a series of trials at a single level, the number of correct trials is plotted against the total number of trials (N). If performance were perfect, the num-

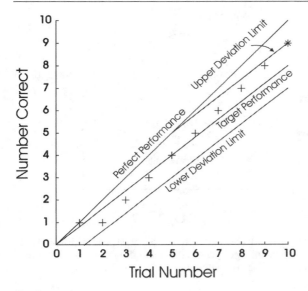

FIG. 11.3. Trials correct versus total trials during a Wald run. Target performance is $p(c) = .8$, and deviations of one trial above or below target are needed to change level. In the example, a new (lower) level is finally called for after the 10th trial.

ber of correct trials would equal the number of trials, and the graph would be a line through the origin with a slope of 1. The expected number of correct trials at the target equals $p(T)N$, the target proportion correct times the number of trials. In the figure, this expectation is shown as a straight line with slope $p(T)$. The experimenter does not require that the observer perform *exactly* at the target—in general, this is not possible—but demands that performance be within a *deviation limit*. In the example, this limit is 1: If the number correct deviates from the expected target by at least 1, the stimulus level is changed. The lines labeled *Deviation limit* are parallel to the target line and bracket a region of performance consistent (by this standard) with the 80% goal.

Hypothetical data are represented in the figure by crosses. The observer is correct on Trial 1, incorrect on Trial 2, and correct on Trials 3 through 10. For the first 9 trials, the total number correct is within 1 of the target number, but on the 10th trial the total of 9 correct is 1 greater than the expected 8. The stimulus level is decreased before Trial 11.

The deviation limit can be set to any value, and both narrow and wide limits have advantages. Narrow limits reject response proportions that are even slightly discrepant from the target, and therefore lead to rapid decisions. Figure 11.4 shows the mean number of trials to reach a decision for $p(T) = .8$ and deviation limits of 1.0, 1.5, and 2.0. The speed advantage of the narrower limit varies with the true $p(c)$ at the level being tested, but can be as great as 5 to 1.

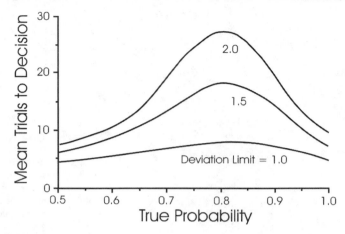

FIG. 11.4.   Number of trials to reach a decision in a Wald test aimed at $p(T) = .8$, as a function of the true probability of a correct response, for three deviation limits. Curves are based on Monte Carlo simulations.

Narrow limit decisions are quick, but they are often wrong. By waiting for a larger discrepancy to occur, the experimenter can be more confident of changing level in the right direction. In Fig. 11.5, the probability that a Wald test decides that the level is too high is shown for each true $p(c)$ value. For a perfect test, this probability would be 0 for all values less than $p(T) = .8$ and 1 for all higher values. The deviation-2.0 limit comes closest to this ideal. If the true proportion correct is .65 and the target .8, a deviation-1.0 test incorrectly decreases the stimulus level about 20% of the time, a deviation-2.0 test only about 5% of the time.

In psychophysical applications, narrow deviation tests are often used for their speed; accuracy derives from the use of many repeated tests as an experimental run proceeds. Hall (1974) showed, on the basis of simulations, that greater accuracy in threshold estimates could be obtained from a larger number of fast, variable, narrow deviation tests than from a smaller number of slow, accurate, wide limit tests.

Target: What Performance Level to Track

In subjective yes-no (matching) experiments the target percentage is almost always 50%, but in objective sensitivity tests the choice is less obvious. Probably the most popular target percentage in application is the 71% tracked by Levitt's (1971) 2/1 rule. This rule is easy to implement, but Green (1990) has

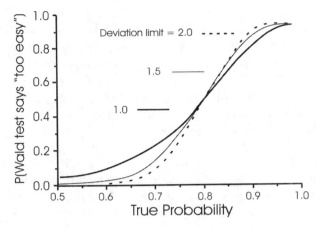

FIG. 11.5. Probability that a Wald test aimed at $p(T) = .8$ yields the result "too high," as a function of the true probability of a correct response, for three deviation limits. Curves are based on Monte Carlo simulations.

argued that the most efficient target percentage is much higher. The inherent variability of a threshold estimate depends on the target percentage because this percentage is influenced by both the slope of the psychometric function and the binomial variance of observer responses. Steep slopes (which occur near the midpoint of the function) and low variance (which occurs near the extremes) are desirable. Thus, Green suggested a "sweet point" that represents a compromise between these two goals. For 2AFC, this optimum occurs at 91%; for procedures in which chance is less than 50%, it is lower. Experimenters wishing not to venture too far from the 2/1 rule can improve the precision of their estimates by using 3/1, 4/1, or some higher criterion for lowering the stimulus level. Kollmeier, Gilkey, and Sieben (1988) showed that the 3/1 (79%) rule was more efficient than the 2/1 (71%) rule in an auditory masked threshold experiment.

Stepping Rules: What Size Change in Level to Make

Having decided to abandon the old stimulus level, we must now select a new one. How large a "step" in the direction determined by our test shall we take: always the same size, a size decreasing during the run, or an adjustable size? If the last, how much prior data should enter into our decision?

Fixed Steps.   The simplest rule is to change the stimulus up or down by a fixed amount; by architectural analogy, this is called a *staircase* procedure (e.g., Blough, 1958; Cornsweet, 1962). To use fixed steps, one must know the appropriate step size beforehand. This is the sort of advance planning from which adaptive procedures are supposed to free us, but sometimes the apparatus confines an experiment to fixed stimulus values, or the experimenter is required to prepare a set of graded stimuli beforehand (as when the stimuli are colored papers with differing spectral characteristics or odorants). When fixed steps are unavoidable, the step size must not be too small (lest performance change insignificantly between steps) or too large (such that one step changes the task from trivially easy to impossibly difficult). In between, experimenters face one of the many tradeoffs of adaptive procedures, that between inaccuracy and tedium.

Step Size Determined by Target Level.   Kaernbach (1991) showed that any point on the psychometric function could be estimated by choosing increasing and decreasing step sizes that are in the appropriate ratio. In general, to reach the target proportion p the ratio of magnitudes must be $p/(1-p)$. For example, a target of 75% is reached by setting the increasing step to be three times the magnitude of a decreasing one.

Decreasing Steps.   In an early paper, Dixon and Mood (1948) suggested that steps in stimulus size be made smaller as an experimental run progresses to take advantage of increasing certainty about threshold location. The Dixon–Mood rule prescribes a step size equal to the initial size divided by the number of steps taken to date in the experimental run. The original application was in research on explosives, where amounts of various constituent chemicals could be chosen arbitrarily; stimulus continua in many behavioral applications have this graded characteristic.

With steps of continually decreasing size, an incorrect decision about the direction of the next step can add to the time to reach a threshold estimate, because recovery from a bad decision takes longer with smaller steps. Most decision rules are designed to be quick rather than highly accurate, so such an incorrect decision—even a string of them—is likely. Some means to recover from steps in the wrong direction by *increasing* step size is appropriate.

Adjustable Steps Determined by Immediately Preceding Trials. The first proposal to address this problem was Parameter Estimation by Sequential Testing (Taylor & Creelman, 1967), or PEST (the field's first marginally clever acronym). PEST rules generally use decreasing step sizes, but switch to increasing ones to recover from apparently incorrect decisions. There are five rules:

1. After each *reversal*, halve the step size. A reversal is a step in the opposite direction from the previous step, for example, an increase in

level following a decrease. A minimum value is specified below which step size is not decreased.

2. A step in the same direction as the last uses the same step size as previously, with the following exceptions.

3. A third step in the same direction calls for a doubled step size, and each successive step in the same direction is also doubled until the next reversal. This rule has its own exceptions.

4. If a reversal follows a doubling of step size, then an extra same-size step is taken after the original two before doubling.

5. A maximum step size is specified, at least 8 or 16 times the size of the minimum step.

These rules are illustrated in Fig. 11.6, which shows the sequence of stimuli used after each of several decisions to change level; the actual number of trials needed to make a decision varies and is not shown. We begin at a level of 35, chosen to be relatively easy for the observer. Testing shows the level to be too high, so it is changed downward by 20 units. This level yields performance that is too low; applying Rule 1, the level is increased by half the previous step size. The figure shows successive applications of the foregoing rules. From this point, the rules applied are as follows: 1, 2, 1, 2, 3, 1, 2, 4, 1, 1, and 1.

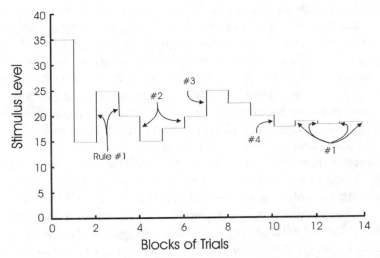

FIG. 11.6.   Example of stimulus levels during a PEST run for successive blocks of trials. The level changes for each block according to the indicated rule.

Adjustable Steps Determined by the Entire History of the Run.
PEST's computation of the next level depends on the past history of the run, but only some of it. In maximum-likelihood methods, a best estimate of the threshold is calculated after each trial from the entire run history, the result of all trials at all levels tested so far. The new level is set to that estimate, and testing is continued.

A maximum-likelihood procedure assumes the underlying psychometric function to have one of the specific forms discussed earlier. This function, which we call $p(x)$, specifies the proportion correct for every stimulus level x. If $p(x)$ describes the data, then the likelihood $L(x)$ of a particular sequence of R correct responses followed by $N - R$ incorrect responses to stimulus x is:

$$L(x) = [p(x)]^R [1 - p(x)]^{N-R} .$$
(11.5)

(The probability of the entire sequence equals the product of probabilities because we assume that trials are independent.) Thus, if a particular theoretical function predicts that $p(c) = .75$ for a stimulus value x, and if in the testing to date there have been four correct trials followed by two incorrect ones, then $L(x) = (.75)^4 (.25)^2 = .0198$ for that sequence. An expression of this form can be written for any sequence of trials and any possible theoretical function. The overall likelihood of the function is the product of $L(x)$ values for all stimulus levels x for which data have been collected.

The experimenter chooses a form for the psychometric function and uses the data to determine which function of that form is correct. Choosing the form of $p(x)$ specifies a *family* of curves whose members differ in mean (threshold) and variance (slope). The likelihood of the data is computed from Equation 11.5 for each member of such a family, and the curve giving the largest value—the maximum likelihood—is selected. The current threshold is then the 80% point (or some other point) on that curve. (For more discussion of maximum-likelihood estimation, see Madigan & Williams, 1987; computational issues are considered in Press, Flannery, Teukolsky, & Vetterling, 1986).

In a seminal paper, Robbins and Monro (1951) showed that this strategy is the optimally efficient way to find the threshold. Among the modern methods that use the maximum-likelihood approach are Pentland's "Best PEST," which assumes the underlying function to be logistic (Lieberman & Pentland, 1982; Pentland, 1980), and Watson and Pelli's (1983) QUEST, which assumes the Weibull function. The calculations these procedures require between trials of an ongoing experiment are well within the capability of laboratory computers.

In both methods, the slope is fixed by the experimenter, and the likelihood calculation gives the probability of the obtained data assuming each possible psychometric function with that slope. QUEST differs from Best PEST in requiring the experimenter to provide some initial guesses, an a priori distribution of likely threshold locations. The program uses a *Bayesian* strategy to successively revise the odds as data are collected. Not having to predict where the threshold might lie is an advantage of adaptive procedures, but the prior estimates have been shown to improve QUEST's precision (Madigan & Williams, 1987).

Suspending the Rules.   Many small choices the experimenter makes are not specified by the adaptive procedure, and it is important that these choices turn out to be truly irrelevant to the outcome. For example, the rules for changing level, whatever they are, are often suspended at the beginning of a run in order to rapidly locate the region of the threshold. Many experimenters begin testing with relatively large stimulus differences. To reach the neighborhood of threshold, they may reduce the stimulus difference by a large step after only one or two correct responses, and revert to normal rules after two incorrect responses. Other experimenters attempt to begin each run at the current best estimate of threshold so that special "run-in" rules are unnecessary. In maximum-likelihood techniques, outlying points (due to inattention or slow learning) can distort threshold estimates. If these points are truly far from threshold, however, they are visited rarely and have only a small influence on threshold estimates late in a run.

Stopping Rules: When to End a Run

As with deciding when to change levels and what to change to, the experimenter can make more or less use of information from the observer in deciding when to stop.

Fixed Number of Trials.   Many experimenters employ runs of fixed length. Run length is, of course, perfectly predictable with this strategy, a significant advantage, but some of the flexibility of adaptive procedures is lost. For example, sufficient information to locate the threshold to the desired degree of accuracy may be available before the run is complete.

Fixed Number of Reversals.   The experimenter can decide to stop the run after a fixed number of reversals, say 5 to 10. This strategy avoids ending a run when the participant has not yet "settled down." The exact number of trials in the run varies, but not dramatically.

Minimum Step Size.   For stepping rules in which the step size varies, one can terminate the run when some minimum-size step is demanded. In the example of Fig. 11.6, for example, data collection might be stopped when a sub-minimum step of 0.3125 is required. This rule leads to considerable unpredictability in run length, but ensures that when the run *is* stopped the stimulus level is near threshold.

Minimum Confidence Interval.   The maximum-likelihood QUEST computations allow a stronger version of the foregoing strategy. After each trial, the prior distribution plus all data thus far collected are used to calculate a confidence interval around the estimated threshold. The run is ended when this interval is less than some minimum. Emerson (1986b) attributed some of QUEST's apparent advantage over other procedures, in simulations, to this stopping rule.

Summary Rules: How to Calculate a Threshold Estimate

Most strategies for calculating a threshold ignore the earliest segment of a run. Many possible summary statistics remain, notably the following:

Average of All Trials.   The simplest approach is to find the mean, or median, of all levels visited.

Average of a Fixed Sample of Trials.   In an adaptive session, the stimuli presented on successive trials are heavily dependent, and those used on more separated trials less so. Kaplan (1975) estimated PEST thresholds by averaging stimulus values obtained every 16 trials and was able to make precise estimates of the auditory threshold of highly trained observers. This has been called the rapid adaptive tracking (RAT) mode for PEST, as opposed to the previously described minimum overshoot and undershoot sequential estimation (MOUSE) mode, in which testing is stopped when a step smaller than some minimum is called for. In the RAT mode, testing is continued by taking the allowed minimum step if smaller steps are requested. RAT is PEST with a long tail.

Average of All Reversals.   Averaging stimulus values at each reversal is equivalent to finding the midpoint between reversals and averaging these, on the assumption that the threshold lies halfway between reversals. Besides omitting early reversals, some experimenters calculate a "trimmed mean" by leaving out the most extreme values, perhaps on the assumption that the observer should be allowed at least one extended lapse in attention.

Final Testing Level or Point on Best-Fitting Psychometric Function. The original PEST package used the final testing level—the one called for by the final minimum-size step—as its estimate of threshold. With a Bayesian or maximum-likelihood estimation run, this is also a logical summary datum because all the data contribute to it. However, the target level could in principle be different from the desired definition of threshold. For example, one might choose a high (say 90%) target but still wish to report the stimulus level corresponding to the 75% point. To do this, one finds the best-fitting psychometric function and calculates the level that leads to 75% correct on that function.

2AFC Threshold Estimation Without Response Bias.   None of these summary measures is completely free of bias. Does such a statistic exist? We illustrate several possible solutions for a 2AFC detection experiment with three levels of intensity; hypothetical data are shown in Table 11.1. The second column gives the proportions correct for each intensity, and these values are plotted in Fig. 11.7a. A typical definition of threshold in this case is the level leading to 75% correct; interpolating between points yields an estimate of 1.57 units. Sometimes 2AFC data are corrected for guessing, via a rearrangement of Equation 11.3:

$$p(x) = \frac{P(x) - \gamma}{1 - \gamma} \ . \tag{11.6}$$

Because .5 is the chance ("guessing") rate, γ is set to .5. Values of the equation thus corrected are shown in the third column of Table 11.1. With this version of the psychometric function, the natural definition of threshold is the 50% point, and interpolation shows that this value is still 1.57 units.

To see that neither approach takes account of response bias, we must distinguish the two types of trials on which a given stimulus level can occur. The non-zero intensity can be presented either in Interval 1 (e.g., the sequence <3,0>) or in Interval 2 (e.g., <0,3>). Column 4 in the table lists the trial types in this format, and Column 5 characterizes the various types of trials by the *difference* in intensity between the two intervals. Remember (from chap. 7) that the difference between the stimulus effects in the two intervals is the optimal decision variable in 2AFC. Column 6 provides the proportion of "2" responses for each possible stimulus sequence, and the overall $p(c)$ values can be seen to be the averages of two quite different numbers at each level of intensity—for example, at Level 2, the proportion cor-

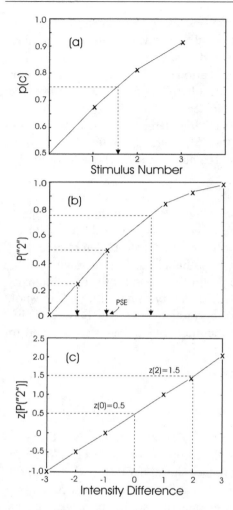

FIG. 11.7. Three treatments of the 2AFC data in Table 11.1. (a) Proportion correct versus stimulus intensity; (b) proportion correct adjusted for the false-alarm rate versus stimulus intensity; and (c) z score corresponding to proportion correct versus the (signed) difference in intensity between Intervals 2 and 1. Only the last approach permits separation of sensitivity from bias.

rect is .93 when the stimulus was in the second interval, but only $1 - .31 = .69$ when it was in the first.

When the proportion of "2" responses is plotted against this difference, as in Fig. 11.7b, a complete psychometric function that increases from .16 to .98 is obtained. This function offers a natural measure of bias, the PSE; it is not even necessary to interpolate (in this case) to find that the PSE is an intensity difference of -1.0 units. To estimate the threshold from curves like this, one normally finds the average difference between the 75% and 25% points. The resulting value is $.5[0.72 - (-2.40)] = 1.56$. This strategy attempts to eliminate the bias in the estimate by averaging it away, but does not succeed. The bot-

TABLE 11.1 *Constructing Bias-Free Psychometric Functions*

Intensity	p(c)	p(c) Corrected	Intensity Pair	Intensity Difference	P("2")	z[P("2")]
3	.91	.82	<0, 3>	3	.98	2.05
2	.81	.62	<0, 2>	2	.93	1.48
1	.67	.34	<0, 1>	1	.84	0.99
			<1, 0>	−1	.50	0.00
			<2, 0>	−2	.31	−0.50
			<3, 0>	−3	.16	−0.99

tom line is the same as in the earlier calculations because each of the two points being averaged is still influenced by response bias.[3]

The best plan, according to detection theory, is to plot the psychometric function in units of z scores rather than proportions, as shown in Fig. 11.7c (Klein, 2001). A natural definition of threshold is the level difference required to obtain a specific value of d' for a stimulus compared with the null difference, $d' = z(x) - z(0)$. Many investigators choose $d' = 1$, the 76% correct point in unbiased 2AFC. In this example, $z(0)$ is found by interpolation to be 0.5 units and $z(2)$ equals 1.48, so $d' = 1$ is obtained when x approximately equals 2. The threshold for detection is therefore estimated to be 2 units, a value that *is* unaffected by response bias. The bias toward "2" in the raw data led to an exaggerated impression of the observer's sensitivity (i.e., an unduly low estimate of threshold).

Evaluation of Tracking Algorithms

A very large number of adaptive packages can be constructed by combining rules for changing levels, target percentage, rules for finding new levels, stopping testing, and computing a threshold. Like Treutwein (1995), who listed 21 separate procedures, we find the goal of deciding on a single method that can be recommended in all circumstances to be beyond our reach. Instead we attempt to set out rationales that might justify choices in particular applications. A necessary preliminary step is to establish some criteria for evaluation.

[3]For subjective judgment UDTR tasks (such as Example 11b), Jesteadt (1980) suggested a similar strategy: Estimate each of two symmetrically located points (e.g., 71% and 29%) on subsets of trials and average them to obtain the PSE. The task is said to reduce response bias and provide the illusion of having a correspondence function, in compensation for a loss of statistical efficiency.

Evaluation Criteria

Statistical Characteristics of Threshold Estimates.   An empirical threshold is a statistic (see Appendix 1), an estimate of a theoretical parameter, and can be evaluated by asking two questions: (a) On the average, is it equal to the parameter? The average discrepancy between a statistic and the corresponding parameter is called *statistical bias* (a usage unrelated to *response bias*). (b) Is its variability small? To find out, comparisons with other, competing measures are made.

The variability of a threshold (or any other) statistic ordinarily decreases as more trials are used in computing it. Taylor and Creelman (1967) suggested that the work accomplished by a procedure could be measured by the *sweat factor*, equal to the product of the number of trials and variance. The relative *efficiency* of a measure is its sweat factor divided into the sweat factor of an alternative index. One interesting basis for comparison is the "ideal" variability, that constrained only by inevitable binomial variance (see Appendix 1).

Computations and Experiments.   Adaptive procedures can be compared by conducting threshold measurements with human or animal observers or by simulating the outcome. More computations than experiments have been done, not only because they can be more easily conducted, but also because they provide a needed baseline for experimental data. The threshold to be expected can sometimes be calculated by enumerating every possible outcome of a series of trials (e.g., Madigan & Williams, 1987). For less tractable calculations, a common approach is simulation using a *Monte Carlo* method: An assumed underlying distribution determines the effective probability of each response at each stimulus "level," and "runs" of many "trials" are presented (Press et al., 1986, ch. 7). The observer imagined by most simulators is *ideal* (see chap. 12), reaching the best possible performance given the limitations of sensory and response variability (Taylor & Creelman, 1967, Appendix). Simulations must mirror all important aspects of the threshold estimation problem. For example, runs must begin at variable starting points because in real experiments we do not know the relation of the initial stimulus to the observer's threshold; assuming knowledge about the starting point produces unrealistically precise estimates (Watson & Fitzhugh, 1990).

We can now evaluate the main classes of adaptive methods, those depending on maximum likelihood, PESTilent rules, or staircases.

Maximum-Likelihood Methods

If adaptive procedures are compared solely in terms of statistical criteria and assessed by simulation, then a choice is not difficult to make: Maximum-likelihood procedures (QUEST and Best PEST) have the greatest efficiency. Because of its use of a priori information, QUEST is the better of the two in efficiency and bias for large stimulus ranges (Emerson, 1986b; Madigan & Williams, 1987). Watson and Pelli (1983) found the efficiency of QUEST to be 84%.

One reason that pure maximum-likelihood methods have not made their competitors obsolete is that they make a number of assumptions. To determine the psychometric function "most likely" to have produced the data, one needs a constrained set of candidates; thus the form of the function must be known. Most methods also require knowledge of the slope so that all the possible functions differ only in threshold. These assumptions are more attractive in well-mapped research areas than in novel domains. A further assumption is that the data result from an ideal observer: Whatever the psychometric function is, it is the same on every trial, with no shifts in threshold, lapses in attention, or loss of memory for stimulus characteristics. It is straightforward to simulate these kinds of nonoptimality, but not to incorporate them into algorithms for choosing the next stimulus level.

Nonparametric Methods Using PEST

PEST packages are nonparametric in that they make no assumptions about the underlying psychometric function. Taylor and Creelman (1967) calculated PEST to have an efficiency of 40% to 50%, which is better than all but the maximum-likelihood procedures. Madigan and Williams (1987) found, in a word-recognition experiment, that PEST was no less efficient in practice than Best PEST or QUEST. PEST-estimated thresholds are biased low for short experimental runs or large stimulus ranges (Emerson, 1986a; Madigan & Willams, 1987).

Taylor, Forbes, and Creelman (1983) reported data suggesting that PEST observers suffer less from sensitivity fluctuations than do participants in the method of constant stimuli. Shifts that do occur can be detected by examining a plot of stimulus level against trials (Hall, 1983; Leek, Hanna, & Marshall, 1991). Such trends are more evident in PEST or UDTR than in a similar plot derived from the sort of continual adjustment made by maximum-likelihood methods.

Nonparametric Staircase Methods

The UDTR method exercises simple staircase control over stimulus intensity. Its major advantage over the other methods is that computation and stimulus control are simple. Computational complexity is an issue of diminishing importance, but it is still true that continuous adjustment of the stimulus variable is not practical in some domains. When experimenters know fairly well what step size to use, and the ballpark of the threshold, UDTR can be a good choice. Kaernbach's (1991) step-size method provides a modest improvement in efficiency for short runs (Rammsayer, 1992). UDTR decision rules are, in general, slightly less efficient than Wald rules for the same target proportion.

A strategy often used in conjunction with UDTR is the interleaving of multiple adaptive tracks. The particular adaptive track to be used on a trial is chosen at random, and its current stimulus is presented. This reduces the predictability of the next stimulus level and aids memory in that if stimuli on one track are at a low, hard-to-remember level, then those on the other track may not be. An apparent disadvantage is that twice as many trials are required, but in compensation one obtains two distinct estimates of threshold.

Two More Choices: Discrimination Paradigm and the Issue of Slope

Two other choices, not logically part of the tracking algorithm, must also be faced. Of the many available discrimination paradigms discussed in earlier chapters, which is to be used? And is the threshold the only important feature of the psychometric function, or should the slope also be estimated?

Discrimination Paradigm

2AFC. Most modern adaptive psychophysical procedures have used the two-alternative forced-choice paradigm, probably because of its reputation for minimal response bias. Although this reputation is deserved, 2AFC is less efficient and more statistically biased than the yes-no paradigm (Kershaw, 1985; Madigan & Williams, 1987; McKee et al., 1985). The inefficiency results from the reduced range: $p(c)$ takes values between .5 and 1 in 2AFC, whereas the yes-no hit rate increases from 0 (or the false-alarm rate γ) to 1. The response-bias-free method described earlier solves this reduced-range problem. The statistical bias arises because the lowest values of the psychometric function near an asymptote can still yield erroneous decisions, whereas the upper values, near the 100% point, are unlikely to give wrong answers. Threshold estimates are therefore systematically too low.

mAFC. Simulations show that offering more than two alternatives per forced-choice trial results in improved efficiency and smaller bias (e.g., McKee et al., 1985). Some of the advantage arises from the increased range of the psychometric function, the lower limit of which is 33.3% in 3AFC and 25% in 4AFC. In addition, *m*AFC designs make it easier to place the target above the midpoint of the psychometric function—a desirable goal (Leek, 2001). Auditory detection experiments confirm that 3AFC and 4AFC are to be preferred over 2AFC (Kollmeier et al., 1988; Shelton & Scarrow, 1984). Of course more presentations per trial mean longer trials; thus, Schlauch and Rose (1990) recommended three alternatives over four.

Yes-No. The task that maximizes the range of the psychometric function is yes-no, which has a minimum response rate of 0%. An additional advantage, in most applications, is that each trial contains a single temporal interval, reducing experiment time. Kaernbach (1990) developed a "single-interval adjustment-matrix" procedure in which different targets are reached by manipulating step size. Both signal and noise trials can occur, and the method aims at a response rate target of $t = H - F$. Stimulus level is lowered 1 unit following a hit, increased $t/(1 - t)$ units following a miss, and increased $1/(1 - t)$ units following a false alarm. For a 75% target, these values are -1, $+3$, and $+4$ units. In simulations and experiments, Kaernbach showed a substantial benefit of this method over Békésy tracking and 2AFC, especially when the time per trial was taken into account.

Slope

Most of our effort in this chapter has been aimed at estimating a single point on a psychometric function, but it is often useful to know more. Knowledge of a complete function is best obtained with the method of constant stimuli; Miller and Ulrich (2001) provided a nonparametric method for estimating the function in some detail. Several investigators have designed adaptive methods with a more modest goal: an accurate estimate of the function's slope. This statistic gives information about the reliability of threshold estimates and is helpful in maximum-likelihood calculations in which a slope must be assumed.

At the end of an adaptive run, response proportions for a number of stimulus levels are available, and in principle slope could be estimated by fitting functions of differing slopes to the data. In fact, however, a set of levels that is well chosen for the goal of estimating threshold is not ideal for estimating slope–typically, the points are too close together. The first step in modifying adaptive methods for slope estimation is to adjust the rules for selecting stimuli.

Consider, for example, the *adaptive probit estimation (APE)* procedure of Watt and Andrews (1981). Four stimuli are selected that are thought to span the major portion of the psychometric function, and a short run using these values is presented. At the end of the run, the observed response proportions are fit by the probit method, and four new values are selected that cover the new estimate of the function. The procedure continues in this adaptive manner.

One way to view the simultaneous estimation of threshold and slope is as a search through a two-dimensional parameter space, and current methods approach the problem from this point of view. In the earliest of these, Hall (1981) used the PEST tracking rules combined with a large initial step to guarantee a dispersion of stimulus values; both the starting point and the initial step are adjusted between runs. The summary psychometric function is chosen by maximum likelihood from a set varying in both mean and variance. King-Smith and Rose (1997) and Kontsevich and Tyler (1999) improved on this basic approach by the use of maximum likelihood and Bayesian methods in stimulus selection as well as for data summary. One cautionary conclusion from this body of work is that the number of trials needed for an accurate assessment of slope is far greater than the number needed for a threshold. Leek, Hanna, and Marshall (1992) recommend 200 trials to find a slope value. Kontsevich and Tyler (1999) estimated that 300 are required, versus 30 for a simple threshold estimate. Clearly one needs a good reason for all this extra work; one compelling rationale would be the existence of theories that make predictions about psychometric function slope.

Summary

Adaptive procedures estimate the stimulus level needed for a fixed level of performance. The stimulus that yields some proportion of correct responding in a forced-choice task (or a specified hit rate in yes-no) is found by systematically varying the stimulus difference during an experimental run.

Procedures differ in the rules by which they decide to change stimulus level, the target performance accuracy, the rules by which the new level is computed, the criterion for ending a run, and the method of computing a threshold from the data. The simplest rules change level after a fixed number of trials, by a fixed step size, stop after a fixed number of trials or reversals of direction, and compute threshold from one or a few points. More complex rules make greater use of the history of the run, the prior judgments of the experimenter, and the expected form of the psychometric function.

Computer simulations show the more complex rules to be more efficient, that is, to produce less variable and less biased estimates, provided that their assumptions are correct. The most popular discrimination paradigm, 2AFC, is inferior to *m*AFC and yes-no, especially at low performance levels. Psychometric function slope can be estimated with appropriate modifications in adaptive procedures, but at considerable experimental cost.

Problems

11.1.   Following are some stimulus values, together with the number of correct and incorrect responses to date at each level in a 2AFC detection experiment:

Stimulus	Number Correct	Number Incorrect
−2.5	0	2
−2.0	0	1
−1.0	1	4
−0.5	2	3
0.0	3	2
0.5	4	1
1.0	4	0
2.0	2	1
2.5	1	0

For each of the following logistic psychometric functions, find the likelihood of these data using the method of Equation 11.5:

(a) $p_1(x) = \dfrac{1}{1 + e^{-x}}$

(b) $p_2(x) = \dfrac{1}{1 + e^{-\left(x + \frac{1}{2}\right)}}$

(c) $p_3(x) = \dfrac{1}{1 + e^{-.6x}}$

Which function is most likely for these data? Plot both the observed data points and the three theoretical curves.

11.2.   A new trial is run at stimulus intensity 0, and the observer is correct. Recompute the likelihoods of the three functions with the new data. Which is now most likely? Plot the new data point on the graph you drew for the previous problem.

11.3.   A string of trials in an adaptive 2AFC experiment leads to the following correct (+) and incorrect (0) responses:

+0+0++0+++0000++++++

(a) Apply the Wald sequential test with $p(T) = .75$ and deviation limit 1.0 until a decision to change level is made. Continue until you run out of data. On what trials is the decision made, and in which direction is the change?
(b) Apply the 2/1 UDTR rule in which $p(T) = .71$ to the same set of responses, and answer the same questions.
(c) Apply the 4/1 UDTR rule in which $p(T) = .84$ to the same set of responses, and answer the same questions.

11.4.   For each of the three rules in Problem 11.3, find the stimulus level after each decision:
(a) using the PEST stepping rules with initial level of 32, initial step of 16, maximum step of 16, and minimum step of 1;
(b) and (c) using the UDTR stepping rules with initial level of 32 and all steps of size 4.

12

Components of Sensitivity

What determines the degree to which two stimuli can be distinguished? Detection theory offers a two-part answer: Sensitivity is high if the difference in the average neural effects of the two are large or if the variability arising from repeated presentations is small. Common measures of accuracy like d' are accordingly expressed as a mean difference divided by a standard deviation. In most of the applications we have considered, changes in sensitivity are equally well interpreted as changes in mean difference or variability, and attributing such effects to one source or the other is both impossible and unnecessary. In the early chapters of this book, we therefore suppressed the role of distribution variances, dealing only with mean differences and standard deviation *ratios*.

When the experimental situation is expanded beyond two stimuli, the locus of a sensitivity effect may become clear. If three stimuli differ along a single dimension—light flashes varying only in luminance, for example—and the extreme stimuli are more discriminable than the adjacent ones, systematic increases in mean effect provide the simplest interpretation. If the perceptibility of a stimulus decreases when another must also be detected, as in uncertain detection designs, it is natural to imagine that variance rather than mean difference has been affected by the demands of attention. Our treatments of these problems in chapters 5 and 8 adopted exactly these interpretations.

In the pure two-stimulus world, disentangling these two contributions to sensitivity requires another approach. A starting point is to ask whether there is variability within a stimulus class itself, and perusal of our several examples reveals the answer to be: sometimes. Absolute auditory detection typifies one case: Every presentation of a weak tone burst is the same, so all the variability must arise from processing. The variance is entirely *internal*.

Recognition memory is quite different: No matter how carefully the stimulus set is constructed, the items in it must differ in familiarity (or whatever the decision variable is). If recognition is represented as a task of distinguishing two distributions of familiarity, *external* variance contributed by the stimulus set combines with the internal variance of the fallible observer.

In this chapter, we examine efforts to partition blame for imperfect sensitivity between external and internal sources, and among components of each of these. We begin with the simplest case, two distributions on one dimension arising from stimulus classes that are nonconstant. The primary question is the relative importance of internal and external variance—the efficiency of the observer compared with the best possible performance. In the second section, we extend the information combination ideas of chapters 6 to 10 to partition variance among components of a stimulus and among observers in a "team." Finally, we discuss hierarchical models in which variance may arise at multiple levels of processing. The application of these ideas in perception is too widespread to cover in a chapter, and a thorough understanding requires sophistication in particular content areas. We intend this introductory presentation to illustrate an important use of detection theory that is not treated elsewhere in the book.

Stimulus Determinants of d' in One Dimension

Example 12a: The Dice Game

The first detection theory problem encountered by many students is a "dice game" described in the first chapter of Green and Swets' (1966) classic treatment. On each trial, three dice are rolled, two conventional dice with 1 to 6 spots, and a third die that contains 0 spots on 3 sides and 3 spots on the other 3 sides. You are given the total number of spots on the three dice and asked to judge the value of the third, critical die. What decision rule should you adopt, and how well can you expect to do?

In this artificial problem, the decision maker is obviously discriminating between *distributions*, not constant stimuli.[1] Possible totals range from 2 to 12 if the third die is 0 and 5 to 15 if it is 3, so values between 5 and 12 could arise from either 0 or 3. Figure 12.1 shows the distributions, which are triangular in shape. The natural decision rule (natural to the reader of *this* book,

---

[1] A given trial does not contain a distribution, but only an event—in this case, a number. We nonetheless use the term *distribution discrimination* for cases in which the possible events in S_1 and S_2 are explicitly varied by the experimenter.

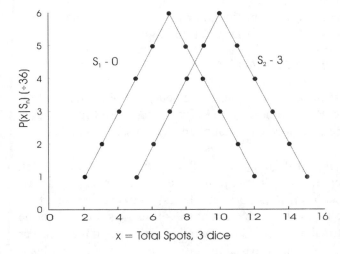

FIG. 12.1. Distribu-
tions of totals for the
dice game. The S_1 dis-
tribution gives the
possible totals for two
conventional dice plus
0, and the S_2 distribution gives the possible
totals for two conven-
tional dice plus 3.

who encounters the problem in chap. 12 rather than chap. 1) is to establish a criterion at some value between 5 and 12.

If presentation probabilities are equal, what is the highest success rate the player can obtain? In this case, the criterion should be at the crossover point of the two distributions, and the decision rule is to respond "0" for totals of 8 or less and "3" for totals of 9 or more. Examination of the distributions reveals that the "hit rate" (correctly saying "3") and the correct rejection rate (correctly saying "0") will each equal 26/36 = .72. This is the best performance level the observer can reach.

Models of optimal performance of this kind are called *ideal observers*. The strategy described for the dice example is ideal in three senses. First, the decision variable is the total number of spots, which is perceived without error. Second, the observed value is compared with a fixed criterion. Third, the criterion is placed at a location that maximizes accuracy as measured by $p(c)$. When either of the first two characteristics is violated by a human observer, as when there is an error in perception or variability in the criterion location, lower than ideal sensitivity results. If only the third characteristic is violated, a point on the ideal ROC is obtained, and performance might or might not be considered ideal depending on the application. Performance that is reliably better than ideal is not possible.

Can we compute a d' for the dice game? The distributions are obviously not normal, but the analogous statistic is easily found. The mean difference is 3, the standard deviation can be shown to be 2.45, and the ratio of these is

1.22. If this *were* a d', it would correspond to a $p(c)_{max}$ of .73, quite close to the true value based on the actual triangular distributions.

Distribution Discrimination

Green and Swets intended the dice game as a pedagogical device, and for that matter so do we. But a number of investigators have used similar tasks to compare actual performance with that of an ideal observer who behaves optimally.

Lee and Janke (1964, 1965), in fact, used numerical distributions as in the dice game, except that the distributions were normal. In a later experiment of this type, Kubovy, Rapoport, and Tversky (1971) found that responding was close to ideal, but that about 6% of responses could not be predicted by the optimal rule. This is what would be expected if the observer shifted the response criterion from trial to trial; we discuss this further later. Lee and his colleagues drew similar conclusions from one-dimensional non-numeric distributions (e.g., grayness of a patch of paper).

Comparisons of Real and Ideal Observers

In measuring detection thresholds, it has been important to find out whether limitations in sensitivity are inherent in the stimulus or derive from shortcomings of the observer. We consider one example from vision and one from hearing, and then we consider how the relative contributions of these two sources of "noise" can be estimated.

Absolute Visual Detection.   An early experiment that compared real and ideal observers predates the development of detection theory. Hecht, Schlaer, and Pirenne (1942) asked how many quanta of light need to be absorbed for a viewer to detect a weak visual stimulus (see Cornsweet, 1970, and Luce & Krumhansl, 1988, for summaries). Figure 12.2a shows the percentage of "yes, I see it" responses as a function of stimulus intensity. Because of uncertainties about how many quanta are filtered out by the optical system of the eye, these data cannot be used directly to determine the minimum number of quanta required for detection. Thus, the stimulus axis is labeled *arbitrary*: The values are only proportional to the number of quanta reaching the receptors and cannot be used directly to find the exact number of quanta being absorbed.

Ideal observers enter the picture because for a fixed light intensity the number of quanta reaching the retina is not constant, but has a Poisson dis-

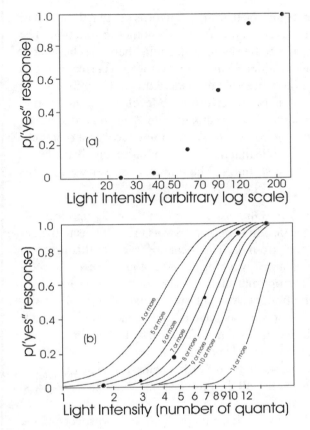

FIG. 12.2. (a) A psychometric function obtained by Hecht, Schlaer, and Pirenne (1942) for visual detection. (b) The same data with theoretical Poisson functions overlaid on them. Each curve corresponds to a different hypothetical number of quanta required for detection, and each has a different slope. The curve that assumes 7 or more quanta to be required for detection provides the best fit. (Adapted with permission from Figs. 4.6 and 4.7 in Cornsweet, 1970.)

tribution. This distribution can be used to predict the shape of the psychometric function. If the data fit the prediction, we could conclude that the only limitation in detecting weak lights lies in the stimulus itself. The Poisson is a family of one-parameter distributions for which the mean and variance are equal, and as a consequence the predicted slope of the psychometric function depends on the number of quanta required for detection. Figure 12.2b shows a family of psychometric functions derived from the Poisson. When the data points obtained by Hecht et al. are superimposed on these theoretical curves, the best fit is the case in which 7 quanta are required for the observer to say "yes."

We know, therefore, that if the observer is ideal the number of quanta needed for detection is 7, but we do not know whether the observer is in fact ideal. As mentioned earlier, one sort of nonoptimality is variation in the observer's response criterion from trial to trial, and it turns out that the effect of such variation is to decrease the slope of the psychometric function. The psychometric function measured by Hecht et al. could, therefore, be steeper than the true function, and the slope of that function would correspond to a higher threshold. Hecht et al. were able to resolve this indeterminacy in a more precise experiment at a single performance level (60%), and found that 8 quanta were sufficient to produce this hit rate. The small difference between this number and the good fit of the 7-quantum, ideal observer theoretical curve led them to conclude that the major limitation on visual detection was the variability of the stimulus. The human observer was, in this case, almost ideal.

Detection of Pure Tones in Noise.   One of the early applications of detection theory, summarized in Green and Swets (1966), was to the detection of tones in noisy backgrounds. The ideal observer for this problem uses all the detail of the stimulus waveform, and is thus termed a *signal-known-exactly* observer. The optimal analysis is to calculate a "cross-correlation" between the observation interval and a remembered copy of the signal. To predict d', one calculates the average output of the cross-correlator for Noise and Signal trials and divides the difference by the standard deviation of this device. Human observers do not meet the prediction made in this way. Their d' values are lower—they have efficiencies less than 1.0—indicating that they are not able to use all the information (e.g., phase) required by the cross-correlation strategy.

If observers are not ideal in solving this problem, what aspects of the waveform *do* they take advantage of? One way to answer this question is to examine the potential performance of some nonoptimal strategies. For example, perhaps the observer simply calculates the energy in the observation interval, discarding other information like phase and frequency. Energy detectors do a much better job of predicting performance. An exact correspondence could be interpreted to mean that human observers are ideal in using the information they collect, even if they do not incorporate other information that could raise their accuracy.

Combining Internal and External Noise.   Because internal noise is never exactly zero, sensitivity in tasks containing stimulus variation is limited by a combination of internal and external noise. The most common approach to modeling this situation is to imagine that the two types of variability are additive, so that the total variance limiting performance is simply the sum of the external and internal contributions.

Suppose, as in Lee and Janke (1964), that an observer is discriminating two distributions of line lengths, with means of 10 and 14 cm and a common standard deviation of 2 cm. The ideal observer's sensitivity can be written as

$$d'_{\text{Ideal}} = (M_2 - M_1)/\sigma_E , \tag{12.1}$$

where M_1 and M_2 are the means of the distributions and σ_E is the external standard deviation. Thus the situation has a d' of 2 and an unbiased $p(c)$ of .84. If actual (unbiased) accuracy is only .76, so that $d' = 1.41$, how much of the variability is internal and how much is external?

We assume that the effective variability is the sum of the external and internal variances and these two components are independent. Then

$$d'_{\text{Observed}} = (M_2 - M_1)/(\sigma_E^2 + \sigma_I^2)^{\frac{1}{2}} , \tag{12.2}$$

where σ_I is the internal standard deviation. The ratio of ideal to observed d' can be used to estimate the ratio of internal to external variance because (combining Eqs. 12.1 and 12.2)

$$\sigma_I^2/\sigma_E^2 = (d'_{\text{Ideal}}/d'_{\text{Observed}})^2 - 1 . \tag{12.3}$$

In the example, this ratio equals 1.0, leading to the conclusion that the amounts of internal and external variance are equal.

What exactly is "internal noise"? In Lee and Janke's length discrimination task, the fault probably lies in the decision rule rather than the encoding process. Two ways in which decision making might be imperfect were raised in the Essay in chapter 2. If the observer is completely inattentive on a proportion γ of trials, performance declines. In this case, the average $p(c)$ of .76 would equal $(.5)\gamma + (.84)(1 - \gamma)$, so $\gamma = .24$—the observer is not attending to about one quarter of the trials. Alternatively, as mentioned earlier, the observer allows the criterion to vary. The equality of internal and external variances implies that the standard deviation of criterion location, like that of the length distribution itself, equals 2 cm.

Our simple model for combining internal and external noise allows us to calculate the observer's *efficiency*, defined as the square of the ratio of ideal to observed d'. A rearrangement of Equation 12.3 shows that in this case it equals $\sigma_E^2/(\sigma_E^2 + \sigma_I^2)$; for the line length example, efficiency equals .5.

Basic Processes in Multiple Dimensions

In one dimension, the ideal observer question is: Does the experimental participant use all the information in the stimulus to make a decision, and if not can the information that is lost be simply described? We now extrapolate these questions to multivariate stimuli. Suppose the stimuli are the horizontal and vertical positions of a dot on a computer screen (as—except for the computer—in Lee, 1963). Depending on the distributions being discriminated, the observer needs to take account of both dimensions. In this section, we show how the multidimensional tools developed in chapters 6 to 10 allow the investigator to determine which stimulus aspects enter into the observer's decision and which are ignored.

We discuss three versions of this approach, all of which exploit the information-combination ideas introduced in chapter 8. First, GRT theorists have extended Lee's distribution-discrimination design to multiple dimensions. Second, components of complex stimuli have been directly manipulated to determine which ones affect overall sensitivity. The central paradigm is the *conditional-on-single-stimulus (COSS)* design of Berg (1989). Finally, Sorkin and his colleagues have extended the analysis to *groups* of observers collaborating on a decision; the "components" of sensitivity are individuals.

The GRT Approach

Ashby and Gott (1988) introduced the "general recognition randomization technique" as a tool for studying pattern classification. Their stimuli were horizontal and vertical line segments joined at the upper left corner. The lengths of both segments were drawn from logistic distributions, and a particular stimulus could be represented as a point in two-dimensional space. Repeated presentations thus defined a bivariate distribution, and the observer's task was to discriminate between two such distributions. In one experiment, for example, the mean of the distribution for Stimulus A was 400 horizontal and 500 vertical units, whereas for Stimulus B it was 500 horizontal and 400 vertical units. The standard deviation for both distributions, and for both segments, was 84. The distributions Ashby and Gott used are shown in Fig. 12.3.

Observers classified items as A or B, and an optimal rule for making this decision is to say "A" if the perceived difference between the vertical and horizontal lengths $v - h$ is greater zero and say "B" if it is not. If this difference is indeed the decision variable (and all three observers adopted rules

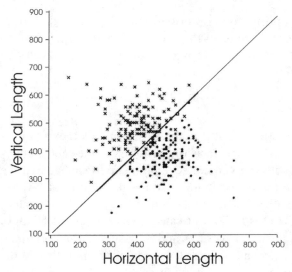

FIG. 12.3. Data reported by Ashby and Gott (1988) for one observer in a classification experiment. Stimuli consisted of a horizontal line segment joined to a vertical line segment. The lengths of both were varied by the experimenters and are represented on the horizontal and vertical axes. xs indicate stimuli that the observer judged to come from a distribution centered at the point (400, 500), •s indicate stimuli judged to come from one centered at (500, 400), and ⌷s indicate response inconsistency. (Adapted with permission from Fig. 7 in Ashby and Gott, 1988.)

very much like it), the problem is converted into a one-dimensional one like that of Lee and Janke (1964). To determine the optimal value of d', we first find the difference between the two means from the Pythagorean Theorem; it is $(100^2 + 100^2)^{1/2} = 141$. The standard deviation of 84 is the same in all directions, so $d' = 141/84 = 1.68$, and the value of $p(c)$ expected for an unbiased observer is .80. All three observers obtained scores close to this (.88, .82, and .79), leading to the conclusion that virtually all the variability limiting a decision in this case was in the stimulus distributions themselves.

   This conclusion is supported by a closer look at Fig. 12.3. The diagonal line is the ideal decision boundary, $v - h = 0$. The xs and •s do not correspond to the *A* versus *B* distributions, but to the responses given by the observer. Virtually all the xs ("A" responses) and •s ("B" responses) were consistent with the use of the ideal boundary, and the "errors" arose from cases in which the *A* distribution produced stimuli in the "B" region and vice versa. Put in terms of the previous section, these observers displayed little internal noise.

   The questions we have asked about classification of two-segment line figures focus on overall performance, and the analysis has been quite similar to that applied by Lee and Janke (1964) to classification of single lines. The two-dimensional distributions also allow us to explore the way in which the two components combine to determine overall performance, and Ashby and Gott used their stimulus set to compare different integration strategies than the optimal one described here. In the GRT framework, this is a question of the shape of the decision bound, a question we explored in chapters 6 and 8. We now turn to a related strategy developed largely in psychoacoustics.

The COSS Approach

Both the horizontal and vertical lengths in the Ashby and Gott experiment are useful in making a classification decision. Let us examine each component separately starting with the vertical segment. The longer this is, the more likely it came from the *A* distribution, so the proportion of "A" responses should increase with vertical length. Figure 12.3 allows us to calculate *P*("A") for vertical lengths in 50-unit categories (200–250, 250–300, etc.) simply by counting the numbers of xs and •s in successive horizontal strips. Proportion of "A" responses increases gradually from 0 to 1 in the manner of a psychometric function. Such functions depend on one component irrespective of others, leading Berg (1989) to call them conditional-on-single-stimulus (COSS) functions. A similar analysis can be applied to the horizontal segments. Because "B" becomes more likely as these lengths increase, it is natural to look at *P*("B") rather than *P*("A"). COSS functions for both segments are given in Table 12.1.

TABLE 12.1   *COSS Functions for the Ashby and Gott (1988) Experiment*

	Vertical		Horizontal	
Lengths	*P("A")*	*z[P("A")]*	*P("B")*	*z[P("B")]*
< 250	0.00		0.00	
250–300	0.09	−1.34	0.00	
300–350	0.03	−1.88	0.11	−1.23
350–400	0.17	−0.95	0.13	−1.13
400–450	0.44	−0.15	0.36	−0.36
450–500	0.72	0.58	0.71	0.55
500–550	0.94	1.56	0.85	1.04
550–600	0.95	1.64	0.92	1.40
> 600	1.00		1.00	

Because the *A* and *B* distributions are approximately normal, converting the COSS functions to *z* scores should produce straight lines, and Fig. 12.4

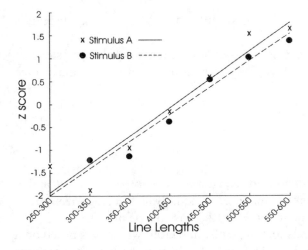

FIG. 12.4. COSS functions constructed from the Ashby and Gott data in Fig. 12.3.

shows that this is roughly true. The slopes of the lines (0.0124 for vertical location and 0.0118 for horizontal) measure the effectiveness of the two components in making the classification judgment and can be used to determine the weighting given to each dimension.

The decision rule assumed by COSS is to compare a weighted sum of component observations with a criterion. Calling the horizontal dimension x_1 and the vertical dimension x_2, the rule is to respond "A" if $a_1 x_1 - a_2 x_2 > c$. (The a_is are weighting constants, and the minus sign arises from the negative relation between the two components.) Solving this relation, we find that x_2 must be greater than y_2, where

$$y_2 = (c + a_1 x_1)/a_2 . \tag{12.4}$$

This equation, which describes a line in the space of Fig. 12.3 with slope a_1/a_2, is the decision bound for the vertical segments. There is an analogous boundary y_1 for the horizontal segment. Berg showed that the values of the weights a_1 and a_2 depend on the variance of these variables in the following way:

$$\frac{a_2^2}{a_1^2} = \frac{\text{var}[y_1] + \sigma_1^2}{\text{var}[y_2] + \sigma_2^2} , \tag{12.5}$$

where σ_1^2 and σ_2^2 are the external variances of the vertical and horizontal length distributions. Intuitively, one component receives more weight than another if the variance of either the stimulus distribution or the observer's responding is smaller.

Applying this analysis is straightforward. It turns out that the slope of each COSS function $[z(P(\text{``A''}|y_i))$ vs $y_i]$ equals the square root of the inverse of var$[y_i]$, so

$$\frac{a_2^2}{a_1^2} = \frac{\frac{1}{.0124^2} + 84^2}{\frac{1}{.0118^2} + 84^2} . \tag{12.6}$$

Because a_1 and a_2 must add to 1, the equation can be solved: $a_1 = .506$ and $a_2 = .494$. The two segments contribute nearly equally to the categorization decision, and the decision bound in Fig. 12.3 has a slope of 1.02.

This conclusion seconds that of Ashby and Gott, which was based on a different analysis of the data. They also found that the best-fitting linear bound for this participant in the space of Fig. 12.3 had a slope of about 1, implying that x_1 and x_2 were equally weighted by their observers. Although this example suggests that GRT and COSS analyses are just two mathematical translations of the same text, more complex perceptual situations reveal advantages for each. GRT is more flexible when the independence assumption is abandoned. As we saw in Part II, dependence between dimensions can be defined in several diagnosable ways.

COSS analysis has an advantage when the number of components contributing to a decision is greater than two. In an auditory experiment analogous to those we have been discussing, Berg, Robinson, and Grantham (1989; summarized in Berg, 1989) presented listeners with a sequence of up to 10 tones, each drawn from a normal distribution with a mean of 1000 Hz and a standard deviation of 100 Hz; or each drawn from a normal distribution with a mean of 1100 Hz and the same standard deviation. The COSS approach assigns each position in the sequence its own weight a_i, and any two positions can be compared using Equation 12.5. The requirement that $\Sigma a_i = 1$ allows just enough equations to estimate all the weights. One finding in the Berg et al. study was that the greatest weight was assigned to the last item in the sequence.

Groups of Observers

In some important real-life situations, groups of individuals (such as juries or committees) must make a decision based on the same evidence. The framework we have been describing for combining "dimensions" within a stimulus can be extended to the problem of combining information from group members. Sorkin and his colleagues (Sorkin & Dai, 1994; Sorkin, Hays, & West, 2001) have studied an experimental situation in which performance of a group can be compared with various ideal models.

In a typical experiment, each member of a team of observers makes a judgment in a visual discrimination task, and the votes are somehow combined into a group response. One kind of model that can be used as a baseline is the *Condorcet group*. In such a group, each individual casts an unbiased vote, the votes are weighted equally, the judgments are treated as independent, and a decision is reached by some type of majority rule. The decision rule may be a simple majority, unanimity, or anything in between. Ideal groups are superior to Condorcet groups in three respects: Individuals make graded judgments rather than binary ones, their judgments are weighted in proportion to their expertise (i.e., their d' values) and summed, and the summed d' statistic is compared with a criterion to make a decision.

Both the ideal group and Condorcet groups with different majority rules predict that performance will increase with the size of the group as shown in Fig. 12.5. Also plotted in the figure are data points from an experiment constructed to remove any artificial constraints on group performance—graded responses were made, a decision was made by consultation rather than strict voting, and the expertise of the group members was known. As can be seen, performance increased with group size in the manner predicted by a Condorcet group operating on the basis of unanimity or near unanimity—a nonoptimal rule. The discrepancy between observed performance and ideal increased with group size, in that efficiency (the square of the d' ratio) dropped from about 90% for a two-person group to 45% for seven people.

Because the analysis follows that developed in the multicomponent stimulus context, tools developed there can be used. For example, Sorkin et al. (2001) were able to apply the COSS method to determine the weights assigned to each individual in a group. Groups in this experiment weighted observations roughly according to the expertise of the contributor, a comforting result. Judgments were essentially uncorrelated and, as predicted, when a correlation was introduced experimentally (by correlating the stimulus arrays), performance dropped.

Decision making by groups has been studied extensively by social psychologists, and some of their findings are illuminated by these results. For example, participants sometimes reduce their efforts when participating in a group, a finding that has been called *social loafing*. In SDT modeling, this result can be understood as a response to high correlations between individ-

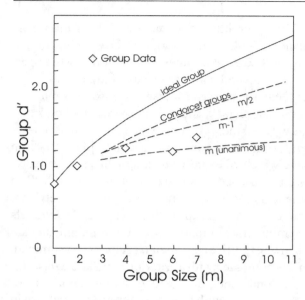

FIG. 12.5. Group performance in a signal-detection task as a measure of group size. Theoretical curves describe the ideal method of combining information and various simpler "Condorcet" rules based on unweighted tallying of individual votes. The data are most consistent with the least optimal Condorcet rules. (Adapted with permission from Fig. 4 in Sorkin, Hays, & West, 2001.)

ual judgments or a low weighting assigned in group decision making. The existence of Condorcet and ideal comparison models allows for more specific hypotheses in the study of such interesting phenomena.

Hierarchical Models

We have been describing models in which observer performance is compared with ideal performance based on stimulus structure, but the general approach can be elevated one step so that the comparison is between different levels of processing. To make this work, we must construct two tasks that use the same stimuli but require different kinds of treatment by the observer. In the simplest case (and the only one we pursue here), one task depends only on low-level mechanisms, the other on both low- and high-level mechanisms.

An example of such an analysis was presented in chapters 5 and 7. Durlach and Braida (1969) postulated that performance in a 2AFC discrimination paradigm in which the same two stimuli (auditory pure tones) were discriminated on every trial was essentially limited only by the sensory noise that arises inevitably from neural coding. Identification accuracy is also limited by this sensory variance, but also by context coding memory noise; a comparison of discrimination and identification allows calculation of the relative magnitude of these two sources of variance. Because Durlach and Braida assumed the two types of noise to be additive, this proportion can be found from a variant of Equation 12.3 (cf. Eq. 5.5). A similar analysis of 2AFC roving discrimination designs, in which the two stimuli to be discriminated vary from trial to trial, proposes that sensory variance, context coding, and time-dependent trace coding combine to determine sensitivity. Each component can be estimated from a suitable data set; see chapters 5 and 7 for more detail.

A similar strategy has been applied to visual search experiments, which are rather more complicated than pure-tone resolution. In a typical visual search design, the observer must determine whether a target (say, a horizontal line segment) is present in an array of distractors (say, vertical line segments). The dependent variable is usually response time, but accuracy may also be measured. An important finding in this literature is that performance is better when target and distractor differ by a single "feature" (as in the line orientation case) than when two features are relevant (as in finding a red horizontal segment in a field of red vertical and green horizontal segments). A common interpretation of this finding (Treisman & Gelade, 1980) is that additional processing is required in "conjunction" conditions to integrate the two features.

Geisler and Chou (1995) asked whether low-level factors might be responsible for such differences in search performance and so ight to measure the low-level baseline for this task. Like Durlach and Braida, they measured 2AFC discrimination (in their case of a field with a target from a field that consisted only of distractors), limited the range of stimuli (using an adaptive procedure like those described in chap. 11), and provided trial-by-trial feedback. Data from these tests were summarized by a discrimination window describing performance over spatial area and stimulus duration; the wider this window, the better the low-level processing.

If low-level mechanisms account substantially for visual search speed and accuracy, then experimental conditions with larger windows should produce faster and more accurate searches. A strong correlation of this type is exactly what Geisler and Chou (1995) observed, and they concluded that the slowness of conjunction searches compared to feature searches, "may be due (at least in part) to low-level factors and not to complex aspects of the attentional mechanisms" (p. 370). It is clear that high-level processes like attention allocation play an important role in visual search, particularly in multiple-fixation conditions, but accounting for more variance with low-level mechanisms makes the task of developing a general model of visual search both more manageable and more accurate.

Essay: Psychophysics Versus Psychoacoustics (etc.)

In the community of auditory researchers, two terms are used to describe research on the discrimination and classification of sounds, and it is useful to draw a distinction between them. *Psychophysics* is the use of theoretically grounded methodology to interpret perceptual measurements of sounds, and *psychoacoustics* is the project of relating those measurements to the sounds' physical characteristics. A similar distinction can be made in other modalities and in cognitive applications as well.

Until the present chapter, this book has been almost entirely about psychophysics: The questions of how to separate sensitivity and bias, compare different discrimination paradigms, and relate classification to discrimination data have been taken up with the most modest of stimulus descriptions. The variability limiting performance has often been partitioned between components (sensory and memory, attention to one or more sources of information), but these components have kept their distance from the stimuli.

In this chapter, we have taken a few steps toward adjusting the balance. Although psychoacoustics has by no means received equal time, the models here do include stimulus factors as explicit contributors to sensitivity or to its limitations. Detection theory unifies approaches with varying degrees of reliance on stimulus factors, and there is a discernable continuum of application, from complete reliance on the stimulus to explain the data to complete indifference to it.

One early line of detection theory research was heavily psychoacoustic. The path of this program moves from the ideal observer models summarized in Green and Swets (1966) to studies of profile analysis (Green, 1988) and Berg's (1989) COSS analysis. The undoubted success of this body of work, however, was obtained at the cost of the introduction of explicit variability into the stimulus sets. Detection of a noise increment in a noise background is well understood in terms of energy detection (Green, 1960), but the background noise is necessary to calculate the variance that limits performance. This kind of ideal observer model would not be able to predict the detectability of a noise burst in silence. Similarly, both GRT and COSS do a good job of accounting for the discriminability of distributions of line-segment pairs, but could not make much progress if only two such pairs were being distinguished. When no external noise is present, detection theoretic models still describe data (such as ROCs) well, but the limiting variance is internal and not identified with any aspect of the experimental situation. This is pure psychophysics, the other end of the continuum.

It is the in-between cases that are the most interesting. In the "ideal group" analyses of Sorkin and his colleagues, experimental results are compared with a baseline that represents the ideal performance under particular stimulus conditions. The discrepancy is then interpreted in terms of nonoptimal decision processes. Kingston and Macmillan's (1995) analysis of the Garner paradigm showed that some tasks are inherently harder than others, so that the degree of "filtering loss" must be understood in terms of ideal observers, not simple performance measures. For that matter, the many comparisons between discrimination paradigms discussed in earlier chapters (and summarized in Figs. 10.1 and 10.2) show that inherent limitations in the decision space can account for a great deal of variance that might otherwise be understood in psychological terms.

The "internal noise" that remains when inherent limitations are factored out can be conveniently (if not precisely) divided into cognitive and sensory components. The cognitive category includes Durlach and Braida's context memory, Geisler and Chou's high-level processes, and explicit attentional manipulations. The sensory category is neural noise, of the sort identified by Hecht et al., in vision and by auditory-nerve-based models in hearing. This category includes processes whose neural substrates are well understood, but of course all internal noise is neurally based. One direction in which progress is being made is in providing a neuroscience explanation of cognitive processes as well. As an example, Patalano, Smith, and Jonides (2001) have shown that different parts of the brain underlie prototype- and exemplar-based strategies in categorization tasks, even within a single participant.

A sign of progress in research using resolution designs is that increasingly complex tasks are being used; another is that detection theory models are keeping pace. The psychophysical and psychoacoustic poles with which we introduced this essay define an increasingly false dichotomy. Psychoacoustics is becoming more modest in its contributions to moderately complex problems, as other components are better understood, but more compelling in complex situations (e.g., sound localization) in which models of the stimulus situation have advanced faster than those of internal processing. Psychophysics is becoming more ambitious, attempting to incorporate cognitive and neural processes into its models. As our understanding of perception deepens, we can expect to see fewer and fewer theories that rely on either one alone and more that draw from many components of sensitivity.

Summary

Detection theory provides strategies for partitioning the variance that limits performance between external and internal sources, and among subcategories of each. An important tool is the distribution discrimination task, in which participants must determine which of two overlapping distributions led to an observation. The magnitude of the external noise arises from the overlap, and the best possible, ideal performance is what would be found if this were the only limiting factor. If accuracy falls below this level, the discrepancy is attributed to internal noise.

When stimulus classes vary on more than one dimension, external variance can be divided among the dimensions. Discrimination can be described with multidimensional detection theory, and the decision bound between the distributions depends on both dimensions. The weighting assigned to a dimension depends on the variability of the distributions along both dimensions (for the ideal observer) and on the slope of the psychometric COSS functions (for real observers). Similar analyses can be applied to features of geometric shapes, frequency regions of noise stimuli, and individuals within a group.

A comparison of different experimental tasks permits division of internal variance to multiple levels of processing, given a theoretical perspective that determines what kind of processing is required for each task. This approach has been successful in such disparate areas as pure-tone resolution and visual search.

Problems

12.1. Draw the ROC curve for the dice game. How is it similar to, and different from, other theoretical ROCs?

12.2. In the original dice game, there are 3 dice, 2 regular and 1 with half 3s and half 0s. Consider two modified games: (a) There are 4 dice, 3 regular and 1 with half 3s and half 0s. (b) There are 3 dice, 2 regular and 1 with half 2s and half 0s. For each game, what are the underlying distributions for total score? How do their means and variances compare with those of the original game? (More difficult:) What is the maximum possible proportion correct?

12.3. Suppose the observer in the Lee and Janke experiment adopts a criterion location of 10 cm on half the trials and 14 cm on the other half. What is the standard deviation of the criterion location? What proportion correct will be obtained?

12.4.   (a) In an auditory experiment, the listener hears two noise bursts. The average intensity of both bursts is 1 for samples of S_1 and 2 for samples of S_2, and both samples have a variance of 1 (see Fig. 12.6.a). The task is to decide which distribution generated the bursts. The experimenter plots COSS functions for the first and second intervals separately and finds that both have slope 1. What weights has the listener assigned to the two bursts? What is the slope of the decision bound?

(b) Same as (a), but now the COSS function for Interval 2 has a slope of 0.5.

12.5.   Same as Problem 12.4, but the two samples have unequal variance: Samples of S_1 have variance 1 and samples of S_2 have variance 4 (see Fig. 12.6b). The experimenter plots COSS functions for the first and second intervals separately and finds that both have slope 1. What weights has the listener assigned to the two bursts? What decision bound does this imply? (b) Same as (a), but now the COSS function for Interval 2 has a slope of 2.

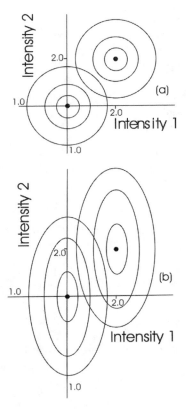

FIG. 12.6.   Distributions of intensity for an experiment in which two noise bursts are presented on each trial. The value of burst 1 is represented on the horizontal axis and that of burst 2 on the vertical axis; the circles and ellipses are equal-likelihood contours, as in previous chapters. Both bursts are drawn either from a distribution with mean = 1 or from a distribution with mean = 2. (a) variance = 1 in both intervals (Problem 12.4), and (b) variance = 1 in interval 1, variance = 4 in interval 2 (Problem 12.5).

IV

Statistics

13

Statistics and Detection Theory

Statistics is commonly divided into two parts. In descriptive statistics, a data set is reduced to a useful measure—a statistic—such as the sample mean or observed proportion. Detection theory includes many possible statistics of sensitivity [d', α, $p(c)$, etc.] and of bias, and this book has been well stocked with (descriptive) statistics.

Inferential statistics, on the other hand, provides strategies for generalizing beyond the data. In chapter 2, for example, we met an observer who was able to correctly recognize 69 of 100 Old faces while producing only 31% false alarms, and thus boasted a d' of 1.0. As a measure of sensitivity for these 200 trials, this value cannot be gainsaid, but how much faith can we have in it as a predictor of future performance? If the same observer were tested again with another set of faces, might d' be only 0.6 or even 0.0?

The statistician views statistics, such as sensitivity measures, as *estimates* of *true* or *population parameters*. In this chapter, we consider how statistics can be used to draw conclusions about parameters. The two primary issues are: (a) How good an estimate have we made? What values, for example, might true d' plausibly have? and (b) Can we be confident that the parameter values, whatever they are, differ from particular values of interest (like 0) or from each other? These two problems are called *estimation* and *hypothesis testing*.

The chapter is in four sections. First, we consider the least processed statistics, hit and false-alarm rates. Second, we examine sensitivity and bias measures. The third section treats an important side issue—the effects of averaging data across stimuli, experimental sessions, or observers. For all these topics, the primary model considered is equal-variance SDT, and the discussion of hypothesis testing is limited to hypotheses about one parameter or the difference between two parameters. The final section shows how

319

the standard statistical technique of logistic regression can be used in testing hypotheses within the basic model of Choice Theory and can be extended to SDT and other models.

Like the rest of the book, most of this chapter should be accessible to the survivor of a one-semester undergraduate statistics course. Relevant concepts from probability and statistics, some of which may be unfamiliar to such a reader, are summarized in Appendix 1.

Hit and False-Alarm Rates

A Single Observed Proportion

To start, consider a face-recognition experiment in which the proportion correct is reported to be .69. Observed proportions vary from sample to sample according to a well-known distribution, the *binomial*. If the true proportion recognized is p, then the observed proportion P varies across separate tests. The expected value of P is the true value p, and the variance of P is $p(1 - p)/N$, where N is the number of trials. We can estimate the variance in our example by using the observed proportion P instead of p; this estimate is $(.69)(.31)/100 = 0.002139$, and the standard error is 0.0462.

When N is fairly large, as in this example, and the products Np and $N(1 - p)$ are not too small, the distribution of P is approximately normal. (For appropriate methods when these conditions are not satisfied, see Darlington & Carlson, 1987.) In a normal distribution, about 95% of scores are within 1.96 standard deviations of the mean, the remaining 5% being equally divided between the two extreme "tails." We can use this fact to construct a *95% confidence interval* around P:

$$p = P \pm z_{.025}[p(1 - p)/N]^{\frac{1}{2}} . \tag{13.1}$$

In our example, approximating p by P,

$$p = .69 \pm (1.96)(0.0462) = .69 \pm .09 .$$

That is, the true proportion is probably between .60 and .78.

The same strategy leads to hypothesis tests about binomial data. The experimenter can test the hypothesis that the true recognition proportion is .5 by simply noting that .5 is not in the 95% confidence interval. Thus, this hypothesis can be rejected "at the .05 level."

Large and small proportions have less variance than intermediate ones: $p(1 - p)$ equals 0.25 when $p = .5$, but is only 0.09 when $p = .9$ and falls to 0.01

when $p = .99$. This suggests that estimation will be most accurate, other things being equal, for large values of d'. Does this mean that, in choosing experimental conditions, one should aim for very high performance levels?

For at least two reasons, the answer is no. The first reason is this: As proportions near 0 or 1, their variability does indeed decrease, but the probability of obtaining an observed proportion of *exactly* 0 or 1 increases. Observed proportions of 0 or 1 can be converted to z scores (or to log odds, the Choice Theory transformation) only by a somewhat arbitrary adjustment and are thus worth avoiding. Techniques for dealing with perfect proportions attained by individuals in a group were introduced in chapter 1 and are discussed further later. The second reason for avoiding very large and small proportions in the first place is also addressed later.

Comparing Two Proportions

Binomial variability also affects comparisons of two data points, each involving only one proportion. To extend the example, suppose a second observer recognizes 89 of 100 faces: Is this a significantly greater proportion than the first observer's .69? An important statistical theorem (see Appendix 1, Equations A1.8 and A1.9) concerns differences between independent variables: The mean of the difference is the difference between the means, and the variance is the *sum* of the variances. In this case, $P_1 - P_2$, the difference in the success rates, has a mean value of $p_1 - p_2$, the true population difference. The variance of P_1 is $p_1(1 - p_1)/N_1$, the variance of P_2 is $p_2(1 - p_2)/N_2$, and if the two proportions are independent—as we assume—the variance of the difference is the sum of these.[1]

Finally, the difference between two normal variables is also normal, so the 95% confidence interval around the observed difference is:

$$P_1 - P_2 = P_1 - P_2 \pm z_{.025}\{[p_1(1 - p_1)/N_1] + [p_2(1 - p_2)/N_2]\}^{1/2}. \quad (13.2)$$

For success rates of .89 and .69, again using observed P values to estimate the p parameters,

$$P_1 - P_2 = .20 \pm (1.96)[0.00098 + 0.00214]^{1/2} = .20 \pm .11$$

---

[1] According to detection theory, H and F are related *across* conditions by an ROC or isobias curve, and this is a form a dependence. But the statistical independence assumed here is that *within* an experimental condition the "yes" rates on S_1 and S_2 trials do not affect each other. This could be false if, for example, the criterion shifts gradually during a set of trials.

The true difference between the two proportions, we can be 95% sure, is between .09 and .31. Because 0 is not in this interval, we can reject the possibility that the two observers' memories are equally good.

(False-Alarm, Hit) Pairs

When S_1 and S_2 trials—New and Old faces, say—are distinguished in an experiment, two proportions (false-alarm and hit rates) are estimated, each with its own binomial variability. The first step in comparing such pairs is to apply the logic of the preceding section to each proportion.

Let us compare a control condition in which $(F, H) = (.31, .69)$ with a condition in which the observer is hypnotized and $(F, H) = (.59, .89)$, assuming that each of the four proportions is based on 100 trials. Both conditions can be represented as points in ROC space. Because F and H are normally distributed, the distribution of the point (F, H) is *bivariate normal* (see Appendix 1). Confidence regions around these points that include 95% of the mass of the bivariate distribution are shown in Fig. 13.1. All points at the edge of region have the same value of likelihood ratio. The regions turn out to be elliptical in shape, with a maximum radius of about 2.5 univariate standard-deviation units. The two contours do not overlap, suggesting that the two points are reliably different.

Figure 13.1 gives an indication of the variation we can expect in detection theory parameters. The confidence region for the ROC point from the control condition includes the points (.23, .77) and (.39, .61), which corre-

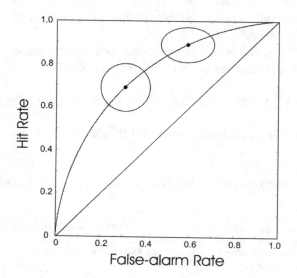

FIG. 13.1 Bivariate binomial (and therefore approximately normal) distributions of (false-alarm, hit) pairs in ROC space. Ellipses indicate regions containing 95% of the distributions.

spond to d' values of 1.48 and 0.56. Also included are the points (.39, .77) and (.23, .61), which correspond to c values of –0.23 and +0.23. Confidence regions based on fewer trials are larger, the radii being inversely related to the square root of the number of trials.

Sensitivity and Bias Measures

Usually our interest is not just in whether two ROC points could have arisen from the same underlying (false-alarm, hit) pair, but in whether the two points reflect the same sensitivity, or the same bias. (In the hypnotic recognition example, both parameters are important.) We consider, in turn, the d' and c parameters of SDT.

Sensitivity

Remember that a sensitivity parameter is computed by subtracting the transformed hit and false-alarm rates: for example, $d' = z(H) – z(F)$. Two separate questions can be asked about this statistic: Is it, on average, equal to true d'? What is its variance? In our discussion of hit and false-alarm rates, we were able to ignore the first, *statistical bias*, issue because observed proportions are accurate—*unbiased*—estimators of population proportions. Things are not so simple with d', and we must answer both questions.[2]

Statistical Bias of d'.   Miller (1996) evaluated the statistical accuracy problem in a straightforward way. Suppose the true hit rate in a yes-no experiment is .69 and the true false-alarm rate is .31, so that true $d' = 1.0$. In an experiment with 16 signal and 16 noise trials, what should we "expect" our estimate of d' to be? The *expected value* is the one obtained, on the average, in experiments of this type. One experiment might yield $H = 12/16 = .75$ and $F = 5/16 = .31$ for an estimated d' of 1.170; in another, perhaps $H = 10/16 = .62$ and $F = 4/16 = .25$, so $d' = 0.979$. The expected value can be calculated from the binomial distributions of H and F. For this particular situation, Miller found it to be 1.064, 6.4% greater than the true value.

Miller conducted this calculation for several values of true d' and the number of trials; some results are shown in Table 13.1, an abbreviated version of Miller's Table 1. Miller simplified his calculations by assuming

---

[2] The terminology is potentially confusing: Statistical bias is conceptually unrelated to response bias, and statistical accuracy (unbiasedness) is unrelated to accuracy as measured by a sensitivity measure. In this chapter, we avoid using the terms *bias* and *accuracy* without a qualifier unless the context makes the usage clear.

TABLE 13.1   *Expected Yes-No d' and Percent Bias*

	Number of Signal and Noise Trials							
	8		32		128		512	
True d'	*d'*	*% bias*	*d'*	*% bias*	*d'*	*% bias*	*d'*	*% bias*
0.5	0.555	11.0	0.514	2.8	0.503	0.6	0.501	0.2
1.0	1.096	9.6	1.029	2.9	1.007	0.7	1.002	0.2
2.0	2.036	1.8	2.084	4.2	2.019	1.0	2.004	0.2
3.0	2.641	−12.0	3.135	4.5	3.048	1.6	3.011	0.4
4.0	2.926	−26.8	3.879	3.0	4.122	3.0	4.032	0.8

equal response bias in all cases and also had to decide what to do about observed hit rates of 1 and false-alarm rates of 0, a problem we examined in chapter 1. Table 13.1 uses the correction in which a frequency of 0 is converted to ½ and a frequency of N is converted to $N − ½$.

Estimation is accurate if the appropriate table entry equals true d'. This is most nearly true for estimates based on large numbers of trials—if true $d' = 1$, for example, the table shows that with 512 trials per stimulus the average observed d' is 1.002, an error of just 0.2%. With fewer trials, unsurprisingly, estimates are less accurate. The most problematic cases—those with the greatest bias—are those in which the number of trials is small and true sensitivity is high; these are the results for which the correction for 0 and 1 cells is most often needed, and any correction leads to some distortion in the estimate.

There are at least two ways in which the pattern in Table 13.1 might affect substantive conclusions. First, sensitivity comparisons involving different numbers of trials entail a constant error. If $d' = 3.14$ in a condition with 32 trials and $d' = 3.05$ in a condition with 128 trials, the apparent difference is entirely attributable to different amounts of bias applied to a true d' of 3.0. Typically, one can avoid such comparisons. The second threat is more insidious: Comparisons of different sensitivity values based on the same number of trials are also contaminated by error. For example, if in two conditions with 32 trials each one measures d' values of 3.14 and 3.88, for a difference of 0.74, the bias pattern implies that the true d' values are 3.0 and 4.0, for a difference of 1.00.

Table 13.1 illustrates a potential distortion in data analysis, but fortunately also contains the information needed to avoid the problem by "correcting" estimates of d' for statistical bias. A d' of 1.10 based on eight trials per stimulus should be adjusted, according to the table, to its most likely true value of 1.00.

Standard Error of d'. The problem of finding the standard error of d' was first solved by Gourevitch and Galanter (1967) using an approximation. We begin with their approach, and then we consider the more exact calculations of Miller (1996).

Because $d' = z(H) - z(F)$, the first step in finding the variance (square of the standard error) of d' is to compute the variances of the transformed proportions. The variance of the difference between the two (independent) variables is then the sum of their variances.

Gourevitch and Galanter showed that observed z scores have an approximately normal distribution, with variance

$$\text{var}[z(p)] = \frac{p(1-p)}{N[\phi(p)]^2} , \tag{13.3}$$

where N is the number of trials and $\phi(p)$ is the height of the normal density function at $z(p)$. As a result,

$$\text{var}(d') = \frac{H(1-H)}{N_2[\phi(H)]^2} + \frac{F(1-F)}{N_1[\phi(F)]^2} , \tag{13.4}$$

where N_2 and N_1 are the number of Signal (S_2) and Noise (S_1) trials.

Values of the function ϕ can be found in Table A5.1 or computed (compare Equations 2.9 and A1.10):

$$\phi(p) = \frac{1}{\sqrt{2\pi}} e^{-\frac{1}{2}z(p)^2} . \tag{13.5}$$

Continuing our example, let us find a 95% confidence interval around d' in the hypnotic condition. The hit and false-alarm rates are .89 and .59, each based on 100 trials. Equation 13.5 reveals that $\phi(.89) = 0.1880$ and $\phi(.59) = 0.3887$. According to Equation 13.4, the variance associated with d' is $0.0277 + 0.0160 = 0.0437$, and the standard error is $(0.0437)^{1/2} = 0.209$. The center of the confidence interval is about 1.00—Table 13.1 shows that the statistical bias is less than 1% when N is approximately 100. The confidence interval extends 1.96 standard errors above and below observed d', that is, $1.00 \pm (1.96)(0.209) = 1.00 \pm 0.41$. We can be 95% confident that true d' is between 0.59 and 1.41, and in particular that it is not 0. The approach can be extended to hypothesis tests about more than two ROC points (Marascuilo, 1970).

An interesting aspect of this example is that the variance associated with the hit rate (0.0277) is substantially greater than the variance associated with the false-alarm rate (0.0160). The general finding, pictured in Fig. 13.2, is that the variance associated with z scores *increases* as proportions approach 0 or 1; this is true even though the variance associated with the proportions themselves *decreases* in this region. Here is the promised second reason to avoid extremely large or small proportions: Even if observations of 0 or 1 can be avoided, the variability associated with d' is large.

Miller (1996) extended the computational approach he used to estimate bias, discussed earlier, to standard errors. Table 13.2 gives an abbreviated version of his Table 2 (again for the ½, $N - ½$ correction) and includes a comparison of the direct calculation with the Gourevitch and Galanter approximation.

A large number of trials lead to a small standard error. The variance should be proportional to $1/N$, and this is almost exactly true for the approximation: For every increase in N by a factor of 4, the variance decreases by that factor. Direct computation shows a much less regular pattern for small N particularly at high sensitivity levels. Because they are exact, Miller's computations are to be preferred to the Gourevitch and Galanter approximation, especially because in some cases the degree of discrepancy is quite large.

The approximation and direct method give exactly the same result, to two decimal places, for the $H = .89$, $F = .59$ running example (letting $N_1 = N_2 = 128$ to allow for more direct reference to Tables 13.1 and 13.2). But

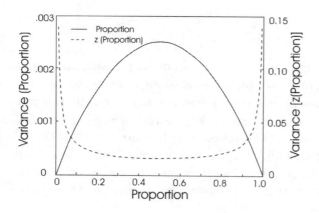

FIG. 13.2. The variance of a proportion p, and of its z transform $z(p)$, as a function of p. Variability of p is greatest when $p = .5$; variability of $z(p)$ is greatest when p is near 0 or 1.

TABLE 13.2   *Variance of Yes-No d'*

Computation	True d'	Number of Signal and Noise Trials			
		8	32	128	512
Direct	0.5	0.491	0.106	0.026	0.0063
	1.0	0.482	0.117	0.027	0.0068
	2.0	0.358	0.173	0.037	0.0090
	3.0	0.168	0.224	0.067	0.0150
	4.0	0.056	0.121	0.141	0.0333
Approximation	0.5	0.402	0.100	0.025	0.0063
	1.0	0.430	0.108	0.027	0.0067
	2.0	0.570	0.142	0.036	0.0089
	3.0	0.929	0.232	0.058	0.0145
	4.0	1.907	0.477	0.119	0.0298
Percent Error in Approximation	0.5	−18.1	−5.7	−3.8	0.0
	1.0	−10.8	−7.7	0.0	−1.5
	2.0	59.2	−36.6	−2.8	−1.1
	3.0	453	3.6	−13.4	−3.3
	4.0	3305	294	−15.6	−10.5

consider two cases with smaller N and higher d'. First, if true $d' = 2$ and $N_1 = N_2 = 32$, the approximation gives a variance of 0.142 (in Table 13.2), so the 95% confidence interval is $2.00 \pm (1.96)(0.142)^{1/2} = 2.00 \pm 0.74$. Direct computation reveals a bias of 4.2% (Table 13.1) and a variance of 0.173 (Table 13.2), so the confidence interval for d' is $2.08 \pm (1.96)(0.173)^{1/2} = 2.08 \pm 0.82$. The discrepancy here is moderate. For a more extreme example, consider the case in which true $d' = 3$ and $N_1 = N_2 = 8$. Now the approximation leads to a confidence interval of 3.00 ± 5.98 (i.e., it is uncertain whether d' is even positive). Direct calculation is more reassuring ($d' = 2.64 \pm 0.80$), but this result is deceptive. The smaller standard error results from the use of an approximation to eliminate undefined values of d'. As Miller (1996) noted, if true d' is high enough, all the data will fall at the maximum value allowed for perfect data, which in this case is $d' = z(7.5/8) - z(0.5/8) = 3.07$. This is not really a precise estimate of anything. Clearly one needs a very good excuse, and considerable caution, to estimate sensitivity from a mere 16 trials.

Response Bias

Bias and sensitivity are much alike statistically, because they are much alike algebraically: d' is the difference between $z(H)$ and $z(F)$, and c is -0.5 times the sum of these terms. The variance of c is found to be just one quarter the variance of d':

$$\begin{aligned}
\text{var}(c) &= \text{var}[-0.5(z(H) + z(F))] \qquad\qquad (13.6)\\
&= 0.25\{\text{var}[z(H)] + \text{var}[z(F)]\}\\
&= 0.25\ \text{var}[z(H) - z(F)]\\
&= 0.25\ \text{var}(d').
\end{aligned}$$

If $H = .89$ and $F = .59$, so that $\text{var}(d') = 0.0437$ as in the previous section, then $\text{var}(c) = 0.0437/4 = 0.0109$.

Chapter 2 describes a number of alternative bias measures. Of those derived from SDT, c has the simplest statistical properties. Indeed these properties are one reason for using c (Banks, 1970).

Comparing Two Conditions

To evaluate differences in sensitivity or response bias between two conditions, we apply Equation 13.4 twice, again using the theorem that the variance of the difference between two independent variables is the sum of their variances. Continuing the example: Before training, an X-ray reader has a hit rate of .89 and a false-alarm rate of .59, each based on 128 trials, for a d' of 1.00 and c of -0.73. After training, the values are $H = .92$, $F = .26$, $d' = 2.04$, and $c = -0.37$. Do the two conditions differ reliably in sensitivity? in bias?

The variance of the first d' is found, by the strategy illustrated in the previous section, to be 0.0437; that of the second, 0.0503. The 95% confidence interval around the difference between the two d' values is $1.04 \pm (1.96)(0.0437 + 0.0503)^{\frac12} = 1.04 \pm 0.60$. Because zero is not in the interval, the two ROC points reflect significantly different sensitivities. The variances of the two c values are 0.0109 and 0.0129, so the confidence interval around the difference is $0.36 \pm (1.96)(0.0238)^{\frac12} = 0.36 \pm 0.30$. Thus, the two response biases can also be reliably distinguished from each other.

Designs Other Than One-Interval

Two-Alternative Forced Choice.   The variance of d' can be calculated for other paradigms. For 2AFC, Bi, Ennis, and O'Mahony (1997) used a method similar to that of Gourevitch and Galanter (1967) to find that

$$\text{var}(d')_{2\text{AFC}} = \frac{H(1-H)}{2N_2\,[\phi(H)]^2} + \frac{F(1-F)}{2N_1\,[\phi(F)]^2} \tag{13.7}$$

This relation is the same as Equation 13.4 except for a factor of 2, which results from the usual analysis of the relation between yes-no and 2AFC (see chap. 7). From this consideration alone, the variance of d' should be *lower* in 2AFC than in yes-no. It is important to realize, however, that higher proportions lead to smaller denominators in Equation 13.7 and thus *higher* variance. The consequence, shown theoretically and empirically by Jesteadt (2004), is that the variance of 2AFC d' is lower than that of yes-no at low performance levels, but higher at high levels. The comparison is illustrated (for unbiased responding) in Table 13.3.

TABLE 13.3   *Variance of d' for Several Paradigms (64 Signal and 64 Noise Trials)*

	Paradigm					
	Yes-No					Same-
True d'	Direct	Approximation	2AFC	3AFC	Oddity	Different
0.5	0.052	0.050	0.026	0.021	0.230	—[a]
1.0	0.056	0.054	0.030	0.022	0.078	0.207
2.0	0.078	0.071	0.053	0.034	0.048	0.097
3.0	0.147	0.116	0.147	0.084	0.060	0.105
4.0	0.157	0.238	0.683	0.364	0.098	0.148

Note: Yes-no "direct" data are from Miller (1996), yes-no approximation uses the Gourevitch and Galanter (1967) formulas, same-different is from Bi (2002), and the others are from Bi et al. (1997).
[a]Table does not contain entries for proportion correct this low.

Same-Different and Other Paradigms.   For the paradigms discussed in chapter 9, d' cannot be expressed as the difference between two transformed variables, so the strategy we have been using for finding the variance of d' fails. Bi et al. (1997) performed the appropriate calculation for 3AFC, oddity, and "duo-trio" (not discussed in this book), as well as 2AFC. Bi (2002) provided a derivation and program for same-different (differencing model only).

Table 13.3 gives a sampling of the results drawn from tables in the two articles. The entries are for symmetric responding (i.e., no response bias) on 128 trials evenly divided among the stimulus alternatives. The two primary patterns concern the number of intervals and the level of performance. In a comparison between 2AFC and 3AFC, variance declines with number of intervals as one might expect. For these paradigms, variance increases monotonically (although very nonlinearly) with performance level. This pattern reflects the increase in the variance of z scores with greater deviation of H and F from .5 (see Fig. 13.2).

Another consequence of this effect is that 2AFC has the highest variance of any paradigm when d' is high, oddity and same-different the lowest. Differences in proportion correct for a given value of d' (see Figs. 10.1 and 10.2) account for this result. Oddity and same-different have the highest variance, however, for $d' = 0.5$ or 1.0.

ROC Curves

Fitting Methods.   Rating experiments (see chap. 3) yield entire ROC curves rather than single points. These curves can usually be summarized by two parameters—for example, the slope of the curve and the distance between the curve and the major diagonal at some fixed point. To find the best-fitting curve, a maximum-likelihood procedure is used (see Appendix 1 and chap. 11). Assuming a specific form for the underlying distribution, the parameter values most likely to have given rise to the observed data points are found. The method provides answers to statistical questions that mirror those about single ROC points: Does an ROC curve differ reliably from chance? Does its slope differ from one? Does its intercept differ from zero?

This approach has been followed by Ogilvie and Creelman (1968) for logistic distributions and by Dorfman and Alf (1969) for normal distributions. Dorfman and Alf's program has been updated and republished in Swets and Pickett (1982). A solution is also available to the more complex problem of testing apparent differences among ROCs (Metz & Kronman, 1980).

Statistics Derived From ROCs.   The statistical properties of A_g, the area under the ROC, are similar to those of a proportion. Pollack and Hsieh (1969) showed that the variance of A_g is less than $A_g(1-A_g)/N$, being largest for low sensitivities (i.e., A_g approximately .5).

The statistical properties of Gaussian ROCs fit to rating data have been studied by Macmillan, Rotello, and Miller (2004). Using simulations, they measured the bias and standard error of ROC area A_z, distance measure d_a, and the slope s. The area measure turns out to be both accurate (low bias) and precise (small standard error); the distance measure is slightly inferior on both counts. Slope, however, can be quite biased and has considerable variability, especially if criteria are spaced close together. A set of tables that permit estimation of bias and standard error can be found at http://www-unix.oit.umass.edu/~caren/Design/Assets/index.htm.

Sensitivity Estimates Based on Averaged Data

According to detection theory, the observer partitions a stable underlying distribution by use of a fixed criterion. This story of the decision process is most convincing if the experiment uses only one pair of stimuli, all data are collected in one session, and the analysis is applied to a single observer. In experiments with multiple stimuli, sessions, or observers—that is, all experiments of real interest—some kind of averaging must be done. The data analysis procedure most consistent with our assumptions is to calculate sensitivity and bias separately for each combination of stimulus, session, and observer, and then average the resulting estimates.

This idealized approach is often not possible because the number of trials contributing to a single estimate is too small, so that sensitivity and bias cannot be computed. An extreme example is provided by class discrimination (e.g., recognition memory experiments), in which a single stimulus occurs only once per observer. Less dramatically, a single session using a roving design often contains few trials per stimulus. And with some observers—infants, for example—a few trials is all one can hope for.

In this section, we consider an alternative strategy: Sensitivity is estimated from data that have been combined across multiple stimuli, sessions, or observers. The resulting statistic is called *pooled sensitivity*. We compare this method with the ideal approach, in which parameter estimates from different subsets of the data are averaged to yield *mean sensitivity*. The questions to be asked fall into two classes. First, how much does pooled sensitivity differ from mean sensitivity, that is, how much statistical bias does the method entail? The second question is one of *efficiency*: How variable are estimates of pooled sensitivity compared with those of mean sensitivity? (The analogous questions about response bias, which we do not consider, can be treated similarly.)

We shall conclude that pooled sensitivity is, under many conditions, an acceptable performance index. This conclusion is tempered, however, by an important caveat. Pooling is most necessary when a proportion of 0 is observed, an event with two possible causes: (a) the true underlying d' is infinite, or at least very large, or (b) the true underlying d' is of moderate size, but the sampling variability associated with the small number of trials leads to no errors. These alternatives can usually be distinguished pragmatically.

If situation (b) seems to obtain, pooled d' is a useful statistic; in situation (a), it is not. Experiments in which observers are able to respond unanimously to stimuli are poorly suited for detection theory analysis—they display ceiling effects—and should be redesigned. Often a simple change from two responses to a rating design solves the problem.

Effects of Averaging on (Statistical) Bias

Participants With the Same Sensitivity but Different Response Biases. Two observers whose sensitivity is the same will produce hit and false-alarm rates that lie on the same ROC curve, but (in general) at different points. If their hit and false-alarm rates are averaged, the resulting point will be halfway along a line connecting the original points. Because ROC curves are concave downward, the average-performance point will be lower than the original points and will yield a lower estimate of sensitivity, as shown in Fig. 13.3.

It is clear from Fig. 13.3 that the decrement in estimated sensitivity will be severe only if the two points are quite discrepant in bias. The exact size of the effect, for $d' = 1$, is shown in Fig. 13.4, in which pooled d' is plotted for all values of the two observers' criteria that yield hit and false-alarm rates between .02 and .98. Large decrements occur only when the criteria are very different, perhaps 1.5 standard deviations apart. The average location of the two criteria on the ROC matters little. Macmillan and Kaplan (1985) showed that the effect illustrated in Fig. 13.4 is not a function of the true underlying sensitivity: Pooled d', as a fraction of true d', depends only on the difference between the criteria divided by d'.

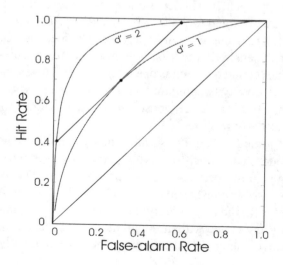

FIG. 13.3 Averaging two points on the same ROC curve yields a point on a curve with a lower value of d'.

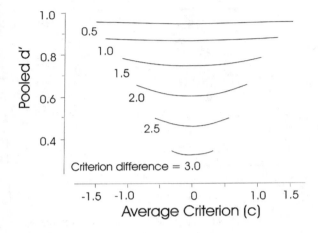

FIG. 13.4 The effect of averaging hit and false-alarm rates from observers with the same d' but different criteria. Apparent d' is less than true d', and the difference essentially depends only on the difference between criteria, not on criterion location. The decline in d' is substantial only when the criteria are substantially different. The means of the two distributions are at 0 and 1.

Clearly it is desirable to average proportions from only those participants whose criteria are similar. A possible procedure is to divide participants into subgroups so that all members of a subgroup have similar response biases. Pooled d' can be computed for each subgroup and the results averaged to estimate d' for the entire group.

Participants With the Same Response Bias but Different Sensitivities. What about ROC points that differ in sensitivity and not in bias? There are many plausible candidates for a bias parameter, and therefore several possible interpretations of "constant-bias." In this section, an observer is said to maintain constant bias if, as d' varies, the criterion is always located the same proportion of the distance between the two distribution means—that is, if c' remains the same.

The relation between proportion correct and d', for a constant relative criterion, is nonlinear, as shown in Fig. 13.5, so pooled d' is generally lower than mean d'. Figure 13.6 shows that this decrement is substantial only if the original d' values differ substantially, say by 2.0. Thus, pooled d' estimates are affected similarly by varying bias (Fig. 13.4) and varying sensitivity. The actual values being averaged have little influence on the effect; it is the difference between them that counts. Macmillan and Kaplan (1985) showed that these conclusions generalize to a variety of absolute criterion locations and to unequal variances.

The error introduced by pooling data from participants with different sensitivities can be reduced by assigning observers to subgroups as described in the previous section. Pooled d' for a subgroup is nearest to true

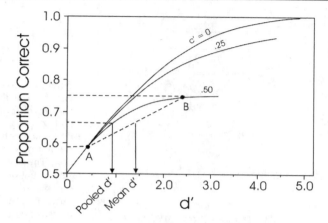

FIG. 13.5   The relation between d' and proportion correct for three different criteria. If two observers operate at Points A and B on the same curve, d' based on average proportion correct is always lower than average d' because the curve is concave downward.

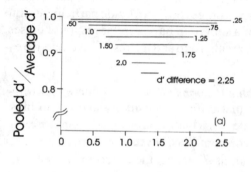

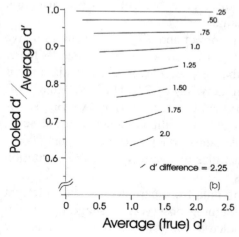

FIG. 13.6   Pooled d' as a fraction of average d' for observers with different d' values but the same bias. Relative criterion c' equals 0 in (a), 0.5 in (b). Pooled d' is less than true d', and the difference essentially depends only on the difference between the two, not on the particular values. The decline in d' is substantial only when the two values of d' are very different.

average d' if the participants in the subgroup have similar sensitivities as well as similar biases.

Effects of Averaging on Variability

We turn now to the issue of variability: In many samples, will similar or widely disparate values of pooled d' be produced? In particular, we calculate the efficiency of pooled d', its variance divided into the variance of mean d', for a constant number of trials.

Figure 13.7 illustrates the result of such a calculation using a technique based on that of Gourevitch and Galanter (1967; see Macmillan & Kaplan, 1985, for details). Values greater than 1.0 imply that the pooled d' estimate is less variable than the mean d' estimate. The figure reveals that pooled d' is *always* less variable than average d', and that its variance decreases as the discrepancy between the participants increases. There is also a tendency, other things being equal, for variance to decrease (relative to that of mean d') as the false-alarm rates of the participants become more extreme.

These effects result from the fact, illustrated in Fig. 13.2, that the variability of a z score computed from an observed proportion increases non-linearly with the absolute value of z. When data from two or more participants are pooled, the z score of the average has less variability than the average of the two z scores. The one important exception to this rule occurs when hit or false-alarm rates equal 0 or 1: The variability of a z score for those proportions is not finite. As previously discussed, whether true d' is infinite must be assessed independently by the experimenter.

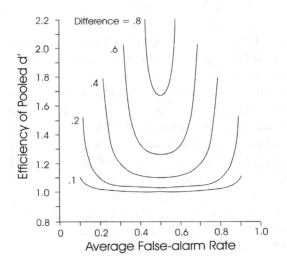

FIG 13.7 The efficiency of pooled d' (its variance divided into that of average d') is shown for two participants with the same hit rate and different false-alarm rates. Pooled d' is always more efficient, especially if the two false-alarm rates are very different or if either of them is extremely large or small.

Hautus (1997) noted a tradeoff between two kinds of statistical bias in estimating d' from a sample of individuals. If individual values of d' are averaged, the smaller number of trials for each subject will increase the likelihood of observing $H = 1$ or $F = 0$, either of which require an adjustment. If the data are pooled before computing d', however, the biases we have just discussed come into play. In a series of simulations, *he* found that pooling led to *less* biased estimates when the number of trials per observer was small, whereas for larger numbers of trials pooled d' was *more* biased than average d'. The crossover point ranged from about 10 to 100 and depended on population values of d' and c, as well as the rule for adjusting perfect scores.

Small-N versus Large-N Experiments

Suppose an experimenter wishes to determine whether a particular variable, say divided versus unitary attention, affects the sensitivity and response bias of an observer in a visual experiment. How many observers should be used? How many trials per observer? The answers to these questions typically interact: If there are few trials per participant many participants are required, whereas with many trials one can justify fewer participants. A large number of observations must be introduced *somehow* to overcome the inherent noisiness of the experimental situation.

SDT is easiest to apply in the few-participant, many-trials case: The observers can be analyzed separately and any individual differences discussed. Such designs are in fact popular for some perceptual problems. To construct confidence intervals and test simple hypotheses for data from one observer is fairly straightforward, as we have seen.

Other content areas or populations of participants require the opposite tack. In recognition memory studies, repeating items to increase the number of trials has effects on memory itself. For various special populations, including infants, the experimenter is limited to a small number of trials, and undefined values of d' and c can be expected. Pooling data can give reasonable estimates of sensitivity and bias, at least under the limited conditions we have explored, but this strategy suppresses the differences among individuals that are generally used to estimate the reliability of dependent variables. How could one conduct a hypothesis test that took the variability among individuals into account?

This may seem to be an area in which detection theory brings more problems (undefined performance indexes) than solutions (good estimates of sensitivity and bias), and a temptation is to fall back on proportion correct despite the problems with this index discussion in chapter 4. A compromise approach was used by Maddox and Estes (1997), who pooled their data to arrive at d' sensitivity estimates, but conducted inferential statistics on values of $H - F$. This latter statistic is equivalent to proportion correct, and thus implies a threshold model, but Maddox and Estes argued that these characteristics need not prevent its use provided that a detection-theoretic index is found for the pooled data. Some degree of distortion is certainly introduced by this approximation, but its degree is unknown.

Entire ROC Curves

Finally, ROC curves obtained by averaging data across participants require special estimation techniques. This problem has been solved by Dorfman and Bernbaum (1986), who used a "jackknife" statistical procedure. Their article provides both a listing of their computer program and an explanation of the jackknife method.

Systematic Statistical Frameworks for Detection Theory

We have seen that statistical treatment of single subjects can be accomplished with reference to the inherent binomial variability in their responses, and SDT parameters can be estimated and evaluated for groups of subjects by prudent averaging. But realistic experimental situations provide further complexities: Performance is measured in several conditions, and multiple hypotheses must be evaluated.

Logistic regression provides a method for estimating parameters and testing hypotheses in experiments with dichotomous (e.g., yes-no) or ordinal (rating) responses. The technique makes use of logistic distributions, and thus merges smoothly with Choice Theory models. Other distributional assumptions, like normality, can be incorporated into *generalized linear models*, which build on logistic regression. Our summary here draws from DeCarlo (1998).

Regression techniques attempt to account for variations in a dependent variable as linear or nonlinear functions of independent variables. For example, psychometric functions are often fit by regression, as we saw in chapter 11. In this and other familiar cases, the data variables involved are typically continuous. In the simplest case, with one independent variable, the dependent variable y is predicted to be a linear function of the independent variable x—that is, $y = mx + b$.

Logistic regression is suitable for detection theory analysis because it is applied to frequency data. Again taking the simplest case, suppose we wish to write the hit and false-alarm rates as a function of trial type (S_2 vs. S_1). The function can be made linear by an appropriate transformation of response rate, which for Choice Theory is logits, or log odds:

$$\ln\left[\frac{P(\text{"yes"}|S_i)}{1-P(\text{"yes"}|S_i)}\right]=-k+dX \, . \tag{13.8}$$

The "dummy" variable X equals 1 for S_2 and 0 for S_1, so this equation reduces to

$$\ln\left(\frac{H}{1-H}\right)=-k+d$$

$$\ln\left(\frac{F}{1-F}\right)=-k \, . \tag{13.9}$$

This is the Choice Theory model, with $k = \ln(b) + \ln(\alpha)$ (the criterion as a distance from the mean of the S_1 distribution) and $d = 2 \ln(\alpha)$.

As a statistical model, Equation 13.8 is *saturated*—that is, it has just as many parameters as data points. Therefore, it is possible to convert the hit and false-alarm rates to measures of sensitivity and bias (as we have been doing throughout the book), but not to perform statistical feats such as goodness-of-fit evaluation. More realistic experiments do allow for such work to be done. Consider the hypothetical face-recognition experiment from chapter 2 in which $H = .69$ and $F = .31$ for an unhypnotized observer (Condition 1) and $H = .89$ and $F = .59$ for a hypnotized observer (Condition 2). Each condition i has a criterion k_i and a sensitivity d_i, and the logistic regression model is:

$$\ln\left[\frac{P(\text{"yes"}|S_i)}{1-P(\text{"yes"}|S_i)}\right]$$
$$=-k_1-(k_2-k_1)X_1+d_1X_2+(d_2-d_1)X_1X_2 \tag{13.10}$$

The terms X_1 and X_2 represent dummy variables: X_1 equals 1 for Old items and 0 for New items; X_2 equals 1 for the hypnotized observer and 0 for the unhypnotized observer.

Equation 13.9 has four parameters, and there are now four data points (hit and false-alarm rates for two conditions), so this too is a saturated model. But it can be used to test the hypotheses that the conditions differ in sensitivity (if not, the last coefficient is 0) or bias (if not, the second coefficient is 0). These reduced models have only three parameters, leaving one degree of freedom for testing goodness of fit. DeCarlo (1998) applied the model and found that the two conditions differ in response bias, but not in sensitivity. His SPSS program is reproduced in the chapter appendix.

To convert the analysis to normal rather than logistic distributions, "link" functions are used to rescale the data. Equation 13.8 becomes

$$z(H) = -k + d \tag{13.11}$$
$$z(F) = -k \, .$$

This is the equal-variance SDT model, with $k = c + d'/2$ (the criterion as a distance from the mean of the S_1 distribution) and $d = d'$.

Logistic regression can also be applied to more complex experiments, such as the rating design and identification. DeCarlo (1998) also described these methods, illustrated their use (with data from chap. 10, as well as other sources), and provided sample SAS programs.

Summary

Single proportions based on moderately large numbers of trials have a normal distribution, which can be used to construct confidence intervals and perform simple hypothesis tests. The bivariate normal distribution characterizes (false-alarm, hit) pairs under similar conditions. The SDT statistics d' and c are also normally distributed, provided that N is reasonably large and ceiling and floor effects on the hit and false-alarm rates are avoided.

Pooling data across stimuli, sessions, or observers may be necessary to avoid observed frequencies of zero. Estimates of sensitivity obtained in this way are biased, but the amount of bias is small unless estimates of very different bias or sensitivity are combined.

Logistic regression models, which are equivalent to those of Choice Theory, permit statistical evaluation of hypotheses about sensitivity and bias parameters. They are particularly valuable for testing hypotheses about the many main and interaction effects involving these parameters that arise in factorial experiments.

Computational Appendix

DeCarlo (1998) provided the following SPSS program for the example in this chapter and others (starting in chap. 2) of face recognition by hypnotized and nonhypnotized subjects:

```
set width = 80 length = none
title "Face recognition".
*(Note—responses are coded yes = 1 and no = 0).
data list list /hypno signal yes count *.
begin data
0 0 0 69
0 0 1 31
0 1 0 31
0 1 1 69
1 0 0 41
1 0 1 59
1 1 0 11
1 1 1 89
end data.
compute sighypno = signal*hypno.
weight by count.
logistic regression yes with hypno signal sighypno
/criteria lcon(0).
*Next is the restricted model without the interaction term.
*The –2log L can be used to test for constant $d_j$.
logistic regression yes with hypno signal
/criteria lcon(0).
```

Problems

13.1.   (a) Find the 95% confidence interval (CI) for a proportion correct of .5 based on 50 trials.
(b) Find the CI for the difference between two proportions of .9 and .5 based on 50 trials each. Use the CI to test the hypothesis that the true difference is 0.
(c) Redo (a) and (b) assuming 200 instead of 50 trials.

13.2.   Find the highest and lowest plausible (95%) true sensitivity and true bias that could lead to the data point $H = .75$, $F = .25$, assuming (a) 50 trials per stimulus, and (b) 200 trials per stimulus. (*Hint*: The 95% confidence area is 2.5 standard deviations wide at $H = .75$ and circular in form.)

13.3.   (a) Find the 95% CI for d' and c for each matrix in Problem 7.1. In each case, test the hypothesis that the underlying parameter is 0.
(b) Find the CI for the difference between matrixes 3 and 4, and use them to test hypotheses about whether the parameters are different.
(c) Multiply all cells by 5, and redo (a) and (b).

13.4.   Suppose $H = .66$, $F = .50$ based on 50 trials each. Find the 95% CI if this is (a) a one-interval experiment (b) 2AFC. Is sensitivity significantly different from 0?

13.5.   Consider the following data points generated by different observers: $(F, H) = (.8, .9), (.5, .8), (.2, .5), (.1, .2)$. (a) Find average and pooled d' and c. (b) Divide into two subgroups, keeping observers with similar criteria together, and find the average of the two pooled d' values. (c) Divide into two subgroups, keeping observers with similar sensitivities together, and find the average of the two pooled d' values.

Appendix 1

Elements of Probability and Statistics

In this appendix, we offer a brief survey of the parts of probability theory and its application to statistics that are most relevant to psychophysics. Because our aim is to make references to these ideas in the body of book more comprehensible, we frequently allude to psychophysical applications. Most concepts would be covered in a one-semester behavioral statistics course, but some ideas (e.g., random variable) are not usually found at that level. We do not believe, of course, that we have exhausted in this brief chapter topics usually covered in a semester or two. Our incomplete discussion is also relatively informal. Hays (1994) provided a thorough treatment of all issues raised here.

Probability

Probability for Finite Sets

Definition of Probability. In the simplest probabilistic situation, an *elementary event* is chosen at random from the *sample space S*. If *A* is a subset of *S*, then the probability that an elementary event that is in *A* will occur is

$$P(A) = n(A)/n(S) \ , \tag{A1.1}$$

where the function *n* counts the number of elementary events in a set. For example, when a fair coin is tossed, the probability of a Head occurring is 1/2, as is the probability of a Tail. When a die is tossed, the probability of a "2" is 1/6, and the probability of an Even outcome ("2," "4," or "6") is 3/6 or 1/2.

Some important characteristics of probabilities are evident from these examples: All probabilities must lie between 0 and 1, and the sum of probabilities for all elementary events must be 1.

An experimenter's choice of stimulus presentation is generally a random event. In the typical two-alternative forced-choice paradigm, each of the two possible orders has probability 1/2. In m-interval forced-choice, the sample space has m elementary events (correct intervals); usually, each is chosen with probability $1/m$.

True Versus Estimated Probabilities.   Participants' responses can also be thought of as probabilisitic events. The *true probability* of saying "yes" in a one-interval experiment is a long-term tendency that cannot be measured exactly. The corresponding *estimated probability* is found from the outcomes of a finite number of observed trials. The "yes" probability, for example, is estimated by dividing the number of "yes" responses by the number of trials. Thus, if a participant responds "yes" on 30 of 50 trials, $P(\text{"yes"}) = .6$.

Probabilities Not Defined by Counting.   Equation A1.1 does not provide an adequate description of all probabilistic situations. In models of discrimination, internal events are assigned probabilities. For example, in high-threshold theory (chap. 4), stimulus S_2 gives rise to state D_2 with probability q and to state D_1 with probability $1 - q$. The two events cannot be decomposed into equiprobable elementary events, but the probabilities of their occurrence have the same characteristics as in the earlier examples.

Probabilities That Depend on More Than One Event

Probabilities can be defined on subsets of the sample space. For example, the estimated hit rate is the probability of saying "yes" when event S_2 occurs: $H = P(\text{"yes"}|S_2)$. We can find such *conditional probabilities* from

$$P(A|B) = P(A \text{ and } B)/P(B) \ . \tag{A1.2}$$

Suppose the data matrix for our experiment is

	"yes"	"no"
S_2	20	5
S_1	10	15

Then $P(\text{"yes"}|S_2) = P(\text{"yes" and } S_2)/P(S_2) = n(\text{"yes" and } S_2)/n(S_2) = 20/25 = .8$. All of these are, of course, estimated probabilities.

In the foregoing matrix, it is clear that the response variable depends, albeit imperfectly, on the stimulus variable, because the probability of saying "yes" is different for the two stimulus events. We say the response and the stimulus are not *independent*. Two events A and B are independent if

$$P(A|B) = P(A) \; , \tag{A1.3}$$

or equivalently

$$P(A \text{ and } B) = P(A) \, P(B) \; . \tag{A1.4}$$

Stimulus presentations on successive trials are independent in most psychophysical experiments: The probability that S_1 is presented on trial n does not depend on whether it was presented on trial $n - 1$.

Discrete Random Variables and Their Probability Functions

Random Variables. The events in a sample space can be assigned numerical values; a mathematical function that accomplishes this is called a *random variable*. As an example, consider a single S_2 trial; let X assign the value 1 to "yes" responses, 0 to "no." Then if p is the hit rate, the random variable X can be summarized as follows:

$x = $ *Value of X*	$p(x) = $ *Probability of This Value*
1	p
0	$1 - p$

The variable X is called a *Bernoulli* random variable; the table describes its *probability function*. The event for which $X = 1$ is often called a "success," that for which $X = 0$ a "failure."

Binomial Random Variables. A Bernoulli random variable counts the number of successes in one trial; the number of successes in N independent trials is given by the *binomial* random variable. The number of hits, when S_2 is presented, is a binomial random variable. Binomial variables differ according to the probability of success on any one trial (the hit rate, in this example) and the number of trials (number of S_2 presentations). The probability of k successes in N trials, when the Bernoulli success rate is p, is given by

$$p(k) = {}_NC_k \, p^k \, (1-p)^{N-k} \, , \tag{A1.5}$$

where ${}_NC_k$ is the binomial coefficient, tabled in most statistics texts.

Equation A1.5 is the probability function of the binomial; the Bernoulli was represented by a table, but is actually a special case of the binomial when $N = 1$. Probability functions can also be represented graphically; Figure A1.1 shows the probability function of the binomial for $N = 1$, 4, and 20, when $p = 1/2$.

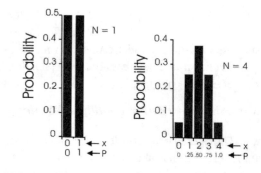

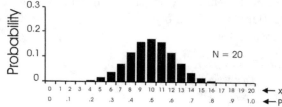

FIG. A1.1. Three binomial probability functions, $N = 1$, 4, and 20, $p = \frac{1}{2}$. Each can be interpreted as the distribution of the number of successes in N trials (X) or the proportion of successes $P = X/N$.

Frequently, we are interested in the proportion rather than the number of successes, that is, in the variable $P = X/N$ rather than X. Figure A1.1 shows values of this *binomial proportion* random variable as well as values of the binomial itself.

Mean and Variance of Random Variables

Definitions.   The *mean*, or *expectation*, of a discrete random variable is given by

$$E(X) = \mu = \Sigma x \, p(x) \, . \tag{A1.6}$$

For the Bernoulli random variable, $E(X) = p$, the success rate.

The *variance* of a random variable, a measure of its spread, is

$$\text{var}(X) = \sigma^2 = \Sigma(x - \mu)^2 \, p(x) \, . \tag{A1.7}$$

The variance of the Bernoulli is $p(1 - p)$. The square root of the variance is called the *standard deviation*, denoted σ.

Mean and Variance of Linear Combinations.   Some simple rules tell us the mean and variance of a constant, of a variable multiplied by a constant, and of the sum (or difference) of two variables. If X and Y are random variables and a is a constant,

$$E(a) = a \tag{A1.8a}$$

$$E(aX) = aE(X) \tag{A1.8b}$$

$$E(X + Y) = E(X) + E(Y) \tag{A1.8c}$$

$$\text{var}(a) = 0 \tag{A1.9a}$$

$$\text{var}\,(aX) = a^2 \, \text{var}(X) \tag{A1.9b}$$

$$\text{var}\,(X + Y) = \text{var}(X) + \text{var}(Y) \, . \tag{A1.9c}$$

Equation A1.9c holds only if X and Y are independent.

These rules allow us to find the mean and variance of any new random variable that is a linear combination (weighted sum) of old ones. For example, it is often useful to express values of a random variable X in standard-deviation units from the mean. The resulting variable, denoted z, equals $(X - \mu)/\sigma$. Application of Equations A1.8 and A1.9 reveals that z has a mean of 0 and a variance of 1.

Mean and Variance of the Binomial.   We can use rules A1.8c and A1.9c to find the mean and variance of the binomial. The binomial random variable, the number of successes in N trials, equals the number of successes on trial 1 plus the number of successes on trial 2, and so on. It is therefore the sum of N Bernoulli random variables. The mean of a binomial is thus the sum of the means of the N Bernoullis, or Np. The variance, similarly, is the sum of the variances of the N Bernoullis, or $Np(1 - p)$.

Rules A1.8b and A1.9b can be used to find the mean and variance of the binomial proportion variable X/N. The mean of this variable is $Np/N = p$, the variance $Np(1-p)/N^2 = p(1-p)/N$. The expected value of an observed proportion does not change as the number of trials increases, but its variance is reduced.

Continuous Random Variables and Their Density Functions

Binomial distributions are *discrete*: Only a countable number of values can occur. Variables that can take on any value on a continuum are called *continuous*. A curve of the sort shown in Fig. A1.2(a) is called a *density function*.

The probability that a random variable takes on a value between two points *a* and *b* is represented by the area under the density function between those two points. (The total area under the curve, therefore, equals 1.) To obtain such probabilities, we use the *distribution function*, which assigns to each value of the random variable the probability of a score below that value. Figure A1.2(b) shows the distribution function corresponding to the density function of Fig. A1.2(a).

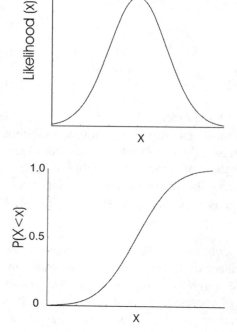

FIG. A1.2.   (a) Density, and (b) distribution functions for a continuous variable. The distribution function gives the area under the density function that is less than each possible value of the random variable.

A particularly important continuous distribution is called the *normal* or *Gaussian* distribution. Its density function (which is in fact the function illustrated in Figure A1.2[a]) is

$$\phi(x) = \frac{1}{\sqrt{2\pi}\sigma} e^{-\frac{1}{2}\left(\frac{x-\mu}{\sigma}\right)^2} . \tag{A1.10}$$

Different normal distributions have different means and variances, but are all the same function of z, namely,

$$\phi(z) = \frac{1}{\sqrt{2\pi}} e^{-\frac{1}{2}z^2} . \tag{A1.11}$$

The distribution function for this normal density is given in Tables A5.1 and A5.2, as is the density itself.

A second distribution of importance in this book is the *logistic*, defined by

$$\lambda(x) = \frac{e^{-(x-\mu)}}{[1 + e^{-(x-\mu)}]^2} . \tag{A1.12}$$

The distribution function, unlike that of the normal, can be calculated directly; the equation is

$$\Lambda(x) = \frac{e^{-(x-\mu)}}{1 + e^{-(x-\mu)}} . \tag{A1.13}$$

The mean and variance of a continuous distribution are not computed from Equations A1.6 and A1.7, but from analogous equations using integrals. No explicit computation is needed to find the mean and variance of the normal distribution plotted in Fig. A1.2; because the plotted variable is z, the mean is 0 and the variance is 1. The logistic distribution has mean 0 and variance $\pi^2/3$.

Bivariate (Two-Dimensional) Distributions

Two Independent Variables.   In several designs analyzed in Parts II and III of this book, we assumed that two variables are simultaneously relevant to the observer's decision. In a same-different design, for example, the magnitudes of sensation in both the first and second intervals (which we call A and B) contribute information. The independence of two dimensions can be represented graphically by placing the A and B axes perpendicular to

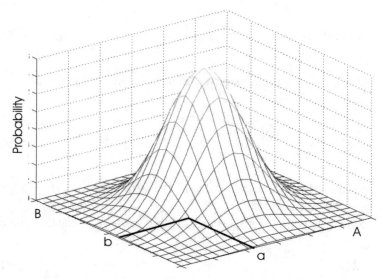

FIG. A1.3.   A bivariate normal distribution. The probability of an observation for which $A < a$ and $B < b$ is the volume above the region to the left of a and below b.

each other, as in Fig. A1.3. Because we have used two dimensions for values of the random variables, probability density must be represented by height above the (A, B) plane and probability by the volume under this bivariate umbrella.

Writing general expressions for probabilities under bivariate distributions is complicated, but we can describe one special (and, fortunately, important) case. Suppose we wish to find the probability that both $A < a$ and $B < b$. Because the variables are independent, Equation A1.4 applies, and

$$P(A < a \text{ and } B < b) = P(A < a) \, P(B < b) \ . \tag{A1.14}$$

This relation is termed the *product rule* (see chap. 6). Both terms on the right side can be found from a table of the (univariate) normal distribution. One way to view the situation described by Equation A1.14 is to first imagine that you are standing along Dimension A; your view of the distribution "in silhouette" is of a univariate normal distribution. Similarly, the view from along B is of another univariate normal distribution. Probabilities from these two distributions are the ones on the right side of Equation A1.14.

The three-dimensional picture in Fig. A1.3 can be reduced to two in another way by constructing an aerial view. To indicate the shape of the distribution in this view, it is common to draw circles (or ellipses, if the variances of *A* and *B* are unequal) to show locations of constant height. This strategy is used extensively in Part II.

Correlation Between Variables.   If the value of one variable does not help in predicting the value of another, we say the variables are not *correlated*. The correlation between two variables is defined by

$$r = [E(XY) - E(X)\,E(Y)]/\sigma_x\,\sigma_y \ . \tag{A1.15}$$

The correlation coefficient measures the degree of linear relation between *X* and *Y* and ranges from −1 (complete negative correlation) to 0 (no correlation) to +1 (complete positive correlation).

Independent variables are always uncorrelated. Uncorrelated variables need not be independent, however, because the dependence may be nonlinear. An important exception: Uncorrelated *normal* variables are always independent.

Statistics

Definitions and Examples

A *statistic* is a function of data—that is, a summary measure that depends on the results of a set of trials. *Statistics* is the mathematical treatment of such measures. An important statistic in psychophysics is *observed proportion*, the number of successes in *N* binomial trials, divided by *N*. When a stimulus variable is being measured (rather than events simply being counted, as in data matrixes), two important statistics are the *sample mean* and *sample variance*:

$$M = \Sigma X/N \tag{A1.16}$$

$$s^2 = \Sigma(X - M)^2/(N - 1) \ .$$

These statistics can be used to estimate from data the theoretical mean and variance of a random variable.

Sampling Distributions

Theoretical means and variances are constant aspects of random variables, but the corresponding sample statistics, which are estimated from data, vary. Distributions of sample statistics are called *sampling distributions*. Important characteristics of sampling distributions are their shape, mean, and variance.

Central Limit Theorem.   An important truth in statistics is the *central limit theorem*, which states that a variable that is the sum of many variables, each with the same distribution, must have an approximately normal distribution. The central limit theorem applies to both the sample mean, which depends on the sum of scores, and the observed number of successes, which is the sum of observed successes on single trials.

Bias and Efficiency.   We can use Equations A1.8 and A1.9 to find the mean and variance of the sampling distributions for M and s. The sample mean has a normal distribution with mean equal to the population (theoretical) mean and variance equal to the population variance divided by N; an observed proportion has a normal distribution with mean equal to the population proportion p and variance equal to $p(1 - p)/N$. Statistics whose expected value equals the corresponding population parameter, like M and s, are called *unbiased*. The *efficiency* of a statistic is reflected in its variance, more efficient measures being less variable. The most important factor influencing efficiency is the number of trials.

Confidence Intervals.   Suppose we observe a proportion of hits H in N trials; what might the true proportion h be? If N is large, H has an approximately normal distribution with mean h and variance $h(1 - h)/N$. Because the distribution is normal, 95% of its area is within 1.96 standard deviations of the mean (see Table A5.2), so we can be 95% sure that H differs from h by no more than 1.96 standard deviations. The *95% confidence interval* around H is therefore

$$H \pm 1.96[h(1 - h)/N]^{\frac{1}{2}} \, . \tag{A1.17}$$

For an approximate calculation, we can use H to estimate h in finding the variance. For example, if H is .8 in 25 trials, the 95% confidence interval is $.8 \pm (1.96)(0.0064)^{\frac{1}{2}}$ or $.8 \pm .16$. We can be confident that the true hit rate lies somewhere between .64 and .96.

The sampling distribution of d' is also approximately normal when N is not too small and d' is not too large, so confidence intervals can be constructed around this statistic using the normal distribution. For small N or large d', the deviation from normality can be substantial. Both cases are discussed in chapter 13.

Hypothesis Testing

Values of Population Parameters.   Sometimes we wish to make yes-or-no decisions about the value of a population statistic; for example, we may wish to know whether a hit rate equals .5. The most common strategy for doing this is to ask whether the value of interest lies within the confidence interval; the hypothesized value is plausible if it does, implausible if not. In the foregoing example, an observed hit rate of .8 based on 25 trials is unlikely to have occurred if the true hit rate was .5, because .5 is not in the 95% confidence interval around .8. We can thus reject the hypothesis that $h = .5$.

Independence.   Another important hypothesis that arises in psychophysics is that of independence. In the data matrix given at the beginning of this appendix, for example, we may ask whether the participant's responses are independent of the stimuli. If they were, then the true matrix would be given by the numbers in parentheses:

	"Yes"	"No"
S_2	20 (15)	5 (10)
S_1	10 (15)	15 (10)

Two similar statistics evaluate the discrepancy between the observed frequencies (O) and the expected frequencies (E):

$$\text{error}_1 = \Sigma \, (|O - E| - 0.5)^2/E \tag{A1.18a}$$

$$\text{error}_2 = \Sigma \, O \ln(O/E) \tag{A1.18b}$$

When N is large and none of the expected frequencies are too small, the distribution of both error functions approximates a known form, called χ^2, which is tabled in most statistics texts. The value of error_1 in this example is 6.75; a table reveals that less than 5% of the area under a χ^2 distribution is greater than this, so we can reject the hypothesis of independence. The estimate provided by error_2 is 8.65, which leads us to the same conclusion.

When matrixes with more than two rows or columns are evaluated, the statistics in Equation A1.18 still apply, with two changes: The 0.5 correction in error_1 is not used, and the distribution approximates a different member of the χ^2 family. Our example has used one *degree of freedom;* larger tables use $(N-1)(M-1)$ degrees of freedom, where N is the number of rows and M the number of columns in the matrix. Tables of χ^2 provide values for many different degrees of freedom.

Model Fitting and Evaluation

In the preceding section we saw how to determine whether two observed proportions are consistent with an independence hypothesis. This is an example of a more general problem, in which data are compared to a theoretical model, and the investigator wishes to know how well the model "fits." Two strategies for fitting models to data are important in this book: least-squares and maximum-likelihood methods.

Least-Squares Fits.   Consider first the case of determining a psychometric function (chap. 11). Using either a fixed or adaptive method, we measure d' for each of several stimulus values. Suppose we believe that the psychometric function should have the shape of a cumulative normal distribution function; we wish to find the particular curve (the particular values of mean and variance) that best describes our data. A *least-squares* technique finds the curve for which the sum of the squared deviations between the points and the curve is as small as possible. A measure of how well the curve fits is the value of this sum, normalized by dividing by the original variance of the d' values. If the curve is a straight line (or has been transformed to a straight line, in this case by using z transformation), this statistic equals one minus the square of the correlation coefficient, which therefore provides an equivalent measure of goodness of fit. This procedure is called *linear regression*.

Maximum-Likelihood Fits.   The least-squares method is concerned with deviations on only one variable (in this case, d' values); the other variable (in this case, stimulus value) is fixed by the experimenter. When the curve to be fitted relates two variables that both contribute variability, a different tack is taken. Consider the problem of fitting ROC curves to points obtained in a rating experiment. An ROC point is determined by a hit and false-alarm rate, each with binomial variability. If we believe that the ROC is a straight line on z coordinates, how can we determine estimates for its intercept and slope? For each possible pair of values for the slope and intercept, we ask, what is the probability that our data would have arisen from this model? One pair will give the largest probability or likelihood; these are the *maximum-likelihood estimates* (MLEs) of the parameters. Because calculating MLEs requires either differential calculus or an iterative computer program, we do not work through an example here; but several MLE programs for ROC estimation are discussed in chapter 3 and are available on the Internet (see Appendix 6).

Maximum-likelihood methods are also used in adaptive procedures (chap. 11), where the parameter being estimated is a point on the observer's psychometric function, and the data consist of all responses made in an experimental run.

Appendix 2

Logarithms and Exponentials

All logarithms are defined with regard to a constant called a *base*. The *logarithm*, or *log*, of a number is the power to which the base must be raised to obtain the number. Thus, if the base is 10 (as in "common" logarithms), $\log(1) = 0$, $\log(10) = 2$, $\log(1000) = 3$, and so on. As this example illustrates, the logarithm of a number increases monotonically with the number itself, but not nearly as quickly.

Natural logarithms, the only kind appearing in this book, have a base of $e = 2.718281828\ldots$ (The ellipsis indicates approximation: e cannot be exactly expressed as a fraction or repeating decimal.) Some of the sensitivity and bias measures in detection theory are defined as the natural logarithm of other measures. Because the logarithm is a monotonic transformation, two measures related by it are equivalent (i.e., have the same isosensitivity or isobias curves).

The easiest way to compute a logarithm is with a hand calculator. On most intermediate (statistical or scientific) calculators, the button relevant to natural logarithms is labeled "ln" ("log" being reserved for common logarithms). If you enter the number 10, and then *ln*, the calculator should display 2.302585.

To calculate measures defined using logs, no deeper knowledge is needed. To follow derivations, however, some other facts are useful:

1. Multiplication of numbers corresponds to addition of their logarithms: $\ln(xy) = \ln(x) + \ln(y)$.

2. Division of numbers corresponds to subtraction of their logarithms: $\ln(x/y) = \ln(x) - \ln(y)$.

3. Raising a number to a constant power corresponds to multiplying its logarithm by that constant: $\ln(x^a) = a\ln(x)$.

357

4. Taking the reciprocal of a number corresponds to taking the negative of its logarithm: $\ln(1/x) = -\ln(x)$.

The function that reverses the effect of the log transformation is the *exponential*, denoted e^x or $\exp(x)$. (It may also be denoted in either of these ways on calculators.) An exponential increases monotonically with its argument, but much faster: $e^0 = 1$, $e^1 = 2.72$, $e^2 = 7.39$, $e^3 = 20.09$, and so on. Using a calculator, entering 1 followed by e^x should produce 2.718281828 (i.e., e).

The properties of exponentials parallel those of logarithms:

1. Multiplication of exponentials corresponds to addition of their exponents: $e^x e^y = e^{x+y}$.

2. Division of exponentials corresponds to subtraction of their exponents: $e^x/e^y = e^{x-y}$.

3. Raising an exponential to a constant power corresponds to multiplying the exponent by that power: $(e^x)^a = e^{xa}$.

4. Taking the reciprocal of an exponential corresponds to negating the exponent: $1/e^x = e^{-x}$.

Because exponentials and logarithms are inverse functions, performing the two operations successively leaves the initial value unchanged: $\ln(e^x) = x$ and $e^{\ln(x)} = x$. Thus, if e^x has been calculated and x is desired, take the logarithm of e^x; if $\ln(x)$ has been calculated but x is desired, take the exponential of $\ln(x)$.

Appendix 3

Flowcharts to Sensitivity and Bias Calculations

The following charts will guide you to the appropriate equations, tables, or computer programs for finding sensitivity and bias. Start with Chart 1, which directs you to other charts depending on the paradigm.

To use the charts, proceed from left to right. Whenever more than one path is available, choose the one that corresponds to your specific application. Each path ends with an outcome in the right-hand column.

Chart 1: Guide to Subsequent Charts

A discrimination experiment measures the ability of an observer to distinguish two stimuli, *A* and *B*. If the experiment has one observation interval, either *A* or *B* is presented on each trial. If it has more than one interval, a sequence of stimuli, each of which is either *A* or *B*, is presented on each trial. Three separate charts are needed to analyze experiments of the second type: one to determine the design of the experiment (Chart 5), one to find the appropriate index of sensitivity (Chart 6), and the last to find the bias index (Chart 7).

Classification experiments measure the ability of an observer to label stimuli (from sets of more than two).

In some designs, the stimuli to be judged are preceded or followed by a specific, constant stimulus on each trial. The appropriate analyses for such experiments are found by ignoring the constant stimulus.

1 interval	2 stimuli	2 responses	sensitivity	Chart 2
	(discrimination)	(yes-no)	bias	Chart 3
		> 2 responses (rating)		Chart 4
	> 2 stimuli (classification)			Chart 8
> 1 interval, 2 stimuli (discrimination)			sensitivity	Charts 5 & 6
			bias	Charts 5 & 7

Chart 2: Yes-No Sensitivity

In this and later charts, two types of decisions are often made, one based on the shape of the assumed underlying distributions, the other on the format of the data. In choosing among the various distributional assumptions, we recommend Gaussian or logistic models. Rectangular-distribution models entail undesirable threshold assumptions (see chap. 4); their only advantage is that they are sometimes simpler to compute.

Data normally are reduced to hit and false-alarm rates (H and F), which should be used whenever they are available. If only proportion correct [$p(c)$] is given, it is necessary to assume that responding is unbiased.

For discussion, see chapters 1 and 4.

Gaussian distributions	distance measure d'	from H and F		Eq. 1.5
		from $p(c)$		Eq. 1.7
	proportion measures	from H and F	$p(c)_{max}$	Eq. 7.5
		from $p(c)$		unchanged
logistic distributions (Choice Theory)		from H and F		Eqs. 4.8, 4.9
		from $p(c)$		Eq. 4.19 (solved for α)
rectangular distributions (threshold theory)	1-threshold model	proportion measure	q	Eq. 4.1
	2-threshold model	proportion measure	$p(c)$	unchanged
logistic distributions for low sensitivity, rectangular distributions for high sensitivity		area measure	A'	Eqs. 4.20, 4.21

Chart 3: Yes-No Response Bias

For discussion, see chapters 2 and 4.

Gaussian distributions	criterion location	c	Eq. 2.1
	relative criterion	c'	Eq. 2.3
	likelihood ratio	β	Eq. 2.6
logistic distributions	criterion location	$\ln(b)$	Eqs. 4.11, 4.12
	relative criterion	b'	Eq. 4.13
	likelihood ratio (equivalent)	β_L	Eq. 4.14
		B''	Eq. 4.23
rectangular distributions	criterion location	yes rate $= \frac{1}{2}(H + F)$	
	relative criterion	error ratio $= (1 - H)/F$	

Chart 4: Rating-Design Sensitivity

To analyze a rating experiment with normal-distribution assumptions, an ROC is fitted to (F, H) pairs, and sensitivity and slope statistics are calculated from the curve. Fitting is best done by a maximum-likelihood computer method (see Appendix 6 for pointers to such programs). The chart assumes that ROC slope and sensitivity are obtained by one of these methods and shows how to obtain other measures.

For discussion, see chapter 3.

Gaussian distributions (any slope)	sensitivity	distance measures	rms standard deviation	d_a	Eqs. 3.4, 3.5
			mean standard deviation	d'_e	Eqs. 3.6, 3.7
		area measure		A_z	Eq. 3.8
	bias (criterion location)	rms standard deviation		c_a	Eq. 3.13
		mean standard deviation		c_e	Eq. 3.14
nonparametric	fit ROC using trapezoidal rule	area measure		A_g	Eq. 3.9

Chart 5: Definitions of Multi-Interval Designs

A discrimination experiment tests the ability to distinguish two stimuli (*A* and *B*), but may use a temporal or spatial sequence of stimuli on each trial. We denote such sequences as bracketed lists; for example, *<AB>* means Stimulus *A* followed by Stimulus *B*. In the lists of possible sequences, the notation "vs" separates sequences with distinct corresponding (correct) responses.

In some designs, the stimulus sequence to be judged is preceded or followed by a specific, constant stimulus on each trial. The appropriate analysis is the same as if these fixed stimuli were not present. Thus, if the only possible sequences are *<AABB>* and *<ABAB>*, the design is the same as if the possibilities were just *<AB>* and *<BA>* (i.e., 2AFC). As another example, if the possible sequences are just *<AAA>* and *<ABA>*, the design is the same as if the possibilities were just *<A>* and *<B>* (i.e., one-interval), and Charts 2 and 3 should be consulted instead of Charts 6 and 7.

Number of intervals	Number of responses	Sequences	Paradigm	Chapter
2	2	*<AB>* vs *<BA>*	2AFC	7
		<AA>, *<BB>* vs *<AB>*, *<BA>*	same-different	9
3	2	*<ABA>*, *<BAB>* vs *<ABB>*, *<BAA>*	ABX	9
	3	*<ABB>* vs *<BAB>* vs *<BBA>*	3AFC	10
		<ABB>, *<BAA>* vs *<BAB>*, *<ABA>* vs *<BBA>*, *<AAB>*	oddity	9
m (*m* ≥ 4)	*m* (*m* ≥ 4)	*<ABB...B>* vs *<BAB...B>* vs...vs *<B...BBA>*	*m*AFC	10

Chart 6: Multi-Interval Sensitivity

Some designs in this chart have two models: one for "independent observations," the other for "differencing." As a rule of thumb, independent observation models are used for fixed designs (only two stimuli in a block of trials) and differencing models for roving designs. As in the one-interval designs, SDT models assume normal distributions. Choice Theory models, however, do not assume logistic distributions in designs other than one-interval, even though a parameter of the one-interval experiment $[\ln(\alpha)]$ is estimated.

For discussion, see chapters 7, 9, and 10.

2AFC	SDT	distance measure	d'	Eqs. 7.2, 7.7
		proportion measure	$p(c)_{max}$	Eq. 7.6
	Choice Theory	distance measure	$\ln(\alpha)$	Eq. 7.3
mAFC	SDT	from $p(c)$	d'	Table A5.7
	Choice Theory	from $p(c)$	$\ln(\alpha)$	Eq. 10.1
		from full matrix	$\ln(\alpha)$	Eq. 10.2
reminder	same as yes-no			Chart 2
same-different (Gaussian)	independent-observation model		from H and F	Eq. 9.7 and Table A5.3
			from $p(c)$	Eq. 9.3
	differencing model			Table 5.4
ABX (Gaussian)	independent-observation model			Table A5.3
	differencing model			Table A5.3
oddity [from $p(c)$ only]	independent-observation model		Gaussian	Table A5.6
	differencing model		Gaussian	Table A5.5
			logistic	Table A5.5

Chart 7: Multi-Interval Bias

For many designs, no bias measures have been developed. (Likelihood ratio is always a possible statistic, but is often difficult to calculate.) If hit and false-alarm rates are available, the yes-no methods of Chart 2 may be used, although the yes-no interpretation (criterion location, likelihood ratio, etc.) cannot be made. We call this a *heuristic use* of these methods.

2AFC	same as yes-no			Chart 3
*m*AFC	Choice Theory			Eq. 10.3
same-different	independent-observation model	heuristic use		Chart 3
		criterion	c_i	Chapter 9
		likelihood ratio	β_i	Chapter 9
	differencing model	heuristic use		Chart 3
		criterion	c_d	Chapter 9
		likelihood ratio	β_d	Eq. 9.10
other 2-response	yes-no methods (heuristic use)			Chart 3
oddity				no method

Chart 8: Classification

All models view sets of more than two stimuli as arranged in a perceptual space. In general, the space may be of any dimension up to one less than the number of stimuli. We consider in this chart only three special (but important) cases: (a) all stimuli are represented on the same dimension, (b) the stimulus set is *feature-complete* (i.e., orthogonally combines values on multiple dimensions), and (c) all stimuli are orthogonal (i.e., each differs from the other on a distinct dimension).

For discussion, see chapters 5 (one dimension) and 10 (more than one dimension).

one dimension	Thurstonian models (Gaussian)	unequal variances	Schönemann & Tucker (1967)
		equal variances	Braida & Durlach (1972)
feature-complete	GRT models	MSDA methods	Kadlec & Townsend (1992a, 1992b)
orthogonal	all stimuli equally discriminable, no bias	SDT	Table A5.7
		Choice Theory	Eqs. 10.1, 10.2
	above assumptions not made	"constant" bias (Choice Theory)	(see chap. 10 for example)
		arbitrary bias	Smith (1982b)

Appendix 4

Some Useful Equations

The equations listed here are taken directly from the text. Only equations useful for computing sensitivity and bias indexes (including all those to which the user of the Appendix 3 flowcharts is directed), or for comparing paradigms, are given. To find out when specific measures are appropriate, see the flowcharts in Appendix 3. For further discussion, refer back to the relevant chapter.

Yes-No Sensitivity

$$d' = z(H) - z(F) \tag{1.5}$$

$$d' = 2\, z[p(c)] \tag{1.7}$$

$$\alpha = \left[\frac{H(1-F)}{(1-H)F} \right]^{\frac{1}{2}} \tag{4.8}$$

$$\ln(\alpha) = \frac{1}{2}\ln\left(\frac{H}{1-H} \right) - \frac{1}{2}\ln\left(\frac{F}{1-F} \right) \tag{4.9}$$

$$p(c)_{max} = \Phi\{[z(H) - z(F)]\,/2\} \tag{7.4}$$

$$p(c)_{max,\,yes\text{-}no} = \Phi(d'/2) \tag{7.5}$$

Yes-No Response Bias

$$c = -\tfrac{1}{2}[z(H) + z(F)] \tag{2.1}$$

$$c' = \frac{c}{d'} = -\frac{1}{2}\frac{[z(H)+z(F)]}{[z(H)-z(F)]} \tag{2.3}$$

$$\beta = e^{cd'} \tag{2.6}$$
$$\ln(\beta) = cd' = -\tfrac{1}{2}[z(H)^2 - z(F)^2]$$

$$b = \left[\frac{(1-H)(1-F)}{HF}\right]^{\frac{1}{2}} \tag{4.11}$$

$$\ln(b) = -\frac{1}{2}\left[\ln\left(\frac{H}{1-H}\right) + \ln\left(\frac{F}{1-F}\right)\right] \tag{4.12}$$

$$b' = \ln(b)/[2\ln(\alpha)] \tag{4.13}$$

$$\beta_L = \frac{H(1-H)}{F(1-F)} \tag{4.14}$$

Rating Experiments

$$s = d'_2/d'_1 \tag{3.1}$$

$$d'_1 = (1/s)z(H) - z(F) \tag{3.2}$$

$$d'_2 = z(H) - sz(F) \tag{3.3}$$

$$d_a = \frac{d'_2}{\left[\tfrac{1}{2}(1+s^2)\right]^{\frac{1}{2}}} = \left(\frac{2}{1+s^2}\right)^{\frac{1}{2}} d'_2 \tag{3.4}$$

$$d_a = \left(\frac{2}{1+s^2}\right)^{\frac{1}{2}}[z(H) - sz(F)] \tag{3.5}$$

$$d'_e = \frac{d'_2}{\tfrac{1}{2}(1+s)} = \frac{2}{(1+s)} d'_2 \tag{3.6}$$

$$d'_e = \left(\frac{2}{1+s}\right)[z(H) - sz(F)] \tag{3.7}$$

$$A_z = \Phi(D_{YN}) = \Phi(d_a/\sqrt{2}) \tag{3.8}$$

$$A_g = \tfrac{1}{2}\Sigma(F_{i+1} - F_i)(H_{i+1} + H_i) \tag{3.9}$$

$$c_2 = \frac{-s}{(1+s)}[z(H) + z(F)] \tag{3.11}$$

$$c_1 = \frac{-1}{(1+s)}[z(H) + z(F)] \tag{3.12}$$

$$c_a = \frac{-\sqrt{2}s}{(1+s^2)^{\frac{1}{2}}(1+s)}[z(H) + z(F)] \tag{3.13}$$

$$c_e = \frac{-2s}{(1+s)^2}[z(H) + z(F)] \tag{3.14}$$

Threshold and "Nonparametric"

$$q = (H - F)/(1 - F) \tag{4.1}$$

$$k = \tfrac{1}{2}[1 - (H + F)] \tag{4.6}$$

$$k' = \left(1 + \frac{F}{1-H}\right)^{-1} \tag{4.7}$$

$$A' = \frac{1}{2} + \frac{(H-F)(1+H-F)}{4H(1-F)} \quad \text{if } H \geq F \tag{4.20}$$

$$A' = \frac{1}{2} - \frac{(F-H)(1+F-H)}{4F(1-H)} \quad \text{if } H \leq F \tag{4.21}$$

$$B'' = \frac{H(1-H) - F(1-F)}{H(1-H) + F(1-F)} \quad \text{if } H \geq F$$

$$B'' = \frac{F(1-F) - H(1-H)}{H(1-H) + F(1-F)} \quad \text{if } H \leq F \tag{4.23}$$

One-Dimensional Classification

$$d'_{\text{discrimination}} = a/\beta \tag{5.3}$$

$$d'_{\text{classification}} = a/(\beta^2 + C^2)^{\frac{1}{2}} \tag{5.4}$$

$$C^2/\beta^2 = (d'_{\text{discrimination}}/d'_{\text{classification}})^2 - 1 \tag{5.5}$$

Forced-Choice

$$d' = \frac{1}{\sqrt{2}}[z(H) - z(F)] \tag{7.2}$$

$$\ln(\alpha) = \frac{1}{2\sqrt{2}}\left[\ln\left(\frac{H}{1-H}\right) - \ln\left(\frac{F}{1-F}\right)\right] \tag{7.3}$$

$$p(c)_{\text{max}} = \Phi\{[z(H) - z(F)]/2\} \tag{7.4}$$

$$p(c)_{\text{max, 2AFC}} = \Phi(d'/\sqrt{2}) \tag{7.6}$$

$$q_{\text{2AFC}} = \frac{p(c)_{\text{2AFC}} - \frac{1}{2}}{1 - \frac{1}{2}} = 2\,p(c)_{\text{2AFC}} - 1 \tag{7.9}$$

$$p(c)_{\text{max, 2AFC}} = \Phi\left(\frac{d_a}{\sqrt{2}}\right) \tag{7.12}$$

$$d'_{\text{roving discrimination}} = \frac{a}{\left[\beta^2 + \dfrac{1}{(AT)^{-1} + (G^2R^2)^{-1}}\right]^{\frac{1}{2}}} \tag{7.14}$$

$$\ln(\alpha) = \frac{1}{\sqrt{2}}\ln\left[\frac{(m-1)p(c)}{1 - p(c)}\right] \tag{10.1}$$

$$\ln(\alpha) = \frac{1}{\sqrt{2m(m-1)}}\ln\left[\prod\frac{P(R_i|S_i)}{P(R_j|S_i)}\right] \tag{10.2}$$

$$\ln\left(\frac{b_i}{b_j}\right) = \frac{1}{m}\ln\left[\prod \frac{P(R_i|S_k)}{P(R_j|S_k)}\right] \tag{10.3}$$

$$q_{m\text{AFC}} = \frac{mp(c)-1}{m-1} \tag{10.5}$$

Same-Different

$$d' = 2z\left(\frac{1}{2}\left\{1 + \left[2\,p(c)_{\text{SD,IO}} - 1\right]^{\frac{1}{2}}\right\}\right) \tag{9.3}$$

$$p(c)_{\text{SD,IO}} = p(c)_{\text{yes-no}}{}^2 + [1 - p(c)_{\text{yes-no}}]^2 \tag{9.4}$$

$$\beta_d = \frac{1}{2}e^{-\frac{1}{4}d'^2}\left(e^{\frac{1}{2}d'k} + e^{-\frac{1}{2}d'k}\right) \tag{9.9}$$

Statistics

$$p = P \pm z_{.025}[p(1-p)/N]^{\frac{1}{2}} \tag{13.1}$$

$$p_1 - p_2 = P_1 - P_2 \pm z_{.025}\{[p_1(1-p_1)/N_1] + [p_2(1-p_2)/N_2]\}^{\frac{1}{2}} \tag{13.2}$$

$$\text{var}[z(p)] = \frac{p(1-p)}{N[\phi(p)]^2} \tag{13.3}$$

$$\text{var}(d') = \frac{H(1-H)}{N_2[\phi(H)]^2} + \frac{F(1-F)}{N_1[\phi(F)]^2} \tag{13.4}$$

$$\text{var}(c) = 0.25\,\text{var}\,(d') \tag{13.6}$$

Appendix 5

Tables

Table A5.1: Normal Distribution (p to z) for Finding d', c, and Other SDT Statistics.

Instructions for the yes-no design:

To find d' *and* c. Look up H in either the p or the p' column and find the corresponding value in the z column; if H is less than .50 (i.e., if it came from the p' column), take the negative of this value. Do the same for F. Then d' is the difference between these values, and c is -0.5 times the sum.

To find β (likelihood ratio). The likelihood ratio β is the ratio of the entries for H and F in the ϕ column. Alternatively, find the z scores corresponding to H and F as above. Then $\ln(\beta)$ is -0.5 times the difference between the *squares* of these values.

TABLE A5.1 *Normal Distribution (p to z)*

p'	p	$z(p) = -z(p')$	ϕ	p'	p	$z(p) = -z(p')$	ϕ
0.001	0.999	3.090	0.0034	0.21	0.79	0.806	0.2882
0.002	0.998	2.878	0.0063	0.22	0.78	0.772	0.2961
0.003	0.997	2.748	0.0091	0.23	0.77	0.739	0.3036
0.004	0.996	2.652	0.0118	0.24	0.76	0.706	0.3109
0.005	0.995	2.576	0.0145	0.25	0.75	0.674	0.3178
0.006	0.994	2.512	0.0170	0.26	0.74	0.643	0.3244
0.007	0.993	2.457	0.0195	0.27	0.73	0.613	0.3306
0.008	0.992	2.409	0.0219	0.28	0.72	0.583	0.3366
0.009	0.991	2.366	0.0243	0.29	0.71	0.553	0.3423
				0.30	0.70	0.524	0.3477
0.01	0.99	2.326	0.0267	0.31	0.69	0.496	0.3528
0.02	0.98	2.054	0.0484	0.32	0.68	0.468	0.3576
0.03	0.97	1.881	0.0680	0.33	0.67	0.440	0.3621
0.04	0.96	1.751	0.0862	0.34	0.66	0.412	0.3664
0.05	0.95	1.645	0.1031	0.35	0.65	0.385	0.3704
0.06	0.94	1.555	0.1191	0.36	0.64	0.358	0.3741
0.07	0.93	1.476	0.1343	0.37	0.63	0.332	0.3776
0.08	0.92	1.405	0.1487	0.38	0.62	0.305	0.3808
0.09	0.91	1.341	0.1624	0.39	0.61	0.279	0.3837
0.10	0.90	1.282	0.1755	0.40	0.60	0.253	0.3863
0.11	0.89	1.227	0.1880	0.41	0.59	0.228	0.3887
0.12	0.88	1.175	0.2000	0.42	0.58	0.202	0.3909
0.13	0.87	1.126	0.2115	0.43	0.57	0.176	0.3928
0.14	0.86	1.080	0.2226	0.44	0.56	0.151	0.3944
0.15	0.85	1.036	0.2332	0.45	0.55	0.126	0.3958
0.16	0.84	0.994	0.2433	0.46	0.54	0.100	0.3969
0.17	0.83	0.954	0.2531	0.47	0.53	0.075	0.3978
0.18	0.82	0.915	0.2624	0.48	0.52	0.050	0.3984
0.19	0.81	0.878	0.2714	0.49	0.51	0.025	0.3988
0.20	0.80	0.842	0.2800	0.50	0.50	0.000	0.3989

TABLE A5.2   *Normal Distribution (z to p)*
Given z, find Φ(z), the proportion less than z .

z	Φ(z)	z	F(z)	z	Φ(z)
.00	.5000000				
.01	.5039894	.31	.6217195	.61	.7290691
.02	.5079783	.32	.6255158	.62	.7323711
.03	.5119665	.33	.6293000	.63	.7356527
.04	.5159534	.34	.6330717	.64	.7389137
.05	.5199388	.35	.6368307	.65	.7421539
.06	.5239222	.36	.6405764	.66	.7453731
.07	.5279032	.37	.6443088	.67	.7485711
.08	.5318814	.38	.6480273	.68	.7517478
.09	.5358564	.39	.6517317	.69	.7549029
.10	.5398278	.40	.6554217	.70	.7580363
.11	.5437953	.41	.6590970	.71	.7611479
.12	.5477584	.42	.6627573	.72	.7642375
.13	.5517168	.43	.6664022	.73	.7673049
.14	.5556700	.44	.6700314	.74	.7703500
.15	.5596177	.45	.6736448	.75	.7733726
.16	.5635595	.46	.6772419	.76	.7763727
.17	.5674949	.47	.6808225	.77	.7793501
.18	.5714237	.48	.6843863	.78	.7823046
.19	.5753454	.49	.6879331	.79	.7852361
.20	.5792597	.50	.6914625	.80	.7881446
.21	.5831662	.51	.6949743	.81	.7910299
.22	.5870604	.52	.6984682	.82	.7938919
.23	.5909541	.53	.7019440	.83	.7967306
.24	.5948349	.54	.7054015	.84	.7995458
.25	.5987063	.55	.7088403	.85	.8023375
.26	.6025681	.56	.7122603	.86	.8051055
.27	.6064199	.57	.7156612	.87	.8078498
.28	.6102612	.58	.7190427	.88	.8105703
.29	.6140919	.59	.7224047	.89	.8132671
.30	.6179114	.60	.7257469	.90	.8159399

TABLE A5.2   *Normal Distribution (z to p)*
(cont.)

z	Φ(z)	z	Φ(z)	z	Φ(z)
.91	.8185887	1.21	.8868606	1.51	.9344783
.92	.8212136	1.22	.8887676	1.52	.9357445
.93	.8238145	1.23	.8906514	1.53	.9369916
.94	.8263912	1.24	.8925123	1.54	.9382198
.95	.8289439	1.25	.8943502	1.55	.9394292
.96	.8314724	1.26	.8961653	1.56	.9406201
.97	.8339768	1.27	.8979577	1.57	.9417924
.98	.8364569	1.28	.8997274	1.58	.9429466
.99	.8389129	1.29	.9014747	1.59	.9440826
1.00	.8413447	1.30	.9031995	1.60	.9452007
1.01	.8437524	1.31	.9049021	1.61	.9463011
1.02	.8461358	1.32	.9065825	1.62	.9473839
1.03	.8484950	1.33	.9082409	1.63	.9484493
1.04	.8508300	1.34	.9098773	1.64	.9494974
1.05	.8531409	1.35	.9114920	1.65	.9505285
1.06	.8554277	1.36	.9130850	1.66	.9515428
1.07	.8576903	1.37	.9146565	1.67	.9525403
1.08	.8599289	1.38	.9162067	1.68	.9535213
1.09	.8621434	1.39	.9177356	1.69	.9544860
1.10	.8643339	1.40	.9192433	1.70	.9554345
1.11	.8665005	1.41	.9207302	1.71	.9563671
1.12	.8686431	1.42	.9221962	1.72	.9572838
1.13	.8707619	1.43	.9236415	1.73	.9581849
1.14	.8728568	1.44	.9250663	1.74	.9590705
1.15	.8749281	1.45	.9264707	1.75	.9599408
1.16	.8769756	1.46	.9278550	1.76	.9607961
1.17	.8789995	1.47	.9292191	1.77	.9616364
1.18	.8809999	1.48	.9305634	1.78	.9624620
1.19	.8829768	1.49	.9318879	1.79	.9632730
1.20	.8849303	1.50	.9331928	1.80	.9640697

TABLE A5.2   *Normal Distribution (z to p)*
(cont.)

z	Φ(z)	z	Φ(z)	z	Φ(z)
1.81	.9648521	2.11	.9825708	2.41	.9920237
1.82	.9656205	2.12	.9829970	2.42	.9922397
1.83	.9663750	2.13	.9834142	2.43	.9924506
1.84	.9671159	2.14	.9838226	2.44	.9926564
1.85	.9678432	2.15	.9842224	2.45	.9928572
1.86	.9685572	2.16	.9846137	2.46	.9930531
1.87	.9692581	2.17	.9849966	2.47	.9932443
1.88	.9699460	2.18	.9853713	2.48	.9934309
1.89	.9706210	2.19	.9857379	2.49	.9936128
1.90	.9712834	2.20	.9860966	2.50	.9937903
1 91	.9719334	2.21	.9864474	2.51	.9939634
1.92	.9725711	2.22	.9867906	2.52	.9941323
1 93	.9731966	2.23	.9871263	2.53	.9942969
1.94	.9738102	2.24	.9874545	2.54	.9944574
1.95	.9744119	2.25	.9877755	2.55	.9946139
1.96	.9750021	2.26	.9880894	2.56	.9947664
1.97	.9755808	2.27	.9883962	2.57	.9949151
1.98	.9761482	2.28	.9886962	2.58	.9950600
1.99	.9767045	2.29	.9889893	2.59	.9952012
2.00	.9772499	2.30	.9892759	2.60	.9953388
2.01	.9777844	2.31	.9895559	2.70	.9965330
2.02	.9783083	2.32	.9898296	2.80	.9974449
2.03	.9788217	2.33	.9900969	2.90	.9981342
2.04	.9793248	2.34	.9903581	3.00	.9986501
2.05	.9798178	2.35	.9906133		
				3.20	.9993129
2.06	.9803007	2.36	.9908625	3.40	.9996631
2.07	.9807738	2.37	.9911060	3.60	.9998409
2.08	.9812372	2.38	.9913437	3.80	.9999277
2.09	.9816911	2.39	.9915758	4.00	.9999683
2.10	.9821356	2.40	.9918025		

TABLE A5.2   *Normal Distribution (z to p)*
(cont.)

z	$\Phi(z)$
4.50	.9999966
5.00	.9999997
5.50	.9999999

Source: Tables A5.1 and A5.2 excerpted from *Tables for Statisticians and Biometricians, Part II*, edited by K. Pearson (1931). Reprinted by permission of the *Biometrika* Trustees.

Table A5.3. Values of d' for Same-Different (Independent-Observation Model) and ABX (Independent-Observation and Differencing Models).

To find d' from H and F, first calculate $z(H) - z(F)$ and find the result in the first column. Then look across to the appropriate design and model. If H and F are not available, assume that the observer is unbiased: Find $p(c)$ in the second column and look across for d'.

| | | | d' | |
| | | Same-Different (Independent Observation) | ABX | |
$z(H) - z(F)$	$p(c)_{unb}$		Independent Observation	Differencing
0.01	0.502	0.16	0.13	0.15
0.02	0.504	0.22	0.19	0.21
0.03	0.506	0.28	0.23	0.26
0.04	0.508	0.32	0.27	0.30
0.05	0.510	0.36	0.30	0.33
0.06	0.512	0.39	0.33	0.36
0.07	0.514	0.42	0.35	0.39
0.08	0.516	0.45	0.38	0.42
0.09	0.518	0.48	0.40	0.45
0.10	0.520	0.51	0.43	0.47
0.11	0.522	0.53	0.45	0.50
0.12	0.524	0.56	0.47	0.52
0.13	0.526	0.58	0.49	0.54
0.14	0.528	0.60	0.51	0.56
0.15	0.530	0.62	0.52	0.58
0.16	0.532	0.64	0.54	0.60
0.17	0.534	0.66	0.56	0.62
0.18	0.536	0.68	0.58	0.64
0.19	0.538	0.70	0.59	0.66
0.20	0.540	0.72	0.61	0.68

TABLE A5.3   *Values of d′ for Same-Different and ABX Models (cont.)*

z(H) − z(F)	p(c)_unb	d′ Same-Different (Independent Observation)	ABX Independent Observation	Differencing
0.21	0.542	0.74	0.62	0.69
0.22	0.544	0.76	0.64	0.71
0.23	0.546	0.78	0.65	0.73
0.24	0.548	0.80	0.67	0.74
0.25	0.550	0.81	0.68	0.76
0.26	0.552	0.83	0.70	0.78
0.27	0.554	0.85	0.71	0.79
0.28	0.556	0.86	0.73	0.81
0.29	0.558	0.88	0.74	0.82
0.30	0.560	0.89	0.75	0.84
0.31	0.562	0.91	0.77	0.85
0.32	0.564	0.93	0.78	0.87
0.33	0.566	0.94	0.79	0.88
0.34	0.567	0.96	0.81	0.90
0.35	0.569	0.97	0.82	0.91
0.36	0.571	0.99	0.83	0.92
0.37	0.573	1.00	0.84	0.94
0.38	0.575	1.01	0.86	0.95
0.39	0.577	1.03	0.87	0.96
0.40	0.579	1.04	0.88	0.98
0.41	0.581	1.06	0.89	0.99
0.42	0.583	1.07	0.90	1.00
0.43	0.585	1.09	0.92	1.02
0.44	0.587	1.10	0.93	1.03
0.45	0.589	1.11	0.94	1.04

TABLE A5.3   *Values of d' for Same-Different and ABX Models (cont.)*

		d'		
		Same-Different	ABX	
$z(H) - z(F)$	$p(c)_{unb}$	(Independent Observation)	Independent Observation	Differencing
0.46	0.591	1.13	0.95	1.06
0.47	0.593	1.14	0.96	1.07
0.48	0.595	1.15	0.97	1.08
0.49	0.597	1.17	0.98	1.09
0.50	0.599	1.18	0.99	1.11
0.51	0.601	1.19	1.01	1.12
0.52	0.603	1.20	1.02	1.13
0.53	0.604	1.22	1.03	1.14
0.54	0.606	1.23	1.04	1.16
0.55	0.608	1.24	1.05	1.17
0.56	0.610	1.25	1.06	1.18
0.57	0.612	1.27	1.07	1.19
0.58	0.614	1.28	1.08	1.20
0.59	0.616	1.29	1.09	1.22
0.60	0.618	1.30	1.10	1.23
0.61	0.620	1.32	1.11	1.24
0.62	0.622	1.33	1.12	1.25
0.63	0.624	1.34	1.13	1.26
0.64	0.626	1.35	1.14	1.27
0.65	0.627	1.36	1.15	1.29
0.66	0.629	1.38	1.16	1.30
0.67	0.631	1.39	1.17	1.31
0.68	0.633	1.40	1.18	1.32
0.69	0.635	1.41	1.19	1.33
0.70	0.637	1.42	1.20	1.34

TABLE A5.3 *Values of d' for Same-Different and ABX Models (cont.)*

| | | d' | | |
| | | Same-Different | ABX | |
$z(H) - z(F)$	$p(c)_{unb}$	(Independent Observation)	Independent Observation	Differencing
0.71	0.639	1.43	1.21	1.35
0.72	0.641	1.45	1.22	1.36
0.73	0.642	1.46	1.23	1.38
0.74	0.644	1.47	1.24	1.39
0.75	0.646	1.48	1.25	1.40
0.76	0.648	1.49	1.26	1.41
0.77	0.650	1.50	1.27	1.42
0.78	0.652	1.51	1.28	1.43
0.79	0.654	1.52	1.29	1.44
0.80	0.655	1.54	1.30	1.45
0.81	0.657	1.55	1.31	1.46
0.82	0.659	1.56	1.32	1.47
0.83	0.661	1.57	1.33	1.49
0.84	0.663	1.58	1.34	1.50
0.85	0.665	1.59	1.35	1.51
0.86	0.666	1.60	1.36	1.52
0.87	0.668	1.61	1.37	1.53
0.88	0.670	1.62	1.38	1.54
0.89	0.672	1.63	1.38	1.55
0.90	0.674	1.65	1.39	1.56
0.91	0.675	1.66	1.40	1.57
0.92	0.677	1.67	1.41	1.58
0.93	0.679	1.68	1.42	1.59
0.94	0.681	1.69	1.43	1.60
0.95	0.683	1.70	1.44	1.61

TABLE A5.3   *Values of d′ for Same-Different and ABX Models (cont.)*

		d′		
		Same-Different	*ABX*	
		(Independent	*Independent*	
z(H) − z(F)	*p(c)*$_{unb}$	*Observation)*	*Observation*	*Differencing*
0.96	0.684	1.71	1.45	1.62
0.97	0.686	1.72	1.46	1.63
0.98	0.688	1.73	1.47	1.64
0.99	0.690	1.74	1.48	1.65
1.00	0.691	1.75	1.48	1.66
1.01	0.693	1.76	1.49	1.68
1.02	0.695	1.77	1.50	1.69
1.03	0.697	1.78	1.51	1.70
1.04	0.698	1.79	1.52	1.71
1.05	0.700	1.80	1.53	1.72
1.06	0.702	1.81	1.54	1.73
1.07	0.704	1.82	1.55	1.74
1.08	0.705	1.83	1.56	1.75
1.09	0.707	1.84	1.57	1.76
1.10	0.709	1.85	1.57	1.77
1.11	0.711	1.87	1.58	1.78
1.12	0.712	1.88	1.59	1.79
1.13	0.714	1.89	1.60	1.80
1.14	0.716	1.90	1.61	1.81
1.15	0.717	1.91	1.62	1.82
1.16	0.719	1.92	1.63	1.83
1.17	0.721	1.93	1.64	1.84
1.18	0.722	1.94	1.64	1.85
1.19	0.724	1.95	1.65	1.86
1.20	0.726	1.96	1.66	1.87

TABLE A5.3   *Values of d' for Same-Different and ABX Models (cont.)*

z(H) − z(F)	$p(c)_{unb}$	Same-Different (Independent Observation)	ABX Independent Observation	Differencing
1.21	0.727	1.97	1.67	1.88
1.22	0.729	1.98	1.68	1.89
1.23	0.731	1.99	1.69	1.90
1.24	0.732	2.00	1.70	1.91
1.25	0.734	2.01	1.71	1.92
1.26	0.736	2.02	1.71	1.93
1.27	0.737	2.03	1.72	1.94
1.28	0.739	2.04	1.73	1.95
1.29	0.741	2.05	1.74	1.96
1.30	0.742	2.06	1.75	1.97
1.31	0.744	2.07	1.76	1.98
1.32	0.745	2.08	1.77	1.99
1.33	0.747	2.09	1.77	2.00
1.34	0.749	2.09	1.78	2.01
1.35	0.750	2.10	1.79	2.02
1.36	0.752	2.11	1.80	2.03
1.37	0.753	2.12	1.81	2.04
1.38	0.755	2.13	1.82	2.05
1.39	0.756	2.14	1.83	2.06
1.40	0.758	2.15	1.83	2.07
1.41	0.760	2.16	1.84	2.08
1.42	0.761	2.17	1.85	2.09
1.43	0.763	2.18	1.86	2.10
1.44	0.764	2.19	1.87	2.11
1.45	0.766	2.20	1.88	2.12

TABLE A5.3   *Values of d' for Same-Different and ABX Models (cont.)*

		d'		
		Same-Different	ABX	
		(Independent	Independent	
$z(H) - z(F)$	$p(c)_{unb}$	Observation)	Observation	Differencing
1.46	0.767	2.21	1.88	2.13
1.47	0.769	2.22	1.89	2.14
1.48	0.770	2.23	1.90	2.15
1.49	0.772	2.24	1.91	2.16
1.50	0.773	2.25	1.92	2.17
1.51	0.775	2.26	1.93	2.18
1.52	0.776	2.27	1.94	2.19
1.53	0.778	2.28	1.94	2.20
1.54	0.779	2.29	1.95	2.21
1.55	0.781	2.30	1.96	2.22
1.56	0.782	2.31	1.97	2.23
1.57	0.784	2.32	1.98	2.24
1.58	0.785	2.33	1.99	2.25
1.59	0.787	2.34	1.99	2.26
1.60	0.788	2.35	2.00	2.27
1.61	0.790	2.36	2.01	2.28
1.62	0.791	2.36	2.02	2.29
1.63	0.792	2.37	2.03	2.30
1.64	0.794	2.38	2.04	2.31
1.65	0.795	2.39	2.04	2.32
1.66	0.797	2.40	2.05	2.33
1.67	0.798	2.41	2.06	2.34
1.68	0.800	2.42	2.07	2.35
1.69	0.801	2.43	2.08	2.36
1.70	0.802	2.44	2.09	2.37

TABLE A5.3   *Values of d' for Same-Different and ABX Models (cont.)*

		d'		
		Same-Different (Independent Observation)	ABX	
			Independent Observation	Differencing
$z(H) - z(F)$	$p(c)_{unb}$			
1.71	0.804	2.45	2.10	2.38
1.72	0.805	2.46	2.10	2.39
1.73	0.806	2.47	2.11	2.40
1.74	0.808	2.48	2.12	2.41
1.75	0.809	2.49	2.13	2.42
1.76	0.811	2.50	2.14	2.43
1.77	0.812	2.51	2.15	2.44
1.78	0.813	2.52	2.15	2.45
1.79	0.815	2.53	2.16	2.46
1.80	0.816	2.53	2.17	2.47
1.81	0.817	2.54	2.18	2.48
1.82	0.819	2.55	2.19	2.49
1.83	0.820	2.56	2.20	2.50
1.84	0.821	2.57	2.20	2.51
1.85	0.823	2.58	2.21	2.52
1.86	0.824	2.59	2.22	2.53
1.87	0.825	2.60	2.23	2.54
1.88	0.826	2.61	2.24	2.55
1.89	0.828	2.62	2.25	2.56
1.90	0.829	2.63	2.25	2.57
1.91	0.830	2.64	2.26	2.58
1.92	0.831	2.65	2.27	2.59
1.93	0.833	2.66	2.28	2.60
1.94	0.834	2.67	2.29	2.61
1.95	0.835	2.67	2.30	2.62

TABLE A5.3   *Values of d' for Same-Different and ABX Models (cont.)*

		d'		
		Same-Different	*ABX*	
		(Independent	*Independent*	
z(H) − z(F)	*p(c)*$_{unb}$	*Observation)*	*Observation*	*Differencing*
1.96	0.836	2.68	2.30	2.63
1.97	0.838	2.69	2.31	2.64
1.98	0.839	2.70	2.32	2.65
1.99	0.840	2.71	2.33	2.67
2.00	0.841	2.72	2.34	2.68
2.01	0.843	2.73	2.35	2.69
2.02	0.844	2.74	2.35	2.70
2.03	0.845	2.75	2.36	2.71
2.04	0.846	2.76	2.37	2.72
2.05	0.847	2.77	2.38	2.73
2.06	0.848	2.78	2.39	2.74
2.07	0.850	2.79	2.40	2.75
2.08	0.851	2.79	2.40	2.76
2.09	0.852	2.80	2.41	2.77
2.10	0.853	2.81	2.42	2.78
2.11	0.854	2.82	2.43	2.79
2.12	0.855	2.83	2.44	2.80
2.13	0.857	2.84	2.45	2.81
2.14	0.858	2.85	2.45	2.82
2.15	0.859	2.86	2.46	2.83
2.16	0.860	2.87	2.47	2.84
2.17	0.861	2.88	2.48	2.85
2.18	0.862	2.89	2.49	2.86
2.19	0.863	2.90	2.50	2.87
2.20	0.864	2.91	2.50	2.88

TABLE A5.3   *Values of d' for Same-Different and ABX Models (cont.)*

		d'		
		Same-Different	*ABX*	
		(Independent	*Independent*	
z(H) − z(F)	$p(c)_{unb}$	*Observation)*	*Observation*	*Differencing*
2.71	0.912	3.37	2.93	3.42
2.72	0.913	3.38	2.94	3.44
2.73	0.914	3.39	2.95	3.45
2.74	0.915	3.40	2.96	3.46
2.75	0.915	3.41	2.97	3.47
2.76	0.916	3.42	2.98	3.48
2.77	0.917	3.43	2.98	3.49
2.78	0.918	3.44	2.99	3.50
2.79	0.918	3.45	3.00	3.51
2.80	0.919	3.45	3.01	3.52
2.81	0.920	3.46	3.02	3.53
2.82	0.921	3.47	3.03	3.55
2.83	0.921	3.48	3.04	3.56
2.84	0.922	3.49	3.04	3.57
2.85	0.923	3.50	3.05	3.58
2.86	0.924	3.51	3.06	3.59
2.87	0.924	3.52	3.07	3.60
2.88	0.925	3.53	3.08	3.61
2.89	0.926	3.54	3.09	3.62
2.90	0.926	3.55	3.10	3.63
2.91	0.927	3.56	3.11	3.65
2.92	0.928	3.56	3.11	3.66
2.93	0.929	3.57	3.12	3.67
2.94	0.929	3.58	3.13	3.68
2.95	0.930	3.59	3.14	3.69

TABLE A5.3   *Values of d′ for Same-Different and ABX Models (cont.)*

| | | *d′* | | |
| | | *Same-Different* | *ABX* | |
z(H) − z(F)	*p(c)*_unb	*(Independent Observation)*	*Independent Observation*	*Differencing*
2.96	0.931	3.60	3.15	3.70
2.97	0.931	3.61	3.16	3.71
2.98	0.932	3.62	3.17	3.72
2.99	0.933	3.63	3.17	3.74
3.00	0.933	3.64	3.18	3.75
3.01	0.934	3.65	3.19	3.76
3.02	0.934	3.66	3.20	3.77
3.03	0.935	3.67	3.21	3.78
3.04	0.936	3.67	3.22	3.79
3.05	0.936	3.68	3.23	3.80
3.06	0.937	3.69	3.24	3.82
3.07	0.938	3.70	3.24	3.83
3.08	0.938	3.71	3.25	3.84
3.09	0.939	3.72	3.26	3.85
3.10	0.939	3.73	3.27	3.86
3.11	0.940	3.74	3.28	3.87
3.12	0.941	3.75	3.29	3.88
3.13	0.941	3.76	3.30	3.90
3.14	0.942	3.77	3.31	3.91
3.15	0.942	3.78	3.32	3.92
3.16	0.943	3.78	3.32	3.93
3.17	0.944	3.79	3.33	3.94
3.18	0.944	3.80	3.34	3.95
3.19	0.945	3.81	3.35	3.96
3.20	0.945	3.82	3.36	3.98

TABLE A5.3 *Values of d' for Same-Different and ABX Models (cont.)*

		d'		
		Same-Different (Independent Observation)	ABX	
			Independent Observation	Differencing
z(H) − z(F)	$p(c)_{unb}$			
3.21	0.946	3.83	3.37	3.99
3.22	0.946	3.84	3.38	4.00
3.23	0.947	3.85	3.39	4.01
3.24	0.947	3.86	3.39	4.02
3.25	0.948	3.87	3.40	4.03
3.26	0.948	3.88	3.41	4.05
3.27	0.949	3.89	3.42	4.06
3.28	0.949	3.90	3.43	4.07
3.29	0.950	3.90	3.44	4.08
3.30	0.951	3.91	3.45	4.09
3.31	0.951	3.92	3.46	4.10
3.32	0.952	3.93	3.47	4.12
3.33	0.952	3.94	3.48	4.13
3.34	0.953	3.95	3.48	4.14
3.35	0.953	3.96	3.49	4.15
3.36	0.954	3.97	3.50	4.16
3.37	0.954	3.98	3.51	4.18
3.38	0.954	3.99	3.52	4.19
3.39	0.955	4.00	3.53	4.20
3.40	0.955	4.01	3.54	4.21
3.41	0.956	4.02	3.55	4.22
3.42	0.956	4.02	3.56	4.23
3.43	0.957	4.03	3.57	4.25
3.44	0.957	4.04	3.57	4.26
3.45	0.958	4.05	3.58	4.27

TABLE A5.3   *Values of d' for Same-Different and ABX Models (cont.)*

$z(H) - z(F)$	$p(c)_{unb}$	d' Same-Different (Independent Observation)	ABX Independent Observation	Differencing
3.46	0.958	4.06	3.59	4.28
3.47	0.959	4.07	3.60	4.29
3.48	0.959	4.08	3.61	4.31
3.49	0.960	4.09	3.62	4.32
3.50	0.960	4.10	3.63	4.33
3.51	0.960	4.11	3.64	4.34
3.52	0.961	4.12	3.65	4.35
3.53	0.961	4.13	3.66	4.37
3.54	0.962	4.14	3.67	4.38
3.55	0.962	4.15	3.67	4.39
3.56	0.962	4.16	3.68	4.40
3.57	0.963	4.16	3.69	4.41
3.58	0.963	4.17	3.70	4.43
3.59	0.964	4.18	3.71	4.44
3.60	0.964	4.19	3.72	4.45
3.61	0.964	4.20	3.73	4.46
3.62	0.965	4.21	3.74	4.47
3.63	0.965	4.22	3.75	4.49
3.64	0.966	4.23	3.76	4.50
3.65	0.966	4.24	3.77	4.51
3.66	0.966	4.25	3.78	4.52
3.67	0.967	4.26	3.79	4.53
3.68	0.967	4.27	3.79	4.55
3.69	0.967	4.28	3.80	4.56
3.70	0.968	4.29	3.81	4.57

TABLE A5.3   *Values of d′ for Same-Different and ABX Models (cont.)*

		d′		
		Same-Different	*ABX*	
		(Independent	*Independent*	
$z(H) - z(F)$	$p(c)_{unb}$	*Observation)*	*Observation*	*Differencing*
3.71	0.968	4.30	3.82	4.58
3.72	0.969	4.31	3.83	4.60
3.73	0.969	4.31	3.84	4.61
3.74	0.969	4.32	3.85	4.62
3.75	0.970	4.33	3.86	4.63
3.76	0.970	4.34	3.87	4.64
3.77	0.970	4.35	3.88	4.66
3.78	0.971	4.36	3.89	4.67
3.79	0.971	4.37	3.90	4.68
3.80	0.971	4.38	3.91	4.69
3.81	0.972	4.39	3.92	4.71
3.82	0.972	4.40	3.93	4.72
3.83	0.972	4.41	3.94	4.73
3.84	0.973	4.42	3.95	4.74
3.85	0.973	4.43	3.95	4.76
3.86	0.973	4.44	3.96	4.77
3.87	0.974	4.45	3.97	4.78
3.88	0.974	4.46	3.98	4.79
3.89	0.974	4.47	3.99	4.81
3.90	0.974	4.48	4.00	4.82
3.91	0.975	4.49	4.01	4.83
3.92	0.975	4.50	4.02	4.84
3.93	0.975	4.50	4.03	4.86
3.94	0.976	4.51	4.04	4.87
3.95	0.976	4.52	4.05	4.88

TABLE A5.3   *Values of d' for Same-Different and ABX Models (cont.)*

$z(H) - z(F)$	$p(c)_{unb}$	Same-Different (Independent Observation)	ABX	
			Independent Observation	Differencing
3.96	0.976	4.53	4.06	4.89
3.97	0.976	4.54	4.07	4.91
3.98	0.977	4.55	4.08	4.92
3.99	0.977	4.56	4.09	4.93
4.00	0.977	4.57	4.10	4.94
4.01	0.978	4.58	4.11	4.96
4.02	0.978	4.59	4.12	4.97
4.03	0.978	4.60	4.13	4.98
4.04	0.978	4.61	4.14	4.99
4.05	0.979	4.62	4.15	5.01
4.06	0.979	4.63	4.16	5.02
4.07	0.979	4.64	4.17	5.03
4.08	0.979	4.65	4.18	5.05
4.09	0.980	4.66	4.19	5.06
4.10	0.980	4.67	4.20	5.07
4.11	0.980	4.68	4.21	5.08
4.12	0.980	4.69	4.22	5.10
4.13	0.981	4.70	4.23	5.11
4.14	0.981	4.71	4.24	5.12
4.15	0.981	4.72	4.25	5.13
4.16	0.981	4.73	4.26	5.15
4.17	0.981	4.74	4.27	5.16
4.18	0.982	4.75	4.28	5.17
4.19	0.982	4.76	4.29	5.19
4.20	0.982	4.77	4.30	5.20

TABLE A5.3   *Values of d' for Same-Different and ABX Models (cont.)*

		d'		
		Same-Different	ABX	
		(Independent	Independent	
$z(H) - z(F)$	$p(c)_{unb}$	Observation)	Observation	Differencing
4.21	0.982	4.78	4.31	5.21
4.22	0.983	4.79	4.32	5.23
4.23	0.983	4.80	4.33	5.24
4.24	0.983	4.81	4.34	5.25
4.25	0.983	4.82	4.35	5.26
4.26	0.983	4.83	4.36	5.28
4.27	0.984	4.84	4.37	5.29
4.28	0.984	4.85	4.38	5.30
4.29	0.984	4.86	4.39	5.32
4.30	0.984	4.87	4.40	5.33
4.31	0.984	4.88	4.41	5.34
4.32	0.985	4.89	4.42	5.36
4.33	0.985	4.90	4.43	5.37
4.34	0.985	4.91	4.44	5.38
4.35	0.985	4.92	4.45	5.40
4.36	0.985	4.93	4.46	5.41
4.37	0.986	4.94	4.47	5.42
4.38	0.986	4.95	4.48	5.43
4.39	0.986	4.96	4.49	5.45
4.40	0.986	4.97	4.50	5.46
4.41	0.986	4.98	4.51	5.47
4.42	0.986	4.99	4.52	5.49
4.43	0.987	5.00	4.53	5.50
4.44	0.987	5.01	4.54	5.51
4.45	0.987	5.02	4.55	5.53

TABLE A5.3   *Values of d' for Same-Different and ABX Models (cont.)*

		d'		
		Same-Different (Independent Observation)	ABX	
			Independent Observation	Differencing
z(H) − z(F)	p(c)_unb			
4.46	0.987	5.03	4.56	5.54
4.47	0.987	5.04	4.57	5.56
4.48	0.987	5.05	4.58	5.57
4.49	0.988	5.06	4.59	5.58
4.50	0.988	5.07	4.60	5.60
4.51	0.988	5.08	4.62	5.61
4.52	0.988	5.09	4.63	5.62
4.53	0.988	5.10	4.64	5.64
4.54	0.988	5.11	4.65	5.65
4.55	0.989	5.12	4.66	5.66
4.56	0.989	5.13	4.67	5.68
4.57	0.989	5.14	4.68	5.69
4.58	0.989	5.15	4.69	5.70
4.59	0.989	5.16	4.70	5.72
4.60	0.989	5.17	4.71	5.73
4.61	0.989	5.18	4.72	5.75
4.62	0.990	5.19	4.73	5.76
4.63	0.990	5.20	4.74	5.77
4.64	0.990	5.21	4.76	5.79
4.65	0.990	5.22	4.77	5.80
4.66	0.990	5.23	4.78	5.81
4.67	0.990	5.25	4.79	5.83
4.68	0.990	5.26	4.80	5.84
4.69	0.990	5.27	4.81	5.86
4.70	0.991	5.28	4.82	5.87

TABLE A5.3 *Values of d′ for Same-Different and ABX Models (cont.)*

| | | d' | | |
| | | Same-Different | ABX | |
$z(H) - z(F)$	$p(c)_{\text{unb}}$	(Independent Observation)	Independent Observation	Differencing
4.71	0.991	5.29	4.83	5.88
4.72	0.991	5.30	4.84	5.90
4.73	0.991	5.31	4.85	5.91
4.74	0.991	5.32	4.87	5.93
4.75	0.991	5.33	4.88	5.94
4.76	0.991	5.34	4.89	5.96
4.77	0.991	5.35	4.90	5.97
4.78	0.992	5.36	4.91	5.98
4.79	0.992	5.37	4.92	6.00
4.80	0.992	5.39	4.93	6.01
4.81	0.992	5.40	4.94	6.03
4.82	0.992	5.41	4.96	6.04
4.83	0.992	5.42	4.97	6.06
4.84	0.992	5.43	4.98	6.07
4.85	0.992	5.44	4.99	6.09
4.86	0.992	5.45	5.00	6.10
4.87	0.993	5.46	5.01	6.11
4.88	0.993	5.47	5.03	6.13
4.89	0.993	5.49	5.04	6.14
4.90	0.993	5.50	5.05	6.16
4.91	0.993	5.51	5.06	6.17
4.92	0.993	5.52	5.07	6.19
4.93	0.993	5.53	5.08	6.20
4.94	0.993	5.54	5.10	6.22
4.95	0.993	5.55	5.11	6.23

TABLE A5.3   *Values of d' for Same-Different and ABX Models (cont.)*

| | | *d'* | | |
| | | *Same-Different* | *ABX* | |
z(H) − z(F)	$p(c)_{unb}$	*(Independent Observation)*	*Independent Observation*	*Differencing*
4.96	0.993	5.56	5.12	6.25
4.97	0.994	5.58	5.13	6.26
4.98	0.994	5.59	5.14	6.28
4.99	0.994	5.60	5.16	6.29
5.00	0.994	5.61	5.17	6.31

Source: Adapted from Kaplan et al. (1978) by permission of The Psychonomic Society, Inc.

Table A5.4. Values of d' for Same-Different (Differencing Model).

H^a	F^b 0.01	0.02	0.03	0.04	0.05	0.06	0.07	0.08	0.09	0.10
0.01	0.00									
0.02	0.71	0.00								
0.03	0.97	0.56	0.00							
0.04	1.16	0.78	0.49	0.00						
0.05	1.31	0.94	0.69	0.45	0.00					
0.06	1.44	1.08	0.84	0.63	0.42	0.00				
0.07	1.55	1.19	0.96	0.77	0.59	0.39	0.00			
0.08	1.65	1.30	1.07	0.88	0.72	0.56	0.38	0.00		
0.09	1.75	1.39	1.16	0.98	0.83	0.68	0.53	0.36	0.00	
0.10	1.83	1.47	1.25	1.07	0.92	0.79	0.65	0.51	0.35	0.00
0.11	1.91	1.55	1.33	1.15	1.01	0.88	0.75	0.63	0.50	0.34
0.12	1.98	1.63	1.40	1.23	1.09	0.96	0.84	0.73	0.61	0.49
0.13	2.05	1.70	1.47	1.30	1.16	1.04	0.92	0.81	0.71	0.60
0.14	2.11	1.76	1.54	1.37	1.23	1.11	1.00	0.89	0.79	0.69
0.15	2.18	1.82	1.60	1.43	1.29	1.17	1.07	0.96	0.87	0.77
0.16	2.24	1.88	1.66	1.49	1.36	1.24	1.13	1.03	0.94	0.85
0.17	2.29	1.94	1.72	1.55	1.41	1.30	1.19	1.09	1.00	0.91
0.18	2.35	1.99	1.77	1.61	1.47	1.35	1.25	1.15	1.06	0.98
0.19	2.40	2.05	1.83	1.66	1.52	1.41	1.30	1.21	1.12	1.04
0.20	2.45	2.10	1.88	1.71	1.58	1.46	1.36	1.26	1.18	1.10
0.21	2.50	2.15	1.93	1.76	1.63	1.51	1.41	1.32	1.23	1.15
0.22	2.55	2.20	1.98	1.81	1.68	1.56	1.46	1.37	1.28	1.20
0.23	2.60	2.24	2.02	1.86	1.72	1.61	1.51	1.42	1.33	1.25
0.24	2.64	2.29	2.07	1.90	1.77	1.66	1.56	1.46	1.38	1.30
0.25	2.69	2.34	2.11	1.95	1.82	1.70	1.60	1.51	1.43	1.35
0.26	2.73	2.38	2.16	1.99	1.86	1.75	1.65	1.56	1.47	1.40
0.27	2.78	2.42	2.20	2.04	1.90	1.79	1.69	1.60	1.52	1.44
0.28	2.82	2.47	2.24	2.08	1.95	1.83	1.73	1.64	1.56	1.49
0.29	2.86	2.51	2.29	2.12	1.99	1.87	1.78	1.69	1.61	1.53
0.30	2.90	2.55	2.33	2.16	2.03	1.92	1.82	1.73	1.65	1.57

$^a H$ = hit rate = P("different" | Different).
$^b F$ = false-alarm rate = P("different" | Same).

TABLE A5.4   *Values of* d′ *for Same-Different (Differencing Model) (cont.)*

H^a	F^b									
	0.01	0.02	0.03	0.04	0.05	0.06	0.07	0.08	0.09	0.10
0.31	2.94	2.59	2.37	2.20	2.07	1.96	1.86	1.77	1.69	1.61
0.32	2.98	2.63	2.41	2.24	2.11	2.00	1.90	1.81	1.73	1.66
0.33	3.02	2.67	2.45	2.28	2.15	2.04	1.94	1.85	1.77	1.70
0.34	3.06	2.71	2.49	2.32	2.1	2.07	1.98	1.89	1.81	1.73
0.35	3.10	2.74	2.52	2.36	2.23	2.11	2.02	1.93	1.85	1.77
0.36	3.14	2.78	2.56	2.40	2.26	2.15	2.05	1.97	1.89	1.81
0.37	3.17	2.82	2.60	2.43	2.30	2.19	2.09	2.00	1.92	1.85
0.38	3.21	2.86	2.64	2.47	2.34	2.23	2.13	2.04	1.96	1.89
0.39	3.25	2.89	2.67	2.51	2.38	2.26	2.17	2.08	2.00	1.93
0.40	3.28	2.93	2.71	2.55	2.41	2.30	2.20	2.12	2.04	1.96
0.41	3.32	2.97	2.75	2.58	2.45	2.34	2.24	2.15	2.07	2.00
0.42	3.36	3.00	2.78	2.62	2.49	2.37	2.28	2.19	2.11	2.04
0.43	3.39	3.04	2.82	2.65	2.52	2.41	2.31	2.22	2.15	2.07
0.44	3.43	3.08	2.86	2.69	2.56	2.45	2.35	2.26	2.18	2.11
0.45	3.47	3.11	2.89	2.73	2.59	2.48	2.38	2.30	2.22	2.15
0.46	3.50	3.15	2.93	2.76	2.63	2.52	2.42	2.33	2.25	2.18
0.47	3.54	3.18	2.96	2.80	2.67	2.55	2.46	2.37	2.29	2.22
0.48	3.57	3.22	3.00	2.83	2.70	2.59	2.49	2.40	2.33	2.25
0.49	3.61	3.25	3.03	2.87	2.74	2.62	2.53	2.44	2.36	2.29
0.50	3.64	3.29	3.07	2.90	2.77	2.66	2.56	2.48	2.40	2.32
0.51	3.68	3.33	3.10	2.94	2.81	2.70	2.60	2.51	2.43	2.36
0.52	3.71	3.36	3.14	2.98	2.84	2.73	2.63	2.55	2.47	2.40
0.53	3.75	3.40	3.18	3.01	2.88	2.77	2.67	2.58	2.50	2.43
0.54	3.78	3.43	3.21	3.05	2.91	2.80	2.70	2.62	2.54	2.47
0.55	3.82	3.47	3.25	3.08	2.95	2.84	2.74	2.65	2.57	2.50
0.56	3.86	3.50	3.28	3.12	2.99	2.87	2.78	2.69	2.61	2.54
0.57	3.89	3.54	3.32	3.15	3.02	2.91	2.81	2.72	2.65	2.57
0.58	3.93	3.58	3.35	3.19	3.06	2.95	2.85	2.76	2.68	2.61
0.59	3.96	3.61	3.39	3.23	3.09	2.98	2.88	2.80	2.72	2.65
0.60	4.00	3.65	3.43	3.26	3.13	3.02	2.92	2.83	2.76	2.68

[a] H = hit rate = P("different" | Different).
[b] F = false-alarm rate = P("different" | Same).

TABLE A5.4 *Values of* d' *for Same-Different (Differencing Model)* (cont.).

H^a	F^b									
	0.01	0.02	0.03	0.04	0.05	0.06	0.07	0.08	0.09	0.10
0.61	4.04	3.68	3.46	3.30	3.17	3.05	2.96	2.87	2.79	2.72
0.62	4.07	3.72	3.50	3.34	3.20	3.09	2.99	2.91	2.83	2.76
0.63	4.11	3.76	3.54	3.37	3.24	3.13	3.03	2.94	2.87	2.79
0.64	4.15	3.80	3.58	3.41	3.28	3.17	3.07	2.98	2.90	2.83
0.65	4.19	3.83	3.61	3.45	3.32	3.20	3.11	3.02	2.94	2.87
0.66	4.23	3.87	3.65	3.49	3.36	3.24	3.15	3.06	2.98	2.91
0.67	4.26	3.91	3.69	3.53	3.39	3.28	3.18	3.10	3.02	2.95
0.68	4.30	3.95	3.73	3.57	3.43	3.32	3.22	3.14	3.06	2.99
0.69	4.34	3.99	3.77	3.61	3.47	3.36	3.26	3.18	3.10	3.03
0.70	4.38	4.03	3.81	3.65	3.51	3.40	3.30	3.22	3.14	3.07
0.71	4.43	4.07	3.85	3.69	3.55	3.44	3.34	3.26	3.18	3.11
0.72	4.47	4.11	3.89	3.73	3.60	3.48	3.39	3.30	3.22	3.15
0.73	4.51	4.16	3.94	3.77	3.64	3.53	3.43	3.34	3.26	3.19
0.74	4.55	4.20	3.98	3.81	3.68	3.57	3.47	3.39	3.31	3.24
0.75	4.60	4.24	4.02	3.86	3.73	3.61	3.52	3.43	3.35	3.28
0.76	4.64	4.29	4.07	3.90	3.77	3.66	3.56	3.47	3.40	3.32
0.77	4.69	4.33	4.11	3.95	3.82	3.70	3.61	3.52	3.44	3.37
0.78	4.73	4.38	4.16	4.00	3.86	3.75	3.65	3.57	3.49	3.42
0.79	4.78	4.43	4.21	4.04	3.91	3.80	3.70	3.62	3.54	3.47
0.80	4.83	4.48	4.26	4.09	3.96	3.85	3.75	3.67	3.59	3.52
0.81	4.88	4.53	4.31	4.15	4.01	3.90	3.80	3.72	3.64	3.57
0.82	4.94	4.58	4.36	4.20	4.07	3.95	3.86	3.77	3.69	3.62
0.83	4.99	4.64	4.42	4.25	4.12	4.01	3.91	3.83	3.75	3.68
0.84	5.05	4.70	4.48	4.31	4.18	4.07	3.97	3.88	3.80	3.73
0.85	5.11	4.76	4.53	4.37	4.24	4.13	4.03	3.94	3.86	3.79
0.86	5.17	4.82	4.60	4.43	4.30	4.19	4.09	4.00	3.93	3.85
0.87	5.24	4.88	4.66	4.50	4.36	4.25	4.16	4.07	3.99	3.92
0.88	5.30	4.95	4.73	4.57	4.43	4.32	4.22	4.14	4.06	3.99
0.89	5.38	5.02	4.80	4.64	4.51	4.39	4.30	4.21	4.13	4.06
0.90	5.46	5.10	4.88	4.72	4.58	4.47	4.37	4.29	4.21	4.14

^aH = hit rate = P("different" | Different).
^bF = false-alarm rate = P("different" | Same).

TABLE A5.4   *Values of* d′ *for Same-Different (Differencing Model) (cont.)*

H^a	F^b									
	0.01	0.02	0.03	0.04	0.05	0.06	0.07	0.08	0.09	0.10
0.91	5.54	5.19	4.97	4.80	4.67	4.56	4.46	4.37	4.29	4.22
0.92	5.63	5.28	5.06	4.89	4.76	4.65	4.55	4.46	4.38	4.31
0.93	5.73	5.38	5.16	4.99	4.86	4.75	4.65	4.56	4.48	4.41
0.94	5.84	5.49	5.27	5.10	4.97	4.86	4.76	4.67	4.60	4.52
0.95	5.97	5.62	5.40	5.23	5.10	4.99	4.89	4.80	4.72	4.65
0.96	6.12	5.77	5.54	5.38	5.25	5.14	5.04	4.95	4.87	4.80
0.97	6.30	5.95	5.73	5.56	5.43	5.32	5.22	5.14	5.06	4.99
0.98	6.55	6.19	5.97	5.81	5.68	5.56	5.47	5.38	5.30	5.23
0.99	6.93	6.58	6.36	6.19	6.06	5.95	5.85	5.77	5.69	5.62

H^a	F^b									
	0.11	0.12	0.13	0.14	0.15	0.16	0.17	0.18	0.19	0.20
0.11	0.00									
0.12	0.34	0.00								
0.13	0.48	0.33	0.00							
0.14	0.58	0.47	0.32	0.00						
0.15	0.67	0.57	0.46	0.32	0.00					
0.16	0.75	0.66	0.56	0.45	0.31	0.00				
0.17	0.83	0.74	0.65	0.55	0.44	0.31	0.00			
0.18	0.89	0.81	0.73	0.64	0.55	0.44	0.31	0.00		
0.19	0.96	0.88	0.80	0.72	0.63	0.54	0.43	0.30	0.00	
0.20	1.02	0.94	0.86	0.79	0.71	0.62	0.53	0.43	0.30	0.00
0.21	1.07	1.00	0.93	0.85	0.78	0.70	0.62	0.53	0.43	0.30
0.22	1.13	1.05	0.98	0.91	0.84	0.77	0.69	0.61	0.52	0.42
0.23	1.18	1.11	1.04	0.97	0.90	0.83	0.76	0.69	0.61	0.52
0.24	1.23	1.16	1.09	1.02	0.96	0.89	0.82	0.75	0.68	0.60
0.25	1.28	1.21	1.14	1.08	1.01	0.95	0.88	0.82	0.75	0.68
0.26	1.33	1.26	1.19	1.13	1.06	1.00	0.94	0.87	0.81	0.74
0.27	1.37	1.30	1.24	1.17	1.11	1.05	0.99	0.93	0.87	0.80
0.28	1.42	1.35	1.28	1.22	1.16	1.10	1.04	0.98	0.92	0.86
0.29	1.46	1.39	1.33	1.27	1.21	1.15	1.09	1.03	0.97	0.92
0.30	1.50	1.44	1.37	1.31	1.25	1.20	1.14	1.08	1.03	0.97

[a] H = hit rate = P("different" | Different).
[b] F = false-alarm rate = P("different" | Same).

TABLE A5.4   *Values of* d' *for Same-Different (Differencing Model) (cont.)*

H^a	F^b									
	0.11	0.12	0.13	0.14	0.15	0.16	0.17	0.18	0.19	0.20
0.31	1.54	1.48	1.42	1.36	1.30	1.24	1.18	1.13	1.07	1.02
0.32	1.59	1.52	1.46	1.40	1.34	1.28	1.23	1.17	1.12	1.07
0.33	1.63	1.56	1.50	1.44	1.38	1.33	1.27	1.22	1.17	1.11
0.34	1.67	1.60	1.54	1.48	1.42	1.37	1.32	1.26	1.21	1.16
0.35	1.71	1.64	1.58	1.52	1.47	1.41	1.36	1.31	1.25	1.20
0.36	1.74	1.68	1.62	1.56	1.51	1.45	1.40	1.35	1.30	1.25
0.37	1.78	1.72	1.66	1.60	1.55	1.49	1.44	1.39	1.34	1.29
0.38	1.82	1.76	1.70	1.64	1.58	1.53	1.48	1.43	1.38	1.33
0.39	1.86	1.80	1.73	1.68	1.62	1.57	1.52	1.47	1.42	1.37
0.40	1.90	1.83	1.77	1.72	1.66	1.61	1.56	1.51	1.46	1.41
0.41	1.93	1.87	1.81	1.75	1.70	1.65	1.60	1.55	1.50	1.45
0.42	1.97	1.91	1.85	1.79	1.74	1.68	1.63	1.59	1.54	1.49
0.43	2.01	1.94	1.88	1.83	1.77	1.72	1.67	1.62	1.58	1.53
0.44	2.04	1.98	1.92	1.86	1.81	1.76	1.71	1.66	1.61	1.57
0.45	2.08	2.02	1.96	1.90	1.85	1.80	1.75	1.70	1.65	1.61
0.46	2.11	2.05	1.99	1.94	1.88	1.83	1.78	1.74	1.69	1.64
0.47	2.15	2.09	2.03	1.97	1.92	1.87	1.82	1.77	1.73	1.68
0.48	2.19	2.12	2.07	2.01	1.96	1.91	1.86	1.81	1.76	1.72
0.49	2.22	2.16	2.10	2.05	1.99	1.94	1.89	1.85	1.80	1.76
0.50	2.26	2.20	2.14	2.08	2.03	1.98	1.93	1.88	1.84	1.79
0.51	2.29	2.23	2.17	2.12	2.06	2.01	1.97	1.92	1.87	1.83
0.52	2.33	2.27	2.21	2.15	2.10	2.05	2.00	1.96	1.91	1.87
0.53	2.36	2.30	2.24	2.19	2.14	2.09	2.04	1.99	1.95	1.90
0.54	2.40	2.34	2.28	2.23	2.17	2.12	2.07	2.03	1.98	1.94
0.55	2.44	2.37	2.32	2.26	2.21	2.16	2.11	2.06	2.02	1.98
0.56	2.47	2.41	2.35	2.30	2.24	2.20	2.15	2.10	2.06	2.01
0.57	2.51	2.45	2.39	2.33	2.28	2.23	2.18	2.14	2.09	2.05
0.58	2.54	2.48	2.42	2.37	2.32	2.27	2.22	2.17	2.13	2.09
0.59	2.58	2.52	2.46	2.41	2.35	2.30	2.26	2.21	2.17	2.12
0.60	2.62	2.56	2.50	2.44	2.39	2.34	2.29	2.25	2.20	2.16

[a]H = hit rate = P("different" | Different).
[b]F = false-alarm rate = P("different" | Same).

TABLE A5.4   Values of d' for Same-Different (Differencing Model)
(cont.)

H^a	F^b									
	0.11	0.12	0.13	0.14	0.15	0.16	0.17	0.18	0.19	0.20
0.61	2.65	2.59	2.53	2.48	2.43	2.38	2.33	2.29	2.24	2.20
0.62	2.69	2.63	2.57	2.52	2.47	2.42	2.37	2.32	2.28	2.24
0.63	2.73	2.67	2.61	2.55	2.50	2.45	2.41	2.36	2.32	2.27
0.64	2.77	2.70	2.65	2.59	2.54	2.49	2.44	2.40	2.35	2.31
0.65	2.80	2.74	2.68	2.63	2.58	2.53	2.48	2.44	2.39	2.35
0.66	2.84	2.78	2.72	2.67	2.62	2.57	2.52	2.48	2.43	2.39
0.67	2.88	2.82	2.76	2.71	2.66	2.61	2.56	2.51	2.47	2.43
0.68	2.92	2.86	2.80	2.75	2.70	2.65	2.60	2.55	2.51	2.47
0.69	2.96	2.90	2.84	2.79	2.74	2.69	2.64	2.59	2.55	2.51
0.70	3.00	2.94	2.88	2.83	2.78	2.73	2.68	2.63	2.59	2.55
0.71	3.04	2.98	2.92	2.87	2.82	2.77	2.72	2.68	2.63	2.59
0.72	3.08	3.02	2.96	2.91	2.86	2.81	2.76	2.72	2.67	2.63
0.73	3.13	3.07	3.01	2.95	2.90	2.85	2.81	2.76	2.72	2.68
0.74	3.17	3.11	3.05	3.00	2.94	2.90	2.85	2.80	2.76	2.72
0.75	3.21	3.15	2.09	3.04	2.99	2.94	2.89	2.85	2.81	2.76
0.76	3.26	3.20	3.14	3.09	3.03	2.98	2.94	2.89	2.85	2.81
0.77	3.30	3.24	3.19	3.13	3.08	3.03	2.98	2.94	2.90	2.86
0.78	3.35	3.29	3.23	2.18	3.13	3.08	3.03	2.99	2.94	2.90
0.79	3.40	3.34	3.28	3.23	3.18	3.13	3.08	3.04	2.99	2.95
0.80	3.45	3.39	3.33	3.28	3.23	3.18	3.13	3.09	3.04	3.00
0.81	3.50	3.44	3.38	3.33	3.28	3.23	3.18	3.14	3.09	3.05
0.82	3.55	3.49	3.44	3.38	3.33	3.28	3.23	3.19	3.15	3.11
0.83	3.61	3.55	3.49	3.44	3.38	3.34	3.29	3.24	3.20	3.16
0.84	3.67	3.61	3.55	3.49	3.44	3.39	3.35	3.30	3.26	3.22
0.85	3.73	3.66	3.61	3.55	3.50	3.45	3.41	3.36	3.32	3.28
0.86	3.79	3.73	3.67	3.61	3.56	3.51	3.47	3.42	3.38	3.34
0.87	3.85	3.79	3.73	3.68	3.63	3.58	3.53	3.49	3.45	3.40
0.88	3.92	3.86	3.80	3.75	3.70	3.65	3.60	3.56	3.51	3.47
0.89	3.99	3.93	3.88	3.82	3.77	3.72	3.67	3.63	3.59	3.55
0.90	4.07	4.01	3.95	3.90	3.85	3.80	3.75	3.71	3.67	3.62

[a]H = hit rate = P("different" | Different).
[b]F = false-alarm rate = P("different" | Same).

TABLE A5.4   *Values of* d′ *for Same-Different (Differencing Model)*
(cont.)

H^a	F^b									
	0.11	0.12	0.13	0.14	0.15	0.16	0.17	0.18	0.19	0.20
0.91	4.16	4.09	4.04	3.98	3.93	3.88	3.84	3.79	3.75	3.71
0.92	4.25	4.19	4.13	4.07	4.02	3.97	3.93	3.88	3.84	3.80
0.93	4.35	4.29	4.23	4.17	4.12	4.07	4.03	3.98	3.94	3.90
0.94	4.46	4.40	4.34	4.29	4.23	4.19	4.14	4.09	4.05	4.01
0.95	4.59	4.52	4.47	4.41	4.36	4.31	4.27	4.22	4.18	4.14
0.96	4.74	4.67	4.62	4.56	4.51	4.46	4.42	4.37	4.33	4.29
0.97	4.92	4.86	4.80	4.75	4.70	4.65	4.60	4.56	4.51	4.47
0.98	5.16	5.10	5.05	4.99	4.94	4.89	4.85	4.80	4.76	4.72
0.99	5.55	5.49	5.43	5.38	5.33	5.28	5.23	5.19	5.14	5.10

H^a	F^b									
	0.21	0.22	0.23	0.24	0.25	0.26	0.27	0.28	0.29	0.30
0.21	0.00									
0.22	0.30	0.00								
0.23	0.42	0.30	0.00							
0.24	0.52	0.42	0.29	0.00						
0.25	0.60	0.51	0.42	0.29	0.00					
0.26	0.67	0.60	0.51	0.41	0.29	0.00				
0.27	0.74	0.67	0.59	0.51	0.41	0.29	0.00			
0.28	0.80	0.73	0.66	0.59	0.51	0.41	0.29	0.00		
0.29	0.86	0.79	0.73	0.66	0.59	0.51	0.41	0.29	0.00	
0.30	0.91	0.85	0.79	0.73	0.66	0.59	0.51	0.41	0.29	0.00
0.31	0.96	0.91	0.85	0.79	0.73	0.66	0.59	0.50	0.41	0.29
0.32	1.01	0.96	0.90	0.85	0.79	0.72	0.66	0.58	0.50	0.41
0.33	1.06	1.01	0.95	0.90	0.84	0.78	0.72	0.66	0.58	0.50
0.34	1.11	1.06	1.00	0.95	0.90	0.84	0.78	0.72	0.65	0.58
0.35	1.15	1.10	1.05	1.00	0.95	0.89	0.84	0.78	0.72	0.65
0.36	1.20	1.15	1.10	1.05	1.00	0.95	0.89	0.84	0.78	0.72
0.37	1.24	1.19	1.14	1.10	1.05	1.00	0.94	0.89	0.84	0.78
0.38	1.28	1.24	1.19	1.14	1.09	1.04	0.99	0.94	0.89	0.84
0.39	1.32	1.28	1.23	1.18	1.14	1.09	1.04	0.99	0.94	0.89
0.40	1.37	1.32	1.27	1.23	1.18	1.14	1.09	1.04	0.99	0.94

$^a H$ = hit rate = P("different" | Different).
$^b F$ = false-alarm rate = P("different" | Same).

TABLE A5.4   *Values of* d' *for Same-Different (Differencing Model)* (cont.)

H^a	F^b									
	0.21	0.22	0.23	0.24	0.25	0.26	0.27	0.28	0.29	0.30
0.41	1.41	1.36	1.32	1.27	1.22	1.18	1.13	1.09	1.04	0.99
0.42	1.45	1.40	1.36	1.31	1.27	1.22	1.18	1.13	1.09	1.04
0.43	1.49	1.44	1.40	1.35	1.31	1.27	1.22	1.18	1.13	1.09
0.44	1.52	1.48	1.44	1.39	1.35	1.31	1.26	1.22	1.18	1.13
0.45	1.56	1.52	1.48	1.43	1.39	1.35	1.31	1.26	1.22	1.18
0.46	1.60	1.56	1.51	1.47	1.13	1.39	1.35	1.31	1.26	1.22
0.47	1.64	1.60	1.55	1.51	1.47	1.43	1.39	1.35	1.31	1.26
0.48	1.68	1.63	1.59	1.55	1.51	1.47	1.43	1.39	1.35	1.31
0.49	1.71	1.67	1.63	1.59	1.55	1.51	1.47	1.43	1.39	1.35
0.50	1.75	1.71	1.67	1.63	1.59	1.55	1.51	1.47	1.43	1.39
0.51	1.79	1.75	1.70	1.66	1.62	1.58	1.55	1.51	1.47	1.43
0.52	1.82	1.78	1.74	1.70	1.66	1.62	1.58	1.55	1.51	1.47
0.53	1.86	1.82	1.78	1.74	1.70	1.66	1.62	1.59	1.55	1.51
0.54	1.90	1.86	1.82	1.78	1.74	1.70	1.66	1.62	1.59	1.55
0.55	1.93	1.89	1.85	1.81	1.78	1.74	1.70	1.66	1.63	1.59
0.56	1.97	1.93	1.89	1.85	1.81	1.78	1.74	1.70	1.66	1.63
0.57	2.01	1.97	1.93	1.89	1.85	1.81	1.78	1.74	1.70	1.67
0.58	2.05	2.01	1.97	1.93	1.89	1.85	1.81	1.78	1.74	1.71
0.59	2.08	2.04	2.00	1.96	1.93	1.89	1.85	1.82	1.78	1.75
0.60	2.12	2.08	2.04	2.00	1.96	1.93	1.89	1.86	1.82	1.78
0.61	2.16	2.12	2.08	2.04	2.00	1.97	1.93	1.89	1.86	1.82
0.62	2.20	2.16	2.12	2.08	2.04	2.00	1.97	1.93	1.90	1.86
0.63	2.23	2.19	2.15	2.12	2.08	2.04	2.01	1.97	1.94	1.90
0.64	2.27	2.23	2.19	2.16	2.12	2.08	2.05	2.01	1.98	1.94
0.65	2.31	2.27	2.23	2.19	2.16	2.12	2.09	2.05	2.02	1.98
0.66	2.35	2.31	2.27	2.23	2.20	2.16	2.13	2.09	2.06	2.02
0.67	2.39	2.35	2.31	2.27	2.24	2.20	2.17	2.13	2.10	2.06
0.68	2.43	2.39	2.35	2.31	2.28	2.24	2.21	2.17	2.14	2.10
0.69	2.47	2.43	2.39	2.35	2.32	2.28	2.25	2.21	2.18	2.15
0.70	2.51	2.47	2.43	2.39	2.36	2.32	2.29	2.25	2.22	2.19

^{a}H = hit rate = P("different" | Different).
^{b}F = false-alarm rate = P("different" | Same).

TABLE A5.4   *Values of* d′ *for Same-Different (Differencing Model)* (cont.)

H^a	F^b									
	0.21	0.22	0.23	0.24	0.25	0.26	0.27	0.28	0.29	0.30
0.71	2.55	2.51	2.47	2.44	2.40	2.36	2.33	2.30	2.26	2.23
0.72	2.59	2.55	2.52	2.48	2.44	2.41	2.37	2.34	2.31	2.27
0.73	2.64	2.60	2.56	2.52	2.49	2.45	2.42	2.38	2.35	2.32
0.74	2.68	2.64	2.60	2.57	2.53	2.49	2.46	2.43	2.39	2.36
0.75	2.72	2.68	2.65	2.61	2.57	2.54	2.50	2.47	2.44	2.41
0.76	2.77	2.73	2.69	2.66	2.62	2.58	2.55	2.52	2.48	2.45
0.77	2.81	2.78	2.74	2.70	2.67	2.63	2.60	2.56	2.53	2.50
0.78	2.86	2.82	2.79	2.75	2.71	2.68	2.64	2.61	2.58	2.55
0.79	2.91	2.87	2.83	2.80	2.76	2.73	2.69	2.66	2.63	2.60
0.80	2.96	2.92	2.88	2.85	2.81	2.78	2.74	2.71	2.68	2.65
0.81	3.01	2.97	2.94	2.90	2.86	2.83	2.80	2.76	2.73	2.70
0.82	3.07	3.03	2.99	2.95	2.92	2.88	2.85	2.82	2.78	2.75
0.83	3.12	3.08	3.04	3.01	2.97	2.94	2.90	2.87	2.84	2.81
0.84	3.18	3.14	3.10	3.07	3.03	2.00	2.96	2.93	2.90	2.87
0.85	3.24	3.20	3.16	3.13	3.09	3.06	3.02	2.99	2.96	2.93
0.86	3.30	3.26	3.22	3.19	3.15	3.12	3.08	3.05	3.02	2.99
0.87	3.36	3.33	3.29	3.25	3.22	3.18	3.15	3.12	3.09	3.05
0.88	3.43	3.40	3.36	3.32	3.29	3.25	3.22	3.19	3.15	3.12
0.89	3.51	3.47	3.43	3.39	3.36	3.33	3.29	3.26	3.23	3.20
0.90	3.58	3.55	3.51	3.47	3.44	3.40	3.37	3.34	3.31	3.27
0.91	3.67	3.63	3.59	3.56	3.52	3.49	3.45	3.42	3.39	3.36
0.92	3.76	3.72	3.68	3.65	3.61	3.58	3.55	3.51	3.48	3.45
0.93	3.86	3.82	3.78	3.75	3.71	3.68	3.65	3.61	3.58	3.55
0.94	3.97	3.93	3.90	3.86	3.82	3.79	3.76	3.73	3.69	3.66
0.95	4.10	4.06	4.02	3.99	3.95	3.92	3.89	3.85	3.82	3.79
0.96	4.25	4.21	4.17	4.14	4.10	4.07	4.04	4.00	3.97	3.94
0.97	4.43	4.39	4.36	4.32	4.29	4.25	4.22	4.19	4.16	4.12
0.98	4.68	4.64	4.60	4.57	4.53	4.50	4.46	4.43	4.40	4.37
0.99	5.06	5.02	4.99	4.95	4.92	4.88	4.85	4.82	4.79	4.76

$^a H$ = hit rate = P("different" | Different).
$^b F$ = false-alarm rate = P("different" | Same).

TABLE A5.4   *Values of* d′ *for Same-Different (Differencing Model)* *(cont.)*

H^a	F^b									
	0.31	0.32	0.33	0.34	0.35	0.36	0.37	0.38	0.39	0.40
0.31	0.00									
0.32	0.29	0.00								
0.33	0.41	0.29	0.00							
0.34	0.50	0.41	0.29	0.00						
0.35	0.58	0.50	0.41	0.29	0.00					
0.36	0.65	0.58	0.50	0.41	0.29	0.00				
0.37	0.72	0.65	0.58	0.50	0.41	0.29	0.00			
0.38	0.78	0.72	0.65	0.58	0.50	0.41	0.29	0.00		
0.39	0.84	0.78	0.72	0.66	0.59	0.51	0.41	0.29	0.00	
0.40	0.89	0.84	0.78	0.72	0.66	0.59	0.51	0.41	0.29	0.00
0.41	0.94	0.89	0.84	0.78	0.72	0.66	0.59	0.51	0.41	0.29
0.42	0.99	0.94	0.89	0.84	0.78	0.72	0.66	0.59	0.51	0.42
0.43	1.04	0.99	0.94	0.89	0.84	0.78	0.73	0.66	0.59	0.51
0.44	1.09	1.04	0.99	0.95	0.90	0.84	0.79	0.73	0.66	0.59
0.45	1.13	1.09	1.04	1.01	0.95	0.90	0.84	0.79	0.73	0.67
0.46	1.18	1.13	1.09	1.04	1.00	0.95	0.90	0.85	0.79	0.73
0.47	1.22	1.18	1.14	1.09	1.05	1.00	0.95	0.90	0.85	0.79
0.48	1.26	1.22	1.18	1.14	1.09	1.05	1.00	0.96	0.91	0.85
0.49	1.31	1.27	1.23	1.18	1.14	1.10	1.05	1.01	0.96	0.91
0.50	1.35	1.31	1.27	1.23	1.19	1.14	1.10	1.06	1.01	0.96
0.51	1.39	1.35	1.31	1.27	1.23	1.19	1.15	1.10	1.06	1.01
0.52	1.43	1.39	1.35	1.31	1.28	1.23	1.19	1.15	1.11	1.06
0.53	1.47	1.43	1.40	1.36	1.32	1.28	1.24	1.20	1.16	1.11
0.54	1.51	1.47	1.44	1.40	1.36	1.32	1.28	1.24	1.20	1.16
0.55	1.55	1.51	1.48	1.44	1.40	1.37	1.33	1.29	1.25	1.21
0.56	1.59	1.56	1.52	1.48	1.45	1.41	1.37	1.33	1.29	1.25
0.57	1.63	1.60	1.56	1.52	1.49	1.45	1.41	1.38	1.34	1.30
0.58	1.67	1.64	1.60	1.56	1.53	1.49	1.46	1.42	1.38	1.34
0.59	1.71	1.68	1.64	1.60	1.57	1.53	1.50	1.46	1.43	1.39
0.60	1.75	1.71	1.68	1.65	1.61	1.58	1.54	1.51	1.47	1.43

$^a H$ = hit rate = P("different" | Different).
$^b F$ = false-alarm rate = P("different" | Same).

TABLE A5.4   *Values of* d′ *for Same-Different (Differencing Model) (cont.)*

H^a	F^b									
	0.31	0.32	0.33	0.34	0.35	0.36	0.37	0.38	0.39	0.40
0.61	1.79	1.75	1.73	1.69	1.65	1.62	1.58	1.55	1.51	1.48
0.62	1.83	1.79	1.76	1.73	1.69	1.66	1.62	1.59	1.56	1.52
0.63	1.87	1.83	1.80	1.77	1.73	1.70	1.67	1.63	1.60	1.56
0.64	1.91	1.88	1.84	1.81	1.78	1.74	1.71	1.68	1.64	1.61
0.65	1.95	1.92	1.88	1.85	1.82	1.78	1.75	1.72	1.68	1.65
0.66	1.99	1.96	1.92	1.89	1.86	1.83	1.79	1.76	1.73	1.69
0.67	2.03	2.00	1.96	1.93	1.90	1.87	1.84	1.80	1.77	1.74
0.68	2.07	2.04	2.01	1.97	1.94	1.91	1.88	1.85	1.81	1.78
0.69	2.11	2.08	2.05	2.02	1.98	1.95	1.92	1.89	1.86	1.83
0.70	2.15	2.12	2.09	2.06	2.03	2.00	1.96	1.93	1.90	1.87
0.71	2.20	2.17	2.13	2.10	2.07	2.04	2.01	1.98	1.95	1.92
0.72	2.24	2.21	2.18	2.15	2.11	2.08	2.05	2.02	1.99	1.96
0.73	2.28	2.25	2.22	2.19	2.16	2.13	2.10	2.07	2.04	2.01
0.74	2.33	2.30	2.27	2.23	2.20	2.17	2.14	2.11	2.08	2.05
0.75	2.37	2.34	2.31	2.28	2.25	2.22	2.19	2.16	2.13	2.10
0.76	2.42	2.39	2.36	2.33	2.30	2.27	2.24	2.21	2.18	2.15
0.77	2.47	2.44	2.41	2.37	2.34	2.31	2.28	2.26	2.23	2.20
0.78	2.52	2.48	2.45	2.42	2.39	2.36	2.33	2.30	2.28	2.25
0.79	2.56	2.53	2.50	2.47	2.44	2.41	2.38	2.36	2.33	2.30
0.80	2.62	2.58	2.55	2.52	2.49	2.47	2.44	2.41	2.38	2.35
0.81	2.67	2.64	2.61	2.58	2.55	2.52	2.49	2.46	2.43	2.40
0.82	2.72	2.69	2.66	2.63	2.60	2.57	2.54	2.51	2.49	2.46
0.83	2.78	2.75	2.72	2.69	2.66	2.63	2.60	2.57	2.54	2.52
0.84	2.83	2.80	2.77	2.74	2.72	2.69	2.66	2.63	2.60	2.57
0.85	2.89	2.86	2.83	2.81	2.78	2.75	2.72	2.69	2.66	2.64
0.86	2.96	2.93	2.90	2.87	2.84	2.81	2.78	2.75	2.73	2.70
0.87	3.02	2.99	2.96	2.93	2.91	2.88	2.85	2.82	2.79	2.77
0.88	3.09	3.06	3.03	3.00	2.98	2.95	2.92	2.89	2.86	2.84
0.89	3.17	3.14	3.11	3.08	3.05	3.02	2.99	2.97	2.94	2.91
0.90	3.24	3.21	3.18	3.16	3.13	3.10	3.07	3.04	3.02	2.99

[a]H = hit rate = P("different" | Different).
[b]F = false-alarm rate = P("different" | Same).

TABLE A5.4   *Values of* d′ *for Same-Different (Differencing Model)*
(cont.)

H^a	F^b 0.31	0.32	0.33	0.34	0.35	0.36	0.37	0.38	0.39	0.40
0.91	3.33	3.30	3.27	3.24	3.21	3.18	3.16	3.13	3.10	3.08
0.92	3.42	3.39	3.36	3.33	3.30	3.28	3.25	3.22	3.19	3.17
0.93	3.52	3.49	3.46	3.43	3.40	3.38	3.35	3.32	3.30	3.27
0.94	3.63	3.60	3.57	3.54	3.52	3.49	3.46	3.43	3.41	3.38
0.95	3.76	3.73	3.70	3.67	3.64	3.62	3.59	3.56	3.54	3.51
0.96	3.91	3.88	3.85	3.82	3.80	3.77	3.74	3.71	3.69	3.66
0.97	4.09	4.07	4.04	4.01	3.98	3.95	3.93	3.90	3.87	3.85
0.98	4.34	4.31	4.28	4.25	4.22	4.20	4.17	4.14	4.12	4.09
0.99	4.73	4.70	4.67	4.64	4.61	4.58	4.56	4.53	4.50	4.48

H^a	F^b 0.41	0.42	0.43	0.44	0.45	0.46	0.47	0.48	0.49	0.50
0.41	0.00									
0.42	0.29	0.00								
0.43	0.42	0.29	0.00							
0.44	0.51	0.42	0.30	0.00						
0.45	0.59	0.51	0.42	0.30	0.00					
0.46	0.67	0.60	0.52	0.42	0.30	0.00				
0.47	0.73	0.67	0.60	0.52	0.42	0.30	0.00			
0.48	0.80	0.74	0.67	0.60	0.52	0.43	0.30	0.00		
0.49	0.86	0.80	0.74	0.68	0.61	0.52	0.43	0.30	0.00	
0.50	0.91	0.86	0.80	0.74	0.68	0.61	0.53	0.43	0.31	0.00
0.51	0.97	0.92	0.86	0.81	0.75	0.68	0.61	0.53	0.43	0.31
0.52	1.07	0.97	0.92	0.87	0.81	0.75	0.69	0.62	0.53	0.44
0.53	1.07	1.02	0.98	0.93	0.87	0.82	0.76	0.69	0.62	0.54
0.54	1.12	1.07	1.03	0.98	0.93	0.88	0.82	0.76	0.70	0.62
0.55	1.17	1.12	1.08	1.03	0.99	0.94	0.88	0.83	0.77	0.70
0.56	1.21	1.17	1.13	1.09	1.04	0.99	0.94	0.88	0.83	0.77
0.57	1.26	1.22	1.18	1.14	1.09	1.05	1.00	0.95	0.90	0.84
0.58	1.31	1.27	1.23	1.19	1.14	1.10	1.05	1.01	0.95	0.90
0.59	1.35	1.31	1.27	1.23	1.19	1.15	1.11	1.06	1.01	0.96
0.60	1.40	1.36	1.32	1.28	1.24	1.20	1.16	1.11	1.07	1.02

[a]H = hit rate = P("different" | Different).
[b]F = false-alarm rate = P("different" | Same).

TABLE A5.4   *Values of* d′ *for Same-Different (Differencing Model)* (cont.)

H^a	F^b									
	0.41	0.42	0.43	0.44	0.45	0.46	0.47	0.48	0.49	0.50
0.61	1.44	1.40	1.37	1.33	1.29	1.25	1.21	1.17	1.12	1.08
0.62	1.49	1.45	1.41	1.38	1.34	1.30	1.26	1.22	1.18	1.13
0.63	1.53	1.49	1.46	1.42	1.39	1.35	1.31	1.27	1.23	1.18
0.64	1.57	1.54	1.50	1.47	1.43	1.39	1.36	1.32	1.28	1.24
0.65	1.62	1.58	1.55	1.51	1.48	1.44	1.41	1.37	1.33	1.29
0.66	1.66	1.63	1.59	1.56	1.53	1.49	1.45	1.42	1.38	1.34
0.67	1.71	1.67	1.64	1.61	1.57	1.54	1.50	1.47	1.43	1.39
0.68	1.75	1.72	1.68	1.65	1.62	1.58	1.55	1.51	1.48	1.44
0.69	1.79	1.76	1.73	1.70	1.66	1.63	1.60	1.56	1.53	1.49
0.70	1.84	1.81	1.78	1.74	1.71	1.68	1.65	1.61	1.58	1.54
0.71	1.88	1.85	1.82	1.79	1.76	1.73	1.69	1.66	1.63	1.59
0.72	1.93	1.90	1.87	1.84	1.81	1.77	1.74	1.71	1.68	1.64
0.73	1.98	1.95	1.92	1.88	1.85	1.82	1.79	1.76	1.73	1.69
0.74	2.02	1.99	1.96	1.93	1.90	1.87	1.84	1.81	1.78	1.74
0.75	2.07	2.04	2.01	1.98	1.95	1.92	1.89	1.86	1.83	1.80
0.76	2.12	2.09	2.06	2.03	2.00	1.97	1.94	1.91	1.88	1.85
0.77	2.17	2.14	2.11	2.08	2.05	2.02	1.99	1.96	1.93	1.90
0.78	2.22	2.19	2.16	2.13	2.10	2.07	2.04	2.01	1.98	1.95
0.79	2.27	2.24	2.21	2.18	2.15	2.13	2.10	2.07	2.04	2.01
0.80	2.32	2.29	2.26	2.24	2.21	2.18	2.15	2.12	2.09	2.06
0.81	2.37	2.35	2.32	2.29	2.26	2.23	2.21	2.18	2.15	2.12
0.82	2.43	2.40	2.37	2.35	2.32	2.29	2.26	2.23	2.21	2.18
0.83	2.49	2.46	2.43	2.40	2.38	2.35	2.32	2.29	2.27	2.24
0.84	2.55	2.52	2.49	2.46	2.44	2.41	2.38	2.35	2.33	2.30
0.85	2.61	2.58	2.55	2.53	2.50	2.47	2.44	2.42	2.39	2.36
0.86	2.67	2.64	2.62	2.59	2.56	2.54	2.51	2.48	2.46	2.43
0.87	2.74	2.71	2.69	2.66	2.63	2.61	2.58	2.55	2.53	2.50
0.88	2.81	2.78	2.76	2.73	2.70	2.68	2.65	2.62	2.60	2.57
0.89	2.96	2.86	2.83	2.80	2.78	2.75	2.73	2.70	2.67	2.65
0.90	2.96	2.94	2.91	2.88	2.86	2.83	2.81	2.78	2.76	2.73

$^a H$ = hit rate = P("different" | Different).
$^b F$ = false-alarm rate = P("different" | Same).

TABLE A5.4   *Values of* d′ *for Same-Different (Differencing Model) (cont.)*

H^a	F^b									
	0.41	0.42	0.43	0.44	0.45	0.46	0.47	0.48	0.49	0.50
0.91	3.05	3.02	3.00	2.97	2.94	2.92	2.89	2.87	2.84	2.82
0.92	3.14	3.12	3.09	3.06	3.04	3.01	2.99	2.96	2.94	2.91
0.93	3.24	3.22	3.19	3.17	3.14	3.11	3.09	3.06	3.04	3.01
0.94	3.36	3.33	3.30	3.28	3.25	3.23	3.20	3.18	3.15	3.13
0.95	3.48	3.46	3.43	3.41	3.38	3.36	3.33	3.31	3.28	3.26
0.96	3.64	3.61	3.58	3.56	3.53	3.51	3.49	3.46	3.44	3.41
0.97	3.82	3.80	3.77	3.75	3.72	3.70	3.67	3.65	3.62	3.60
0.98	4.07	4.04	4.02	3.99	3.97	3.94	3.92	3.90	3.87	3.85
0.99	4.45	4.43	4.40	4.38	4.35	4.33	4.31	4.28	4.26	4.24

H^a	F^b									
	0.51	0.52	0.53	0.54	0.55	0.56	0.57	0.58	0.59	0.60
0.51	0.00									
0.52	0.31	0.00								
0.53	0.44	0.31	0.00							
0.54	0.54	0.44	0.31	0.00						
0.55	0.63	0.54	0.45	0.32	0.00					
0.56	0.71	0.63	0.55	0.45	0.32	0.00				
0.57	0.78	0.71	0.64	0.55	0.45	0.32	0.00			
0.58	0.84	0.78	0.72	0.64	0.56	0.46	0.32	0.00		
0.59	0.91	0.85	0.79	0.72	0.65	0.56	0.46	0.33	0.00	
0.60	0.97	0.92	0.86	0.80	0.73	0.65	0.57	0.47	0.33	0.00
0.61	1.03	0.98	0.92	0.87	0.80	0.74	0.66	0.57	0.47	0.33
0.62	1.08	1.04	0.99	0.93	0.87	0.81	0.74	0.67	0.58	0.47
0.63	1.14	1.09	1.05	0.99	0.94	0.88	0.82	0.75	0.67	0.58
0.64	1.19	1.15	1.10	1.05	1.00	0.95	0.89	0.83	0.76	0.68
0.65	1.25	1.20	1.16	1.11	1.06	1.01	0.96	0.90	0.84	0.77
0.66	1.30	1.26	1.22	1.17	1.12	1.07	1.02	0.97	0.91	0.84
0.67	1.35	1.31	1.27	1.23	1.18	1.14	1.09	1.03	0.98	0.92
0.69	1.45	1.42	1.38	1.34	1.30	1.25	1.21	1.16	1.11	1.06
0.70	1.51	1.47	1.43	1.39	1.35	1.31	1.27	1.22	1.17	1.12

$^a H$ = hit rate = P("different" | Different).
$^b F$ = false-alarm rate = P("different" | Same).

TABLE A5.4   *Values of* d′ *for Same-Different (Differencing Model)* *(cont.)*

H^a	F^b 0.51	0.52	0.53	0.54	0.55	0.56	0.57	0.58	0.59	0.60
0.71	1.56	1.52	1.48	1.45	1.41	1.37	1.32	1.28	1.24	1.19
0.72	1.61	1.57	1.54	1.50	1.46	1.42	1.38	1.34	1.30	1.25
0.73	1.66	1.62	1.59	1.55	1.52	1.48	1.44	1.40	1.36	1.31
0.74	1.71	1.68	1.64	1.61	1.57	1.54	1.50	1.46	1.42	1.37
0.75	1.76	1.73	1.70	1.66	1.63	1.59	1.55	1.52	1.48	1.44
0.76	1.82	1.78	1.75	1.72	1.68	1.65	1.61	1.57	1.54	1.50
0.77	1.87	1.84	1.80	1.71	1.74	1.70	1.67	1.63	1.60	1.56
0.78	1.92	1.89	1.86	1.83	1.80	1.76	1.73	1.69	1.66	1.62
0.79	1.98	1.95	1.92	1.88	1.85	1.82	1.79	1.75	1.72	1.68
0.80	2.03	2.00	1.97	1.94	1.91	1.88	1.85	1.81	1.78	1.74
0.81	2.09	2.06	2.03	2.00	1.97	1.94	1.91	1.87	1.84	1.81
0.82	2.15	2.12	2.09	2.06	2.03	2.00	1.97	1.94	1.90	1.87
0.83	2.21	2.18	2.15	2.12	2.09	2.06	2.03	2.00	1.97	1.94
0.84	2.27	2.24	2.21	2.19	2.16	2.13	2.10	2.07	2.04	2.00
0.85	2.34	2.31	2.28	2.25	2.22	2.19	2.16	2.13	2.10	2.07
0.86	2.40	2.37	2.35	2.32	2.29	2.26	2.23	2.20	2.17	2.14
0.87	2.47	2.45	2.42	2.39	2.36	2.33	2.31	2.28	2.25	2.22
0.88	2.55	2.52	2.49	2.46	2.44	2.41	2.38	2.35	2.33	2.30
0.89	2.62	2.60	2.57	2.54	2.52	2.49	2.46	2.43	2.41	2.38
0.90	2.70	2.68	2.65	2.63	2.60	2.57	2.55	2.52	2.49	2.46
0.91	2.79	2.77	2.74	2.71	2.69	2.66	2.64	2.61	2.58	2.56
0.92	2.89	2.86	2.84	2.81	2.78	2.76	2.73	2.71	2.68	2.65
0.93	2.99	2.97	2.94	2.92	2.89	2.86	2.84	2.81	2.79	2.76
0.94	3.11	3.08	3.06	3.03	3.01	2.98	2.96	2.93	2.91	2.88
0.95	3.24	3.21	3.19	3.16	3.14	3.11	3.09	3.07	3.04	3.02
0.96	3.39	3.37	3.34	3.32	3.29	3.27	3.25	3.22	3.20	3.17
0.97	3.58	3.55	3.53	3.51	3.48	3.46	3.44	3.41	3.39	3.36
0.98	3.83	3.80	3.78	3.76	3.73	3.71	3.69	3.66	3.64	3.62
0.99	4.21	4.19	4.17	4.15	4.12	4.10	4.08	4.06	4.03	4.01

[a]H = hit rate = P("different" | Different).
[b]F = false-alarm rate = P("different" | Same).

TABLE A5.4   *Values of* d' *for Same-Different (Differencing Model) (cont.)*

H^a	F^b									
	0.61	0.62	0.63	0.64	0.65	0.66	0.67	0.68	0.69	0.70
0.61	0.00									
0.62	0.34	0.00								
0.63	0.48	0.34	0.00							
0.64	0.59	0.48	0.34	0.00						
0.65	0.69	0.60	0.49	0.35	0.00					
0.66	0.77	0.70	0.60	0.50	0.35	0.00				
0.67	0.85	0.78	0.70	0.61	0.50	0.36	0.00			
0.68	0.93	0.86	0.79	0.71	0.62	0.51	0.36	0.00		
0.69	1.00	0.94	0.87	0.80	0.72	0.63	0.52	0.37	0.00	
0.70	1.07	1.01	0.95	0.89	0.81	0.73	0.64	0.52	0.37	0.00
0.71	1.14	1.08	1.03	0.96	0.90	0.82	0.74	0.65	0.53	0.38
0.72	1.20	1.15	1.10	1.04	0.98	0.91	0.84	0.75	0.66	0.54
0.73	1.27	1.22	1.17	1.11	1.05	0.99	0.92	0.85	0.76	0.67
0.74	1.33	1.28	1.23	1.18	1.13	1.07	1.01	0.94	0.86	0.78
0.75	1.39	1.35	1.30	1.25	1.20	1.15	1.09	1.02	0.95	0.88
0.76	1.46	1.41	1.37	1.32	1.27	1.22	1.16	1.10	1.04	0.97
0.77	1.52	1.48	1.43	1.39	1.34	1.29	1.24	1.18	1.12	1.06
0.78	1.58	1.54	1.50	1.46	1.41	1.36	1.31	1.26	1.20	1.14
0.79	1.64	1.60	1.56	1.52	1.48	1.43	1.39	1.34	1.28	1.22
0.80	1.71	1.67	1.63	1.59	1.55	1.50	1.46	1.41	1.36	1.31
0.81	1.77	1.73	1.70	1.66	1.62	1.58	1.53	1.49	1.44	1.39
0.82	1.84	1.80	1.76	1.73	1.69	1.65	1.60	1.56	1.51	1.47
0.83	1.90	1.87	1.83	1.80	1.76	1.72	1.68	1.64	1.59	1.55
0.84	1.97	1.94	1.90	1.87	1.83	1.79	1.75	1.71	1.67	1.63
0.85	2.04	2.01	1.97	1.94	1.90	1.87	1.83	1.79	1.75	1.71
0.86	2.11	2.08	2.05	2.02	1.98	1.95	1.91	1.87	1.83	1.79
0.87	2.19	2.16	2.13	2.09	2.06	2.02	1.99	1.95	1.91	1.87
0.88	2.27	2.24	2.21	2.17	2.14	2.11	2.07	2.04	2.00	1.96
0.89	2.35	2.32	2.29	2.26	2.23	2.19	2.16	2.13	2.09	2.05
0.90	2.43	2.41	2.38	2.35	2.32	2.28	2.25	2.22	2.18	2.15

[a] H = hit rate = P("different" | Different).
[b] F = false-alarm rate = P("different" | Same).

TABLE A5.4   *Values of d′ for Same-Different (Differencing Model) (cont.)*

H^a	F^b 0.61	0.62	0.63	0.64	0.65	0.66	0.67	0.68	0.69	0.70
0.91	2.53	2.50	2.47	2.44	2.41	2.38	2.35	2.32	2.28	2.25
0.92	2.63	2.60	2.57	2.54	2.51	2.48	2.45	2.42	2.39	2.36
0.93	2.73	2.71	2.68	2.65	2.62	2.60	2.57	2.54	2.50	2.47
0.94	2.85	2.83	2.80	2.77	2.75	2.72	2.69	2.66	2.63	2.60
0.95	2.99	2.96	2.94	2.91	2.89	2.86	2.83	2.80	2.77	2.74
0.96	3.15	3.12	3.10	3.07	3.05	3.02	2.99	2.97	2.94	2.91
0.97	3.34	3.32	3.29	3.27	3.24	3.22	3.19	3.16	3.14	3.11
0.98	3.59	3.57	3.55	3.52	3.50	3.47	3.45	3.42	3.40	3.37
0.99	3.99	3.97	3.94	3.92	3.90	3.87	3.85	3.83	3.80	3.78

H^a	F^b 0.71	0.72	0.73	0.74	0.75	0.76	0.77	0.78	0.79	0.80
0.71	0.00									
0.72	0.38	0.00								
0.73	0.55	0.39	0.00							
0.74	0.68	0.56	0.40	0.00						
0.75	0.79	0.69	0.57	0.40	0.00					
0.76	0.89	0.80	0.70	0.58	0.41	0.00				
0.77	0.99	0.91	0.82	0.71	0.59	0.42	0.00			
0.78	1.08	1.00	0.92	0.83	0.73	0.60	0.43	0.00		
0.79	1.16	1.10	1.02	0.94	0.85	0.74	0.61	0.44	0.00	
0.80	1.25	1.19	1.12	1.04	0.96	0.87	0.76	0.63	0.45	0.00
0.81	1.33	1.27	1.21	1.14	1.07	0.98	0.89	0.78	0.64	0.46
0.82	1.41	1.36	1.30	1.24	1.17	1.09	1.00	0.91	0.79	0.66
0.83	1.50	1.44	1.39	1.33	1.26	1.19	1.12	1.03	0.93	0.81
0.84	1.58	1.53	1.48	1.42	1.36	1.29	1.22	1.14	1.06	0.95
0.85	1.66	1.61	1.56	1.51	1.45	1.39	1.33	1.25	1.17	1.08
0.86	1.75	1.70	1.65	1.60	1.55	1.49	1.43	1.36	1.29	1.21
0.87	1.83	1.79	1.74	1.70	1.64	1.59	1.53	1.47	1.40	1.33
0.88	1.92	1.88	1.84	1.79	1.74	1.69	1.64	1.58	1.51	1.44
0.89	2.01	1.97	1.93	1.89	1.84	1.79	1.74	1.69	1.63	1.56
0.90	2.11	2.07	2.03	1.99	1.95	1.90	1.85	1.80	1.74	1.68

$^a H$ = hit rate = P("different" | Different).
$^b F$ = false-alarm rate = P("different" | Same).

TABLE A5.4   *Values of* d′ *for Same-Different (Differencing Model)*
(cont.)

H^a	F^b 0.71	0.72	0.73	0.74	0.75	0.76	0.77	0.78	0.79	0.80
0.91	2.21	2.18	2.14	2.10	2.06	2.01	1.97	1.92	1.86	1.81
0.92	2.32	2.29	2.25	2.21	2.17	2.13	2.08	2.04	1.99	1.94
0.93	2.44	2.41	2.37	2.33	2.29	2.25	2.21	2.17	2.12	2.07
0.94	2.57	2.54	2.50	2.47	2.43	2.39	2.35	2.31	2.27	2.22
0.95	2.71	2.68	2.65	2.62	2.58	2.54	2.51	2.47	2.42	2.38
0.96	2.88	2.85	2.82	2.79	2.75	2.72	2.68	2.65	2.61	2.56
0.97	3.08	3.05	3.02	2.99	2.96	2.93	2.89	2.86	2.82	2.78
0.98	3.35	3.32	3.29	3.26	3.23	3.20	3.17	3.14	3.10	3.07
0.99	3.75	3.73	3.70	3.68	3.65	3.62	3.59	3.56	3.53	3.50

H^a	F^b 0.81	0.82	0.83	0.84	0.85	0.86	0.87	0.88	0.89	0.90
0.81	0.00									
0.82	0.47	0.00								
0.83	0.67	0.48	0.00							
0.84	0.84	0.69	0.50	0.00						
0.85	0.98	0.86	0.71	0.51	0.00					
0.86	1.12	1.01	0.89	0.74	0.53	0.00				
0.87	1.24	1.15	1.04	0.92	0.75	0.55	0.00			
0.88	1.37	1.28	1.19	1.08	0.95	0.79	0.57	0.00		
0.89	1.49	1.42	1.33	1.23	1.12	0.99	0.82	0.59	0.00	
0.90	1.62	1.55	1.47	1.38	1.28	1.17	1.03	0.86	0.62	0.00

[a]H = hit rate = P("different" I Different).
[b]F = false-alarm rate = P("different" I Same).

TABLE A5.4 *Values of d′ for Same-Different (Differencing Model)*
(cont.)

H^a	F^b									
	0.81	0.82	0.83	0.84	0.85	0.86	0.87	0.88	0.89	0.90
0.91	1.75	1.68	1.61	1.53	1.44	1.34	1.22	1.08	0.90	0.65
0.92	1.88	1.82	1.75	1.68	1.60	1.50	1.40	1.28	1.13	0.95
0.93	2.02	1.96	1.90	1.83	1.76	1.67	1.58	1.47	1.35	1.20
0.94	2.17	2.11	2.06	1.99	1.93	1.85	1.77	1.67	1.56	1.43
0.95	2.33	2.28	2.23	2.17	2.11	2.04	1.96	1.88	1.78	1.67
0.96	2.52	2.47	2.42	2.37	2.31	2.25	2.18	2.10	2.02	1.92
0.97	2.74	2.70	2.65	2.61	2.55	2.50	2.43	2.36	2.29	2.20
0.98	3.03	2.99	2.95	2.90	2.86	2.80	2.75	2.69	2.62	2.54
0.99	3.46	3.43	3.39	3.35	3.31	3.27	3.22	3.16	3.11	3.04

H^a	F^b								
	0.91	0.92	0.93	0.94	0.95	0.96	0.97	0.98	0.99
0.91	0.00								
0.92	0.69	0.00							
0.93	1.00	0.73	0.00						
0.94	1.28	1.07	0.79	0.00					
0.95	1.54	1.37	1.16	0.85	0.00				
0.96	1.80	1.67	1.50	1.27	0.95	0.00			
0.97	2.10	1.98	1.84	1.67	1.43	1.07	0.00		
0.98	2.46	2.36	2.24	2.10	1.92	1.67	1.27	0.00	
0.99	2.97	2.89	2.79	2.68	2.54	2.36	2.10	1.67	0.00

[a]H = hit rate = P("different" | Different).
[b]F = false-alarm rate = P("different" | Same).

Source: Adapted from Kaplan et al. (1978) by permission of The Psychonomic Society, Inc.
Hit and false-alarm rates, defined here in terms of the "different" response, were defined by Kaplan et al.
in terms of the "same" response. In addition, some entries are slightly different because of an improved
algorithm.

TABLE A5.5   Values of d' for Oddity, Gaussian Model (M = Number of Intervals).

p(c)	M												
	3	4	5	6	7	8	9	10	11	12	16	24	32
0.01	–	–	–	–	–	–	–	–	–	–	–	–	–
0.02	–	–	–	–	–	–	–	–	–	–	–	–	–
0.03	–	–	–	–	–	–	–	–	–	–	–	–	–
0.04	–	–	–	–	–	–	–	–	–	–	–	–	–
0.05	–	–	–	–	–	–	–	–	–	–	–	0.32	0.51
0.06	–	–	–	–	–	–	–	–	–	–	–	0.46	0.62
0.07	–	–	–	–	–	–	–	–	–	–	0.27	0.58	0.72
0.08	–	–	–	–	–	–	–	–	–	–	0.41	0.67	0.80
0.09	–	–	–	–	–	–	–	–	–	0.24	0.51	0.75	0.88
0.10	–	–	–	–	–	–	–	0.00	0.28	0.38	0.60	0.82	0.95
0.11	–	–	–	–	–	–	–	0.29	0.40	0.48	0.68	0.88	1.01
0.12	–	–	–	–	–	–	0.27	0.40	0.50	0.56	0.74	0.94	1.07
0.13	–	–	–	–	–	0.20	0.39	0.49	0.57	0.63	0.81	1.00	1.12
0.14	–	–	–	–	–	0.34	0.48	0.57	0.64	0.70	0.86	1.05	1.17
0.15	–	–	–	–	0.23	0.44	0.56	0.64	0.71	0.76	0.92	1.10	1.22
0.16	–	–	–	–	0.36	0.52	0.63	0.70	0.76	0.82	0.97	1.15	1.27
0.17	–	–	–	0.17	0.46	0.59	0.69	0.76	0.82	0.87	1.02	1.20	1.31
0.18	–	–	–	0.32	0.53	0.66	0.75	0.81	0.87	0.92	1.06	1.24	1.35
0.19	–	–	–	0.42	0.60	0.72	0.80	0.87	0.92	0.96	1.11	1.28	1.40
0.20	–	–	0.00	0.51	0.67	0.77	0.85	0.91	0.97	1.01	1.15	1.32	1.44
0.21	–	–	0.28	0.58	0.72	0.82	0.90	0.96	1.01	1.05	1.19	1.36	1.47
0.22	–	–	0.40	0.64	0.78	0.87	0.94	1.00	1.05	1.10	1.23	1.40	1.51
0.23	–	–	0.49	0.70	0.83	0.92	0.99	1.04	1.09	1.14	1.27	1.44	1.55
0.24	–	–	0.56	0.76	0.88	0.96	1.03	1.09	1.13	1.18	1.31	1.47	1.59
0.25	–	0.00	0.63	0.81	0.92	1.01	1.07	1.13	1.17	1.21	1.34	1.51	1.62
0.26	–	0.29	0.69	0.86	0.97	1.05	1.11	1.16	1.21	1.25	1.38	1.54	1.65
0.27	–	0.42	0.75	0.91	1.01	1.09	1.15	1.20	1.25	1.29	1.41	1.58	1.69
0.28	–	0.51	0.80	0.95	1.05	1.13	1.19	1.24	1.28	1.32	1.45	1.61	1.72
0.29	–	0.59	0.86	1.00	1.09	1.17	1.23	1.28	1.32	1.36	1.48	1.64	1.75
0.30	–	0.66	0.90	1.04	1.13	1.20	1.26	1.31	1.36	1.39	1.52	1.68	1.79

TABLE A5.5   *Values of* d' *for Oddity, Gaussian Model*
(M = Number of Intervals)   *(cont.)*

						M							
p(c)	3	4	5	6	7	8	9	10	11	12	16	24	32
0.31	–	0.72	0.95	1.08	1.17	1.24	1.30	1.35	1.39	1.43	1.55	1.71	1.82
0.32	–	0.79	1.00	1.12	1.21	1.28	1.33	1.38	1.42	1.46	1.58	1.74	1.85
0.33	0.00	0.84	1.04	1.16	1.24	1.31	1.37	1.42	1.46	1.50	1.61	1.77	1.88
0.34	0.27	0.90	1.09	1.20	1.28	1.35	1.40	1.45	1.49	1.53	1.64	1.80	1.91
0.35	0.43	0.95	1.13	1.24	1.32	1.38	1.44	1.48	1.52	1.56	1.68	1.83	1.94
0.36	0.55	1.00	1.17	1.27	1.35	1.41	1.47	1.51	1.55	1.59	1.71	1.86	1.97
0.37	0.64	1.05	1.21	1.31	1.39	1.45	1.50	1.55	1.59	1.62	1.74	1.89	2.00
0.38	0.73	1.10	1.25	1.35	1.42	1.48	1.53	1.58	1.62	1.65	1.77	1.92	2.03
0.39	0.81	1.14	1.29	1.38	1.46	1.51	1.57	1.61	1.65	1.68	1.80	1.95	2.06
0.40	0.88	1.19	1.32	1.42	1.49	1.55	1.60	1.64	1.68	1.72	1.83	1.98	2.09
0.41	0.95	1.23	1.36	1.45	1.52	1.58	1.63	1.67	1.71	1.75	1.86	2.01	2.12
0.42	1.01	1.27	1.40	1.49	1.56	1.61	1.66	1.70	1.74	1.78	1.89	2.04	2.14
0.43	1.07	1.32	1.44	1.52	1.59	1.64	1.69	1.74	1.77	1.81	1.92	2.07	2.17
0.44	1.13	1.36	1.47	1.56	1.62	1.68	1.72	1.77	1.80	1.84	1.95	2.10	2.20
0.45	1.19	1.40	1.51	1.59	1.65	1.71	1.75	1.80	1.83	1.87	1.98	2.13	2.23
0.46	1.25	1.44	1.55	1.62	1.69	1.74	1.79	1.83	1.86	1.90	2.00	2.15	2.26
0.47	1.31	1.48	1.58	1.66	1.72	1.77	1.82	1.86	1.89	1.93	2.03	2.18	2.28
0.48	1.36	1.52	1.62	1.69	1.75	1.80	1.85	1.89	1.92	1.96	2.06	2.21	2.31
0.49	1.41	1.56	1.65	1.73	1.78	1.83	1.88	1.92	1.95	1.99	2.09	2.24	2.34
0.50	1.47	1.60	1.69	1.76	1.82	1.86	1.91	1.95	1.98	2.02	2.12	2.27	2.37
0.51	1.52	1.64	1.73	1.79	1.85	1.90	1.94	1.98	2.01	2.04	2.15	2.29	2.40
0.52	1.57	1.68	1.76	1.83	1.88	1.93	1.97	2.01	2.04	2.07	2.18	2.32	2.42
0.53	1.62	1.72	1.80	1.86	1.91	1.96	2.00	2.04	2.07	2.10	2.21	2.35	2.45
0.54	1.67	1.76	1.83	1.89	1.95	1.99	2.03	2.07	2.10	2.13	2.24	2.38	2.48
0.55	1.72	1.80	1.87	1.93	1.98	2.02	2.06	2.10	2.13	2.16	2.27	2.41	2.51
0.56	1.77	1.84	1.90	1.96	2.01	2.05	2.09	2.13	2.16	2.19	2.30	2.44	2.54
0.57	1.82	1.88	1.94	1.99	2.04	2.09	2.12	2.16	2.19	2.22	2.32	2.47	2.56
0.58	1.87	1.92	1.98	2.03	2.07	2.12	2.16	2.19	2.22	2.26	2.35	2.50	2.59
0.59	1.92	1.96	2.01	2.06	2.11	2.15	2.19	2.22	2.26	2.29	2.38	2.52	2.62
0.60	1.98	2.00	2.05	2.10	2.14	2.18	2.22	2.25	2.29	2.32	2.41	2.55	2.65

TABLE A5.5   *Values of* d' *for Oddity, Gaussian Model*
(M = Number of Intervals)   *(cont.)*

p(c)	3	4	5	6	7	8	9	10	11	12	16	24	32
0.61	2.03	2.04	2.09	2.13	2.17	2.22	2.25	2.29	2.32	2.35	2.44	2.58	2.68
0.62	2.08	2.08	2.12	2.17	2.21	2.25	2.29	2.32	2.35	2.38	2.47	2.61	2.71
0.63	2.13	2.12	2.16	2.20	2.24	2.28	2.32	2.35	2.38	2.41	2.50	2.64	2.74
0.64	2.18	2.17	2.20	2.24	2.28	2.32	2.35	2.38	2.41	2.44	2.54	2.67	2.77
0.65	2.23	2.21	2.24	2.27	2.31	2.35	2.38	2.42	2.44	2.47	2.57	2.70	2.80
0.66	2.29	2.25	2.27	2.31	2.35	2.38	2.42	2.45	2.48	2.51	2.60	2.73	2.83
0.67	2.34	2.29	2.31	2.35	2.38	2.42	2.45	2.48	2.51	2.54	2.63	2.77	2.86
0.68	2.39	2.34	2.35	2.38	2.42	2.45	2.48	2.52	2.54	2.57	2.66	2.80	2.89
0.69	2.45	2.38	2.39	2.42	2.45	2.49	2.52	2.55	2.58	2.61	2.70	2.83	2.92
0.70	2.50	2.42	2.43	2.46	2.49	2.52	2.55	2.59	2.61	2.64	2.73	2.86	2.96
0.71	2.56	2.47	2.47	2.50	2.53	2.56	2.59	2.62	2.65	2.67	2.76	2.89	2.99
0.72	2.62	2.52	2.51	2.54	2.57	2.60	2.63	2.66	2.68	2.71	2.80	2.93	3.02
0.73	2.68	2.56	2.56	2.58	2.61	2.64	2.67	2.69	2.72	2.74	2.83	2.96	3.05
0.74	2.74	2.61	2.60	2.62	2.65	2.67	2.70	2.73	2.76	2.78	2.87	3.00	3.09
0.75	2.80	2.66	2.64	2.66	2.69	2.71	2.74	2.77	2.79	2.82	2.90	3.03	3.12
0.76	2.86	2.71	2.69	2.70	2.73	2.75	2.78	2.81	2.83	2.86	2.94	3.07	3.16
0.77	2.92	2.76	2.74	2.75	2.77	2.80	2.82	2.85	2.87	2.90	2.98	3.11	3.20
0.78	2.99	2.81	2.78	2.79	2.81	2.84	2.86	2.89	2.91	2.94	3.02	3.14	3.23
0.79	3.06	2.87	2.83	2.84	2.86	2.88	2.91	2.93	2.96	2.98	3.06	3.18	3.27
0.80	3.13	2.92	2.88	2.89	2.90	2.93	2.95	2.97	3.00	3.02	3.10	3.22	3.31
0.81	3.20	2.98	2.94	2.94	2.95	2.97	3.00	3.02	3.04	3.06	3.14	3.26	3.35
0.82	3.28	3.04	2.99	2.99	3.00	3.02	3.04	3.06	3.09	3.11	3.19	3.31	3.40
0.83	3.35	3.10	3.05	3.04	3.05	3.07	3.09	3.11	3.13	3.16	3.23	3.35	3.44
0.84	3.44	3.17	3.11	3.10	3.10	3.12	3.14	3.16	3.18	3.21	3.28	3.40	3.49
0.85	3.52	3.24	3.17	3.15	3.16	3.17	3.19	3.22	3.23	3.26	3.33	3.45	3.53
0.86	3.61	3.31	3.23	3.21	3.22	3.23	3.25	3.27	3.29	3.31	3.38	3.50	3.58
0.87	3.71	3.38	3.30	3.27	3.28	3.29	3.31	3.33	3.35	3.37	3.44	3.55	3.64
0.88	3.81	3.46	3.37	3.34	3.34	3.36	3.37	3.39	3.41	3.42	3.49	3.61	3.69
0.89	3.91	3.55	3.44	3.41	3.41	3.42	3.44	3.45	3.47	3.49	3.55	3.67	3.75
0.90	4.03	3.64	3.52	3.49	3.48	3.49	3.51	3.52	3.54	3.55	3.62	3.73	3.81

TABLE A5.5 *Values of* d' *for Oddity, Gaussian Model (M = Number of Intervals)* *(cont.)*

						M							
p(c)	*3*	*4*	*5*	*6*	*7*	*8*	*9*	*10*	*11*	*12*	*16*	*24*	*32*
0.91	4.15	3.73	3.61	3.57	3.56	3.57	3.58	3.59	3.61	3.63	3.69	3.80	3.88
0.92	4.29	3.84	3.71	3.66	3.65	3.65	3.66	3.67	3.69	3.71	3.77	3.87	3.95
0.93	4.44	3.96	3.81	3.76	3.74	3.74	3.75	3.76	3.78	3.79	3.85	3.95	4.03
0.94	4.61	4.09	3.93	3.87	3.85	3.85	3.85	3.86	3.88	3.89	3.95	4.04	4.12
0.95	4.80	4.24	4.07	4.00	3.97	3.97	3.97	3.98	3.99	4.00	4.05	4.15	4.23
0.96	5.03	4.42	4.23	4.15	4.12	4.11	4.11	4.11	4.12	4.13	4.18	4.27	4.35
0.97	5.32	4.64	4.43	4.34	4.30	4.28	4.27	4.28	4.29	4.29	4.34	4.43	4.50
0.98	5.70	4.94	4.70	4.59	4.54	4.51	4.50	4.50	4.50	4.51	4.55	4.63	4.70
0.99	6.31	5.42	5.12	4.99	4.92	4.88	4.86	4.85	4.85	4.86	4.88	4.95	5.02

Source: Reprinted from Craven (1992) by permission of the author and the Psychonomic Society.

TABLE A5.6 Values of $p(c)$ given d' for Oddity (Differencing and Independent-Observation Model, Normal), and for mAFC.

	$m = 3$			$m = 4$			$m = 5$		
		Oddity			Oddity			Oddity	
d'	AFC	$\varepsilon^2=0$	$\varepsilon^2=1$	AFC	$\varepsilon^2=0$	$\varepsilon^2=1$	AFC	$\varepsilon^2=0$	$\varepsilon^2=1$
0.0	0.333	0.333	0.333	0.250	0.250	0.250	0.200	0.200	0.200
0.1	0.362	0.334	0.335	0.277	0.252	0.252	0.224	0.201	0.202
0.2	0.391	0.337	0.338	0.304	0.255	0.257	0.249	0.205	0.208
0.3	0.422	0.342	0.345	0.333	0.261	0.267	0.278	0.211	0.219
0.4	0.452	0.348	0.353	0.362	0.268	0.279	0.305	0.221	0.234
0.5	0.483	0.356	0.364	0.393	0.279	0.295	0.334	0.232	0.252
0.6	0.512	0.365	0.376	0.424	0.292	0.313	0.366	0.246	0.275
0.7	0.543	0.376	0.392	0.456	0.306	0.334	0.397	0.261	0.299
0.8	0.574	0.389	0.409	0.487	0.321	0.358	0.428	0.279	0.327
0.9	0.604	0.403	0.427	0.520	0.339	0.385	0.461	0.299	0.357
1.0	0.633	0.418	0.446	0.552	0.360	0.413	0.495	0.321	0.391
1.1	0.663	0.434	0.468	0.583	0.381	0.443	0.528	0.344	0.426
1.2	0.690	0.452	0.491	0.615	0.404	0.474	0.559	0.368	0.460
1.3	0.716	0.468	0.513	0.644	0.426	0.506	0.591	0.394	0.496
1.4	0.741	0.488	0.536	0.673	0.449	0.538	0.624	0.420	0.533
1.5	0.765	0.507	0.560	0.702	0.475	0.571	0.653	0.447	0.568
1.6	0.788	0.525	0.584	0.729	0.499	0.603	0.683	0.475	0.604
1.7	0.810	0.546	0.608	0.755	0.524	0.635	0.711	0.503	0.639
1.8	0.831	0.565	0.631	0.779	0.549	0.665	0.738	0.531	0.673
1.9	0.848	0.585	0.654	0.802	0.575	0.696	0.764	0.559	0.705
2.0	0.865	0.605	0.677	0.823	0.600	0.725	0.788	0.586	0.735
2.1	0.881	0.624	0.699	0.842	0.624	0.751	0.810	0.614	0.764
2.2	0.896	0.645	0.722	0.860	0.649	0.777	0.831	0.640	0.791
2.3	0.909	0.663	0.742	0.877	0.672	0.801	0.851	0.667	0.816
2.4	0.921	0.682	0.763	0.893	0.695	0.824	0.869	0.692	0.839
2.5	0.931	0.700	0.781	0.907	0.716	0.845	0.885	0.716	0.859
2.6	0.941	0.718	0.799	0.919	0.738	0.864	0.900	0.740	0.878
2.7	0.949	0.734	0.816	0.930	0.758	0.882	0.914	0.762	0.895
2.8	0.957	0.751	0.832	0.941	0.778	0.898	0.925	0.783	0.910
2.9	0.963	0.766	0.847	0.949	0.796	0.911	0.936	0.803	0.924
3.0	0.968	0.783	0.862	0.957	0.813	0.924	0.945	0.821	0.935

TABLE A5.6   *Values of* p(c) *given* d' *for Oddity (Differencing and Independent-Observation Model, Normal), and for mAFC   (cont.)*

| | m = 3 | | | m = 4 | | | m = 5 | | |
| | Oddity | | | Oddity | | | Oddity | | |
d'	AFC	$\varepsilon^2 = 0$	$\varepsilon^2 = 1$	AFC	$\varepsilon^2 = 0$	$\varepsilon^2 = 1$	AFC	$\varepsilon^2 = 0$	$\varepsilon^2 = 1$
3.1	0.974	0.795	0.874	0.963	0.829	0.935	0.953	0.839	0.945
3.2	0.978	0.810	0.886	0.969	0.845	0.945	0.961	0.855	0.954
3.3	0.981	0.823	0.897	0.974	0.859	0.953	0.967	0.870	0.962
3.4	0.985	0.836	0.908	0.978	0.873	0.961	0.972	0.884	0.969
3.5	0.987	0.848	0.917	0.982	0.885	0.967	0.977	0.897	0.974
3.6	0.990	0.858	0.926	0.985	0.896	0.973	0.981	0.908	0.979
3.7	0.991	0.870	0.934	0.988	0.907	0.978	0.984	0.919	0.983
3.8	0.993	0.879	0.941	0.990	0.916	0.982	0.987	0.928	0.985
3.9	0.995	0.889	0.947	0.992	0.925	0.985	0.990	0.938	0.989
4.0	0.995	0.898	0.953	0.993	0.933	0.988	0.992	0.945	0.991
4.1	0.996	0.906	0.959	0.995	0.941	0.990	0.993	0.952	0.993
4.2	0.997	0.914	0.964	0.996	0.948	0.992	0.995	0.958	0.994
4.3	0.998	0.921	0.968	0.997	0.953	0.994	0.996	0.964	0.995
4.4	0.998	0.927	0.972	0.997	0.959	0.995	0.996	0.969	0.996
4.5	0.999	0.934	0.976	0.998	0.964	0.996	0.997	0.973	0.997
4.6	0.999	0.939	0.978	0.998	0.968	0.997	0.998	0.977	0.998
4.7	0.999	0.946	0.981	0.999	0.972	0.997	0.998	0.980	0.998
4.8	0.999	0.950	0.983	0.999	0.976	0.998	0.999	0.983	0.999
4.9	0.999	0.955	0.986	0.999	0.979	0.998	0.999	0.986	0.999
5.0	1.000	0.959	0.987	0.999	0.981	0.999	0.999	0.988	0.999

Note: $\varepsilon^2 = 0$ is the differencing model, $\varepsilon^2 = 1$ is independent-observations. Table from Versfeld et aI. (1996) with permission of the authors and The Psychonomic Society.

TABLE A5.7   *Values of d' for m-Interval Forced Choice or Identification.*

$p\backslash m$	2	3	4	5	6	7	8	9	10	11	12	16	24	32	256	1000
.01	-3.29	-2.42	-2.02	-1.77	-1.59	-1.46	-1.35	-1.26	-1.18	-1.12	-1.06	-0.88	-0.65	-0.50	0.35	0.80
.02	-2.90	-2.08	-1.69	-1.45	-1.28	-1.14	-1.04	-0.95	-0.88	-0.81	-0.75	-0.58	-0.35	-0.21	0.64	1.08
.03	-2.66	-1.86	-1.48	-1.24	-1.07	-0.94	-0.84	-0.75	-0.68	-0.62	-0.56	-0.39	-0.16	-0.02	0.82	1.26
.04	-2.48	-1.69	-1.32	-1.09	-0.92	-0.79	-0.69	-0.61	-0.53	-0.47	-0.41	-0.24	-0.02	0.12	0.96	1.40
.05	-2.33	-1.56	-1.19	-0.96	-0.80	-0.67	-0.57	-0.49	-0.41	-0.35	-0.29	-0.12	0.10	0.24	1.07	1.51
.06	-2.20	-1.44	-1.08	-0.85	-0.69	-0.57	-0.47	-0.38	-0.31	-0.25	-0.19	-0.02	0.19	0.34	1.16	1.60
.07	-2.09	-1.34	-0.98	-0.76	-0.60	-0.48	-0.38	-0.29	-0.22	-0.16	-0.11	0.07	0.28	0.42	1.25	1.68
.08	-1.99	-1.25	-0.90	-0.68	-0.52	-0.39	-0.29	-0.21	-0.14	-0.08	-0.03	0.14	0.36	0.50	1.32	1.76
.09	-1.90	-1.17	-0.82	-0.60	-0.44	-0.32	-0.22	-0.14	-0.07	-0.01	0.05	0.22	0.43	0.57	1.39	1.82
.10	-1.81	-1.09	-0.75	-0.53	-0.37	-0.25	-0.15	-0.07	0.00	0.06	0.11	0.28	0.50	0.63	1.45	1.89
.11	-1.73	-1.02	-0.68	-0.46	-0.31	-0.19	-0.09	-0.01	0.06	0.12	0.18	0.34	0.56	0.69	1.51	1.94
.12	-1.66	-0.96	-0.62	-0.40	-0.25	-0.13	-0.03	0.05	0.12	0.18	0.24	0.40	0.61	0.75	1.57	2.00
.13	-1.59	-0.89	-0.56	-0.34	-0.19	-0.07	0.03	0.11	0.18	0.24	0.29	0.46	0.67	0.80	1.62	2.05
.14	-1.53	-0.83	-0.50	-0.29	-0.13	-0.01	0.08	0.16	0.23	0.29	0.34	0.51	0.72	0.85	1.67	2.10
.15	-1.47	-0.78	-0.45	-0.23	-0.08	0.04	0.13	0.21	0.28	0.34	0.39	0.56	0.77	0.90	1.71	2.14
.16	-1.41	-0.72	-0.39	-0.18	-0.03	0.09	0.18	0.26	0.33	0.39	0.44	0.60	0.81	0.95	1.76	2.19
.17	-1.35	-0.67	-0.35	-0.13	0.02	0.13	0.23	0.31	0.37	0.43	0.49	0.65	0.86	0.99	1.80	2.23
.18	-1.29	-0.62	-0.30	-0.09	0.06	0.18	0.27	0.35	0.42	0.48	0.53	0.69	0.90	1.04	1.84	2.27
.19	-1.24	-0.57	-0.25	-0.04	0.11	0.22	0.32	0.39	0.46	0.52	0.57	0.73	0.94	1.08	1.88	2.31
.20	-1.19	-0.53	-0.21	0.00	0.15	0.26	0.36	0.44	0.50	0.56	0.61	0.77	0.98	1.12	1.92	2.35

TABLE A5.7 *Values of d' for m-Interval Forced Choice or Identification* (*cont.*)

$p\backslash m$	2	3	4	5	6	7	8	9	10	11	12	16	24	32	256	1000
.21	-1.14	-0.48	-0.16	0.04	0.19	0.31	0.40	0.48	0.54	0.60	0.65	0.81	1.02	1.16	1.96	2.39
.22	-1.09	-0.44	-0.12	0.08	0.23	0.35	0.44	0.52	0.58	0.64	0.69	0.85	1.06	1.19	1.99	2.42
.23	-1.04	-0.40	-0.08	0.12	0.27	0.38	0.48	0.56	0.62	0.68	0.73	0.89	1.10	1.23	2.03	2.46
.24	-1.00	-0.35	-0.04	0.16	0.31	0.42	0.52	0.59	0.66	0.72	0.77	0.93	1.13	1.27	2.06	2.49
.25	-0.95	-0.31	0.00	0.20	0.35	0.46	0.55	0.63	0.70	0.75	0.80	0.96	1.17	1.30	2.10	2.52
.26	-0.91	-0.27	0.04	0.24	0.38	0.50	0.59	0.66	0.73	0.79	0.84	1.00	1.20	1.34	2.13	2.56
.27	-0.87	-0.23	0.08	0.28	0.42	0.53	0.62	0.70	0.77	0.82	0.87	1.03	1.24	1.37	2.17	2.59
.28	-0.82	-0.20	0.11	0.31	0.46	0.57	0.66	0.74	0.80	0.86	0.91	1.07	1.27	1.40	2.20	2.62
.29	-0.78	-0.16	0.15	0.35	0.49	0.60	0.69	0.77	0.83	0.89	0.94	1.10	1.30	1.43	2.23	2.65
.30	-0.74	-0.12	0.19	0.38	0.53	0.64	0.73	0.80	0.87	0.92	0.97	1.13	1.33	1.47	2.26	2.68
.31	-0.70	-0.08	0.22	0.42	0.56	0.67	0.76	0.84	0.90	0.96	1.01	1.16	1.37	1.50	2.29	2.71
.32	-0.66	-0.05	0.26	0.45	0.59	0.70	0.79	0.87	0.93	0.99	1.04	1.20	1.40	1.53	2.32	2.74
.33	-0.62	-0.01	0.29	0.48	0.63	0.74	0.83	0.90	0.97	1.02	1.07	1.23	1.43	1.56	2.35	2.77
.34	-0.58	0.02	0.32	0.52	0.66	0.77	0.86	0.93	1.00	1.05	1.10	1.26	1.46	1.59	2.38	2.80
.35	-0.55	0.06	0.36	0.55	0.69	0.80	0.89	0.96	1.03	1.08	1.13	1.29	1.49	1.62	2.41	2.83
.36	-0.51	0.09	0.39	0.58	0.72	0.83	0.92	1.00	1.06	1.12	1.16	1.32	1.52	1.65	2.44	2.86
.37	-0.47	0.13	0.42	0.62	0.76	0.86	0.95	1.03	1.09	1.15	1.19	1.35	1.55	1.68	2.46	2.89
.38	-0.43	0.16	0.46	0.65	0.79	0.89	0.98	1.06	1.12	1.18	1.23	1.38	1.58	1.71	2.49	2.91
.39	-0.40	0.20	0.49	0.68	0.82	0.93	1.01	1.09	1.15	1.21	1.26	1.41	1.61	1.74	2.52	2.94
.40	-0.36	0.23	0.52	0.71	0.85	0.96	1.04	1.12	1.18	1.24	1.29	1.44	1.64	1.77	2.55	2.97

TABLE A5.7 *Values of d' for m-Interval Forced Choice or Identification   (cont.)*

p^m	2	3	4	5	6	7	8	9	10	11	12	16	24	32	256	1000
.41	-0.32	0.26	0.55	0.74	0.88	0.99	1.07	1.15	1.21	1.27	1.31	1.47	1.67	1.79	2.58	3.00
.42	-0.29	0.30	0.59	0.77	0.91	1.02	1.10	1.18	1.24	1.30	1.34	1.50	1.69	1.82	2.60	3.02
.43	-0.25	0.33	0.62	0.80	0.94	1.05	1.13	1.21	1.27	1.33	1.37	1.53	1.72	1.85	2.63	3.05
.44	-0.21	0.36	0.65	0.84	0.97	1.08	1.16	1.24	1.30	1.35	1.40	1.55	1.75	1.88	2.66	3.08
.45	-0.18	0.39	0.68	0.87	1.00	1.11	1.19	1.27	1.33	1.38	1.43	1.58	1.78	1.91	2.68	3.10
.46	-0.14	0.43	0.71	0.90	1.03	1.14	1.22	1.30	1.36	1.41	1.46	1.61	1.81	1.94	2.71	3.13
.47	-0.11	0.46	0.74	0.93	1.06	1.17	1.25	1.33	1.39	1.44	1.49	1.64	1.84	1.96	2.74	3.16
.48	-0.07	0.49	0.77	0.96	1.09	1.20	1.28	1.35	1.42	1.47	1.52	1.67	1.86	1.99	2.77	3.18
.49	-0.04	0.52	0.81	0.99	1.12	1.23	1.31	1.38	1.45	1.50	1.55	1.70	1.89	2.02	2.79	3.21
.50	0.00	0.56	0.84	1.02	1.15	1.26	1.34	1.41	1.47	1.53	1.58	1.73	1.92	2.05	2.82	3.24
.51	0.04	0.59	0.87	1.05	1.18	1.29	1.37	1.44	1.50	1.56	1.61	1.75	1.95	2.08	2.85	3.26
.52	0.07	0.62	0.90	1.08	1.21	1.32	1.40	1.47	1.53	1.59	1.63	1.78	1.98	2.10	2.87	3.29
.53	0.11	0.65	0.93	1.11	1.24	1.35	1.43	1.50	1.56	1.62	1.66	1.81	2.01	2.13	2.90	3.32
.54	0.14	0.69	0.96	1.14	1.27	1.38	1.46	1.53	1.59	1.65	1.69	1.84	2.03	2.16	2.93	3.34
.55	0.18	0.72	0.99	1.17	1.30	1.41	1.49	1.56	1.62	1.67	1.72	1.87	2.06	2.19	2.95	3.37
.56	0.21	0.75	1.02	1.20	1.33	1.44	1.52	1.59	1.65	1.70	1.75	1.90	2.09	2.22	2.98	3.40
.57	0.25	0.78	1.06	1.23	1.37	1.47	1.55	1.62	1.68	1.73	1.78	1.93	2.12	2.24	3.01	3.42
.58	0.29	0.82	1.09	1.27	1.39	1.50	1.58	1.65	1.71	1.76	1.81	1.96	2.15	2.27	3.04	3.45
.59	0.32	0.85	1.12	1.30	1.43	1.53	1.61	1.68	1.74	1.79	1.84	1.99	2.18	2.30	3.06	3.48
.60	0.36	0.89	1.15	1.33	1.46	1.56	1.64	1.71	1.77	1.82	1.87	2.02	2.21	2.33	3.09	3.51

TABLE A5.7   Values of d' for m-Interval Forced Choice or Identification   (cont.)

$p\backslash m$	2	3	4	5	6	7	8	9	10	11	12	16	24	32	256	1000
.61	0.40	0.92	1.19	1.36	1.49	1.59	1.67	1.74	1.80	1.85	1.90	2.05	2.23	2.36	3.12	3.53
.62	0.43	0.95	1.22	1.39	1.52	1.62	1.70	1.77	1.83	1.88	1.93	2.07	2.26	2.39	3.15	3.56
.63	0.47	0.99	1.25	1.42	1.55	1.65	1.73	1.80	1.86	1.91	1.96	2.10	2.29	2.42	3.18	3.59
.64	0.51	1.02	1.29	1.46	1.58	1.68	1.77	1.83	1.89	1.94	1.99	2.14	2.32	2.45	3.20	3.62
.65	0.54	1.06	1.32	1.49	1.62	1.72	1.80	1.87	1.92	1.98	2.02	2.17	2.35	2.48	3.23	3.64
.66	0.58	1.09	1.35	1.52	1.65	1.75	1.83	1.90	1.96	2.01	2.05	2.20	2.38	2.51	3.26	3.67
.67	0.62	1.13	1.39	1.56	1.68	1.78	1.86	1.93	1.99	2.04	2.09	2.23	2.42	2.54	3.29	3.70
.68	0.66	1.16	1.42	1.59	1.72	1.81	1.89	1.96	2.02	2.07	2.12	2.26	2.45	2.57	3.32	3.73
.69	0.70	1.20	1.46	1.63	1.75	1.85	1.93	2.00	2.05	2.10	2.15	2.29	2.48	2.60	3.35	3.76
.70	0.74	1.24	1.49	1.66	1.79	1.89	1.96	2.03	2.09	2.14	2.18	2.33	2.51	2.63	3.38	3.79
.71	0.78	1.28	1.53	1.70	1.82	1.92	2.00	2.06	2.12	2.17	2.22	2.36	2.54	2.67	3.42	3.82
.72	0.82	1.31	1.57	1.73	1.86	1.95	2.03	2.10	2.16	2.21	2.25	2.39	2.58	2.70	3.45	3.85
.73	0.87	1.35	1.60	1.77	1.89	1.99	2.07	2.13	2.19	2.24	2.29	2.43	2.61	2.73	3.48	3.89
.74	0.91	1.39	1.64	1.81	1.93	2.03	2.10	2.17	2.23	2.28	2.32	2.46	2.65	2.77	3.51	3.92
.75	0.95	1.43	1.68	1.85	1.97	2.06	2.14	2.21	2.26	2.31	2.36	2.50	2.68	2.80	3.54	3.95
.76	1.00	1.47	1.72	1.89	2.01	2.10	2.18	2.24	2.30	2.35	2.40	2.54	2.72	2.84	3.58	3.99
.77	1.05	1.52	1.76	1.93	2.05	2.14	2.22	2.28	2.34	2.39	2.43	2.57	2.75	2.87	3.61	4.02
.78	1.09	1.56	1.81	1.97	2.09	2.18	2.26	2.32	2.38	2.43	2.47	2.61	2.79	2.91	3.65	4.06
.79	1.14	1.61	1.85	2.01	2.13	2.22	2.30	2.36	2.42	2.47	2.51	2.65	2.83	2.95	3.69	4.09
.80	1.19	1.65	1.89	2.05	2.17	2.26	2.34	2.40	2.46	2.51	2.55	2.69	2.87	2.99	3.73	4.13

TABLE A5.7   Values of d' for m-Interval Forced Choice or Identification   (cont.)

$p^{\backslash m}$	2	3	4	5	6	7	8	9	10	11	12	16	24	32	256	1000
.81	1.24	1.70	1.94	2.10	2.22	2.31	2.38	2.45	2.50	2.55	2.60	2.73	2.91	3.03	3.77	4.17
.82	1.29	1.75	1.99	2.14	2.26	2.35	2.43	2.49	2.55	2.60	2.64	2.78	2.96	3.07	3.81	4.21
.83	1.35	1.80	2.04	2.19	2.31	2.40	2.47	2.54	2.59	2.64	2.68	2.82	3.00	3.12	3.85	4.25
.84	1.41	1.85	2.09	2.24	2.36	2.45	2.52	2.59	2.64	2.69	2.73	2.87	3.05	3.16	3.89	4.29
.85	1.47	1.91	2.14	2.29	2.41	2.50	2.57	2.64	2.69	2.74	2.78	2.92	3.09	3.21	3.94	4.34
.86	1.53	1.97	2.20	2.35	2.46	2.55	2.63	2.69	2.74	2.79	2.83	2.97	3.14	3.26	3.99	4.39
.87	1.59	2.03	2.25	2.41	2.52	2.61	2.68	2.74	2.80	2.85	2.89	3.02	3.20	3.31	4.04	4.44
.88	1.66	2.09	2.32	2.47	2.58	2.67	2.74	2.80	2.86	2.90	2.95	3.08	3.25	3.37	4.09	4.49
.89	1.73	2.16	2.38	2.53	2.64	2.73	2.80	2.86	2.92	2.96	3.01	3.14	3.31	3.43	4.15	4.54
.90	1.81	2.23	2.45	2.60	2.71	2.80	2.87	2.93	2.98	3.03	3.07	3.20	3.37	3.49	4.20	4.60
.91	1.90	2.31	2.53	2.67	2.78	2.87	2.94	3.00	3.05	3.10	3.14	3.27	3.44	3.56	4.27	4.67
.92	1.99	2.39	2.61	2.75	2.86	2.95	3.02	3.08	3.13	3.18	3.22	3.35	3.52	3.63	4.34	4.73
.93	2.09	2.49	2.70	2.84	2.95	3.03	3.10	3.16	3.22	3.26	3.30	3.43	3.60	3.71	4.42	4.81
.94	2.20	2.59	2.80	2.94	3.05	3.13	3.20	3.26	3.31	3.35	3.39	3.52	3.69	3.80	4.50	4.90
.95	2.33	2.71	2.92	3.06	3.16	3.24	3.31	3.37	3.42	3.46	3.50	3.63	3.79	3.91	4.60	4.99
.96	2.48	2.85	3.05	3.19	3.29	3.37	3.44	3.50	3.55	3.59	3.63	3.75	3.92	4.03	4.72	5.11
.97	2.66	3.02	3.22	3.35	3.45	3.53	3.60	3.65	3.70	3.75	3.78	3.91	4.07	4.18	4.86	5.25
.98	2.90	3.25	3.44	3.57	3.67	3.75	3.81	3.87	3.91	3.95	3.99	4.11	4.27	4.38	5.05	5.44
.99	3.29	3.62	3.80	3.92	4.01	4.09	4.15	4.20	4.25	4.29	4.32	4.44	4.59	4.69	5.36	5.73

Note: Equal detectability and unbiased responding is assumed.
Source: Reprinted from Hacker and Ratcliff (1979) by permission of the authors and The Psychonomic Society, Inc.

Appendix 6

Software for Detection Theory

In this appendix, we describe software available for doing the calculations presented in the book. The descriptions take two forms. First, the listing of a program for basic calculation (e.g., of yes-no d', c, and β) is given. Second, we list Web sites from which useful programs can be downloaded. We thank the colleagues who have agreed to publicize their sites in this way.

Listing

The following Pascal program calculates d', c, and β for frequency data typed into a keyboard. It consists of a subroutine for estimating z and a simple driver program. To deal with perfect proportions, it uses the $\frac{1}{2}$, $N - \frac{1}{2}$ rule described in chapter 1.

```pascal
program sdtdrive;

{————————————————————————}
function z(p:real):real; {Odeh & Evans}

var y:real;

begin
    y: = sqrt(-2*1n(p));
    z: = -y + (((((0.0000453642210148*y + 0.0204231210245)*y +
        0.342242088547)*y + 1)*y + 0.322232431088)/
(((((0.0038560700634*y + 0.10353775285)*y + 0.531103462366)*y +
        0.588581570495)*y + 0.099348462606)
```

```
end;
{————————————}

var Nhits,Nmisses,Nfa,Ncr,N2,N1:        real;
    hitrate,farate,dp,cr,beta,zh,zf:     real;
    answer:                              char;
    adjustment:                          boolean;

begin
     writeln('Follow all responses by <RETURN>');
     writeln;
     repeat
           adjustment: = false;
           writeln;
           write('# of hits: ');
           read(Nhits);
           write('      # of misses: ');
           readln(Nmisses);
           write ('# of false alarms: ');
           read (Nfa);
           write('   # of correct rejections: ');
           readln(Ncr);
           N2: = Nhits + Nmisses;
           N1: = Nfa + Ncr;
           if (Nhits < = 0) or (Nmisses < 0) or (Nfa < 0) or (Ncr < = 0)
                 then writeln('Bad data')
                 else begin
           if Nmisses = 0 then begin
                 Nmisses: = 1\2;
                 Nhits: = Nhits – 1\2;
                 adjustment: = true
                 end;
           if Nfa = 0 then begin
                 Nfa: = 1\2;
                 Ncr: = Ncr – 1\2;
                 adjustment: = true
                 end;
           hitrate: = Nhits/N2;
           farate: = Nfa/N1;
```

```
            zh: = z(hitrate);
            zf: = z(farate);
            dp: = zh – zf;
            cr: = –0.5*(zh + zf);
            beta: = exp(–0.5*(zh*zh – zf*zf));
            writeln;
            if adjustment = true
                    then writeln('Data have been adjusted');
            writeln('H = ',hitrate:4:3,' F = ',farate:6:3);
            writeln('d" = ',dp:4:3,', c = ',
                    cr:6:3,', beta = ',beta:6:3)
            end;
        writeln;
        write('Continue? (y/n) ');
        readln(answer);
        until answer = 'n'

end.
```

Web Sites

A useful site for exploring the use of spreadsheets in detection theory (Bob Sorkin):
http://www.psych.ufl.edu/~sorkin

A program for finding d' and other statistics for data collected under a wide variety of paradigms is $d'plus$ (Macmillan & Creelman, 1997), available at http://psych.utoronto.ca/~creelman

Programs for fitting ROCs using maximum-likelihood techniques can be found at sites maintained by Lew Harvey and Charles Metz:
http://psych.colorado.edu/~lharvey
http://www.xray.bsd.uchicago.edu/krl/KRL_ROC/software_index.htm

The MSDA method (Helena Kadlec) is available at
http://castle.uvic.ca/psyc/kadlec/research.htm

Several programs permitting statistical evaluation of SDT data (Larry DeCarlo) are at
http://www.columbia.edu/~ld208

Information about the statistical accuracy and efficiency of ROC parameters (Caren Rotello) is located at
http://www-unix.oit.umass.edu/~caren/Design/Assets/index.htm

Appendix 7

Solutions to Selected Problems

Most answers were obtained using the tables (Appendix 5). Answers that have been found by interpolation carry asterisks.

Chapter 1

1.1.   If the person being tested tells a lie, a hit occurs if the polygraph responds positively, a miss if it responds negatively. If the person tells the truth, a false alarm is a positive response, a correct rejection a negative one.

1.2.

Problem	H	F	H – F	$p(c)$	$p(c)$*
(a)	.6	.47	.13	.57	.57
(b)	.55	.17	.38	.69	.62
(c)	.45	.83	-.38	.31	.38

$p(c)$ is always greater than $H - F$ except when both equal 1 (draw a graph of Eq. 1.3).

1.3.   (a) Computationally, "base rates" do not affect the calculation of conditional probabilities or $p(c)$, but do affect $p(c)$*. Experimentally, the likelihood of a "yes" response may well depend on these rates. For more detail, see chapter 2. (b) Yes, $p(c)$ for S_2 trials is simply the hit rate.

1.4.   (a) 1.99, 4.65, 0 (b) 1.03, 1.28, 2.93.

1.5.   (a) 1.68 assuming no bias. (b) $H = .65$ and $d' = 2.03$.

1.7.

Problem	d' from H and F	d' from p(c)
(a)	0.336*	0.330*
(b)	1.080	0.992
(c)	−1.080	−0.992

1.8.   Yes, in both cases, because the implied ROCs of both d' and $p(c)$ are symmetric.

1.9.   For (.2, .6), $d' = 1.095$. For (.2, .91) and (.03, .6), $d' = 2.190$.

1.11.   If no cell contains a frequency of 0, the largest d' is 1.85, and the smallest −1.11. If a 0 cell does occur, the largest d' is 2.31 and the smallest is −1.47.

Chapter 2

2.1. $p(c) = .65$ before training, .785, .656, and .825 after.

2.2.   (a) When $d' = 1, p(c) = .69$; when $d' = 2, p(c) = .84$. (b) All are equal to 0.

2.3.

H	F	d'	c	c'	ln(β)	β
.6	.47	0.328	−0.089	−0.271	−0.029	0.971
.55	.17	1.08	0.414	0.383	0.447	1.564
.45	.83	−1.08	0.414	−0.383	−0.447	0.639

2.4.   (a) A vertical line. (b) A line with slope −1.

2.5.

H	F	d'	c	c'	β
.6	.2	1.095	0.2945	0.269	1.38
.79	.08	2.211	0.2995		
.71	.05	2.141		0.268	
.83	.11	2.181			1.35

2.6.　The false-alarm rate $[1 - P(\text{"truth"}|\text{truth})]$, c, and d' are as follows:

Experimental group	F	c	d'
Interrogators	.34	−0.215	1.254
Sheriffs	.44	−0.294	0.890
Clinical psychologists	.36	−0.098	0.911
Academic psychologists	.42	0.013	0.378

Trained interrogators are probably a bit better at detecting lying than sheriffs or clinical psychologists, and academic psychologists are—as a group—terrible. It is also interesting that sheriffs and interrogators showed a strong bias toward stating that a person was lying, whereas the psychologists were relatively neutral.

2.7.　(a) Payoff matrix is 10, 0, 0, 10.

2.7.　(b–d)

$P(S_2)$	LR	c	H	F
.5	1	0	.69	.31
.25	3	1.1	.27	.05
.1	9	2.2	.04	.003

2.10.　(a) $(F, H) = (.07, .69), (.31, .93)$; (b) $(F, H) = (.02, .5), (.5, .98)$.

Chapter 3

3.1.　(a) The first (F, H) point is $(.12, .52)$, the second $(.36, .80)$. (If you did not get the second point, remember that the response categories have to be in order of confidence from one alternative to the other.) (b) $d' = 1.225$ and 1.200.

3.2.　If three "not sure" responses of each type are assigned to "sure in tune" and the rest to "sure out of tune," estimated d' will drop to 1.064.

3.3.　Condition 1: $d_a = 2.44$, $s = 0.59$, $A_z = .96$, $c_a = 0.453, -0.147, -0.287$. Condition 2: $d_a = 1.15$, $s = 0.76$, $A_z = .79$, $c_a = 1.00, -0.433, -0.299$.

3.4.   1.235, 0.850, 0.152.

3.5. $d_a = 2.21$, $d'_e = 2.22$, $A_z = .94$.

3.6.   For low-frequency words, $A_g = .879$ and $A_z = .90$. For high-frequency words, $A_g = .752$ and $A_z = .76$.

Chapter 4

4.1.   (a) .75, .6, 0. (b) .2. u always equals F.

4.2.   (a) You can't. (b) On the upper limb, because it is above the minor diagonal.

4.3. $p(c) = .7$ at $F = 0$, and .5 at $F = 1$; $d' = 2.098$ at $F = .01$, and 0.186 at $F = .99$.

4.4.   (a) For (.4, .9), $p(c) = .75$, yes rate $= .65$, error ratio $= (1 - H)/F = 0.25$. For (.2, .9), $p(c) = .85$, yes rate $= .55$, error ratio $= 0.5$. (b) The y-intercept (lowest hit rate) is $2p(c) - 1$, and the x-intercept (highest false-alarm rate) is $2[1 - p(c)]$.

4.5.   .83, .75, .875

4.6.   .8, .2.

4.8.   (a) Area under two-limbed curve $= .75$, $A' = .835$, $d' = 1.366$, $A_z = .833$. (b) Area under two-limbed curve $= .695$, $A' = .842$, $d' = 2.073$, $A_z = .93$.

4.9.   It is systematically greater than for the low-threshold area. It will be similar to the area for high-threshold theory if the point is near the minor diagonal and similar to that for double high-threshold theory if accuracy is high.

4.10.   (a) For both points, $d' = 1.095$, $\ln(\alpha) = 0.90$, $p(c) = .7$, and $A' = .646$. (b) .667, .5. (c) $H = .75$, $F = .25$.

4.13.   For both points, $\beta_L = 0.429$ and $B'' = -0.4$. $\ln(b) = -1.52$ and -0.675.

Chapter 5

5.1. (a) $d' = 3$ (b) $p(c) = .93$. No additivity in $p(c)$ itself, need to convert to z scores: $z[p(c)_{AB}] + z[p(c)_{BC}] = z[p(c)_{AC}]$. This is the same as Equation 5.1 (except for a factor of 2). (c) $d' = 5$, so $p(c)$ is very close to 1.0.

5.2. No. If $s = 0.5$ for both comparisons, then the standard deviations of distributions A, B, and C can be set to 1, 2, and 4, with means at 0, 1, and 5. Then $d'_1(AC) = 5$, not 3.

5.3. (a) 1.168 for 1 presentation, 1.621 for 2, and 2.320 for 4. Criterion is .915 above mean of New distribution. (b) $d' = 1.15$, $p(c) = .72$.

5.4. For each stimulus, find $P(\text{"higher"})$ and convert this to a z score. These scores are -0.842, -0.253, 0.253, 0.842, and 1.282. The PSE is at 999.5 Hz. The jnd is $\frac{1}{2}(1000.7 - 998.3) = 1.2$ Hz (interpolate to find the values for which $z = \pm 0.675$).

5.5. 84% and 16%.

5.6.

Stimulus	d'	cumulative d'
10		0
	0.440	
15		0.440
	0.457	
20		0.897
	0.638	
25		1.535
	0.392	
30		1.927
	0.675	
35		2.602
	0.524	
40		3.126

Criteria are at cumulative $d' = 1.282$ and 2.602, which correspond to approximately 23 and 35 cm.

5.7.

Stimulus	d'	cumulative d'
10		0
	0.363	
15		0.363
	0.348	
20		0.711
	0.616	
25		1.327
	0.318	
30		1.645
	0.674	
35		2.319
	0.362	
40		2.681

Criterion is at cumulative $d' = 1.645$, which corresponds to 30 cm.

5.9.

Stimulus	d'	cumulative d'
1 (the letter E)		0
	0.599	
2		0.599
	1.029	
3		1.628
	1.095	
4		2.723
	0.440	
5 (the letter F)		3.163

Criterion is at cumulative $d' = 1.881$, about stimulus 3.

5.10.

Stimulus pair	SDT	threshold
1, 2	.62	.502
2, 3	.70	.545
3, 4	.71	.580
4, 5	.59	.505
1, 3	.79	.568
2, 4	.86	.745
3, 5	.78	.625

5.11.   0.599, 2.60, and 5.40.

5.12.   Distribution means are 0, 0.589, 1.366, and 2.732. Criteria are 0.253, 0.842, and 2.208.

Chapter 6

6.1.   .98, .50, .16, .31.

6.2.   (a) .92. (b) no change; .84. (c) ignore sound; .98.

6.3.   (a) Same decision axis; .5. (b) Decision axis and criterion are both rotated clockwise compared to Fig. 6.5, by an angle less than 45°.

6.4.   (a). (i) and (ii) .5, (iii) and (iv) .25. (b) (i) no (ii) no (iii) yes. (c) (i) and (ii) .07, (iii) .07 × .07, (iv) .93 × .93. (d) (i) no (ii) yes (iii) yes.

6.5.   .61, .63, .65.

6.6.   Additivity holds in both cases although the d' calculation in (b) is heuristic only.

Chapter 7

7.1.

Matrix	2AFC d'	yes-no d'	c	$p(c) = (H + 1 - F)/2$	$p(c)^*$
A	0.358	0.506	0	.6	.6
B	0.536	0.758	−0.903	.6	.6
C	0.568	0.803	1.244	.575	.575
D	−0.132	−0.187	0.346	.467	.556

7.2.

Matrix	$p(c)_{max}$
A	.6
B	.65
C	.66
D	.46

Notes: (a) $p(c)_{max}$ is the same for 2AFC and yes-no—it depends only on H and F. (b) $p(c)_{max}$ is actually *smaller* than $p(c)$ for Matrix D; it is a "maximum" in that it represents a point that is maximally different from chance.

7.3.   (a) $d' = 1.33$ in both cases, $c = 0.25$ for item recognition and −0.33 for source discrimination. (b) $p(c)_{max} = .83$, so the prediction is exactly right. (c) See if you can account for this by assuming decisional separability.

7.4.   (a) $d_a = 1.19$ independent of s. (b) $d'_2 = 0.94$ assuming $s = 0.5$, but 1.88 assuming $s = 2$. An advantage of d_a is that it can be predicted from $p(c)_{max}$ without knowing s.

Chapter 8

8.1. $d'_{1000} = 1.478$, $d'_{1200} = 1.079$, identification $d' = 1.830$.

8.2. $p(c) = .86, .71$.

8.3.   (a) Yes: $p(c)$ in the uncertain condition is about .59.

8.4.   Unbiased $p(c)$ for S_1 detection is .84, for S_2 detection and S_{12} recognition, .69.

8.7.   Matrix 1 supports the independent-observation model, and Matrix 2 the integration model.

8.9. $p(c) = .79$.

8.10. $p(c) = .71$.

8.11. $p(c) = .91$.

Chapter 9

9.1.

H	F	independent-observation	differencing
		d'	
.6	.4	1.19	1.43
.9	.7	1.49	2.15
.2	.05	1.54	1.58
.6	.667	–0.70	–0.92

9.2. $p(c)_{\text{yes-no}} = .9; p(c)_{\text{same-different}} = .82$ according to the threshold and independent-observation models, .75 for differencing.

9.3.   Threshold model makes no obvious prediction; if $p(c)$ is still .82, then H is .69. According to the independent-observations model, $H = .57$ and $p(c) = .76$; according to the differencing model, $H = .44$ and $p(c) = .70$.

9.4. S_1 versus S_2, $H = .55$, $F = .25$. S_2 versus S_3, $H = .65$, $F = .15$. Same trials count more heavily. Overall, $p(c) = .686$, but average of H and $1 - F$ is .70.

9.5. $p(c)_{\text{yes-no}} = .974*, p(c)_{\text{2AFC}} = .997$.

9.7.

matrix	independent-observation	differencing
	d'	
A	1.01	1.12
B	1.26	1.41
C	1.30	1.45
D	–0.59	–0.66

9.8.   Entries are $p(c)$:

Design	d'	
	1	2
yes-no	.69	.84
2AFC	.76	.92
ABX	.600	.788
same-different	.573	.732
oddity	.45	.68

9.9.   Entries are $p(c)$:

Design	d'	
	1	2
ABX	.583	.747
same-different	.55	.675*
oddity	.42	.60

Results for other paradigms are the same as in Problem 9.8.

Chapter 10

10.1. $p(c)_{25AFC} = .20$; $p(c)_{5AFC} = .49$; $p(c)_{2AFC} = .76$.

10.2.   Entries in last three columns are $p(c)$.

m	SDT	Choice Theory	Boundary
3	.62	.60	.56
4	.54	.50	.42
8	.37	.30	.13
32	.16	.088	.00013
1,000	.015*	.003	$(.75)^{999}$

10.3. $d' = 2.160$ for any pair; points in representation form an equilateral triangle.

10.4.   (a) 0.777; (b) 0.817, −0.217, −0.327, −0.597

10.5. $p(c)_{\text{yes-no}} = p(c)_{3\text{AFC}} = .78$ at $d' = 1.56$.
$p(c)_{\text{yes-no}} = p(c)_{4\text{AFC}} = .88$ at $d' = 2.32$.
$p(c)_{\text{yes-no}} = p(c)_{8\text{AFC}} = .96$ at $d' = 3.44$.

10.7. Hit rates are .915 and .790, false-alarm rates are .47 and .28. Both are reliably different, so there is no marginal response invariance. Values of d' are 1.45 and 1.39 (not different), criteria are 1.37 and 0.81 (different). This pattern implies PS, but not DS.

Chapter 11

11.1. (a) .0000558 (most likely). (b) .000046. (c) .000035.

11.2. (a) .000028. (b) .000029 (most likely). (c) .000017.

11.3. (a) Trials 4 (+), 12 (+), 14 (+), 18 (–).

(b) Trials 2 (+), 4 (+), 6 (–), 7 (+), 9 (–), 11 (+), 12 (+), 13 (+), 14 (+), 16 (–), 18 (–), 20 (–).

(c) Trials 2(+), 4(+), 7(+), 11(+), 12(+), 13(+), 14(+), 18(–).

11.4. After 20 trials, level is (a) 80, (b) 40, (c) 56.

Chapter 12

12.1. Smooth and symmetric, like a normal-normal curve. Consists of points corresponding to the possible cutpoint decision rules, and line segments connecting them corresponding to a mixture of two adjacent criteria.

12.2. (a) Mean difference unchanged, both variances increase. Best $p(c)$ is .70. (b) Mean difference decreases by 0.5, variances unchanged. Best $p(c)$ is .65.

12.4. (a) a_1 and a_2 both equal 0.5, decision bound has slope $-a_2/a_1 = -1$. (b) $a_1 = 0.4$, $a_2 = 0.6$, decision bound has slope $-a_2/a_1 = -1.5$.

12.5. (a) $a_1 = 0.4$, $a_2 = 0.6$, decision bound has slope $-a_2/a_1 = -1.5$. (a) $a_1 = 0.33$, $a_2 = 0.67$, decision bound has slope $-a_2/a_1 = -2$.

Chapter 13

Note: Statistically significant results are indicated by $.

13.1.   (a) .38 to .62.
         (b) .25 to .55$.
         (c) .44 to .56; .33 to .47$.

13.2. $d'_{max} = 2.160$, $d'_{min} = 0.716$ ($c = 0$ for both); $c_{max} = 0.361$, $c_{min} = -0.361$ ($d' = 1.438$ for both).

13.3.   (a) matrix 1: $d' = 0.506 \pm 0.786$, $c = 0 \pm 0.393$
              matrix 2: $d' = 0.758 \pm 0.946$, $c = -0.903 \pm 0.473$$
              matrix 3: $d' = 0.803 \pm 1.118$, $c = 1.244 \pm 0.559$$
              matrix 4: $d' = -0.187 \pm 1.467$, $c = -0.347 \pm 0.735$$
         (b) $d'_3 - d'_4 = 0.990 \pm 1.847$; $c_1 - c_2 = 0.903 \pm 0.615$$

13.4.   (a) 0.412 ± 0.499
         (b) 0.291 ± 0.353

13.5.   (a) average $= 0.641$, pooled $= 0.516$
         (b) 0.651
         (c) 0.511

Glossary

The number in parentheses following each entry gives the chapter in which the term is introduced. Part I is denoted by I, Appendix 1 by A1, and so on.

2AFC (7). Two-alternative forced-choice.

α (4). Sensitivity measure for Choice Theory.

A' (area under the ROC) (4). An estimate of the area under the ROC based on a single point in ROC space. A measure of sensitivity.

A_g (minimum area under the ROC) (4). An estimate of the area under the ROC based on more than one point in ROC space. A measure of sensitivity.

A_z (3). Area under an SDT ROC (i.e., one that is linear on z coordinates). A measure of sensitivity.

absolute identification (5). A classification experiment in which the number of responses equals the number of stimuli.

absolute judgment (5). Same as absolute identification.

ABX (9). A discrimination design in which three stimuli are presented on each trial, and the observer must decide whether the third matches the first or the second.

accuracy (1, 13). (a) Same as sensitivity. (b) In statistics, the degree to which the expected value of an estimator equals the parameter being estimated.

adaptation level theory (5). A theory that states that judgments in identification are relative to a central point, the adaptation level.

adaptive probit estimation (APE) (11). An adaptive procedure that estimates psychometric function slope as well as threshold.

447

adaptive procedure (11). A method for estimating empirical thresholds by choosing stimuli in reaction to the observer's previous responses.

area theorem (7). The equivalence between area under the yes-no ROC and the proportion correct obtainable by an unbiased observer in 2AFC.

attention operating characteristic (AOC) (8). In a divided attention paradigm, accuracy on one task versus accuracy on another as attention is shifted from one to the other.

β (2). In SDT, likelihood ratio for two Gaussian distributions. A measure of response bias.

β_L (4). In Choice Theory, likelihood ratio for two logistic distributions. A measure of response bias.

β_d (9). Likelihood ratio for the differencing model of the same-different paradigm. A measure of response bias.

β_i (9). Likelihood ratio for the independent-observation model of the same-different paradigm. A measure of response bias.

b (4). In Choice Theory, $\ln(b)$ is the location of the criterion in standard deviation units from the equal-bias point. A measure of response bias.

b' (4). In Choice Theory, the relative criterion. A measure of response bias.

B'', B'_H (4). Bias measures based on the geometry of ROC space.

Bayesian (11). Referring to the result that the odds in favor of a hypothesis before an observation is made, multiplied by the likelihood ratio of the observation, equal the odds after the observation.

Békésy audiometry (11). An adaptive procedure, psychophysically informal, in which the observer provides a continuous detection response to a continuously changing stimulus.

Bernoulli random variable (A1). A random variable that can take on only two values, 0 and 1.

bias, response (2). See response bias.

bias, statistical (A1, 11, 13). The average amount by which an estimate differs from the parameter being estimated.

binomial distribution (13, A1). Distribution of a binomial random variable.

binomial proportion distribution (A1). Distribution of the proportion of successes in N trials (i.e., of a binomial random variable divided by N).

binomial random variable (A1). A random variable that is the sum of N Bernoulli random variables; the number of successes in N trials.

bivariate distribution (13, A1). Probability distribution of two variables.

boundary theorem (10). A generalization of the area theorem that predicts a lower bound on mAFC performance, given 2AFC performance.

c (2). In SDT, the location of the criterion in z units from the equal-bias point. A measure of response bias.

c' (2). In SDT, the relative criterion. A measure of response bias.

c_a (3). Criterion location in units of the root mean square standard deviation.

c_d (9). Criterion location for the differencing model for the same-different task. A measure of response bias.

c_e (3). Criterion location in units of the average standard deviation.

c_i (9). Criterion location for the independent-observation model for the same-different task. A measure of response bias.

categorical perception hypothesis (5). The hypothesis that sensitivity in classification is the same as in discrimination and/or that discrimination sensitivity reaches a peak at an intermediate point on a continuum.

categorization (5). A classification experiment in which the number of responses is less than the number of possible stimuli.

category scaling (5). Categorization, usually with stimuli that vary along a single continuum.

central limit theorem (A1). The result that the sum of many independent variables, each with the same distribution, has a normal distribution.

channels (8). Theoretical analyzers of multidimensional stimuli, often assumed independent.

choice axiom (4). Basic tenet of Choice Theory. States that the odds of choosing one stimulus over a second are unaffected by the availability of other possible stimuli.

Choice Theory (4). (a) A theory of choice behavior, derived from the choice axiom, in which responses are determined by the strengths of corresponding stimuli and by response biases. (b) For the yes-no experiment, equivalent to a version of detection theory in which underlying distributions are assumed to be logistic.

city-block metric (1, 8). A distance measure for multidimensional stimuli equal to the sum of the distances on each dimension.

classification experiment (5). An experiment in which one stimulus, from a set of more than two, is presented on each trial.

comparison design (7). Discrimination paradigm with two intervals that can be represented with two underlying distributions.

comparison stimulus (5). Stimulus that varies from trial to trial, in an experiment that contains a standard stimulus (which does not).

complete correspondence experiment (Intro). See correspondence experiment.

compound detection (6). Detection of a multidimensional stimulus.

conditional-on-single-stimulus analysis (12). Method for determining the sensitivity of single components in multidimensional stimuli.

conditional probability (A1). A probability defined on a subset of a sample space; the probability of one event given that another occurs.

Condorcet group (12). Group that makes decisions by counting unweighted votes.

confidence interval (13, A1). Interval within which, with some degree of confidence, a population parameter falls.

constant-ratio rule (10). The assertion that the ratio of response frequencies in a stimulus–response matrix is unchanged by the addition or removal of items to the stimulus set.

context coding (5). Perceptual process in which stimuli being judged are compared with the context provided by previous trials.

context variance (5). Variability in perceptual process contributed by context coding.

continuous random variable (A1). A random variable that can take on any value in a (finite or infinite) interval.

correct rejection (1). In a yes-no experiment, a response of "no" to S_1 (the stimulus class for which "no" is correct).

correct rejection rate (1). The proportion of correct rejections on S_1 trials.

correction for guessing (4). A formula for computing q, the "corrected" hit rate. Equivalent to high-threshold theory.

correlation (A1). The tendency for two variables to covary.

correspondence experiment (Intro). An experiment in which each possible stimulus class is associated with one "correct" response from among a finite set. The determination of which response is correct may be rigidly set by the experimenter or may be limited to a class of possibilities.

COSS (12). See conditional-on-single-stimulus analysis.

criterion (1). The point on a decision axis that divides one response from another. See also decision boundary.

criterion variability (2). Setting the criterion at different locations on different trials in the same experimental condition.

cumulative sensitivity (5). Sensitivity to the difference between a stimulus and an endpoint stimulus, sometimes inferred from sensitivities to adjacent stimulus pairs.

d' (1). Sensitivity measure for SDT, assuming equal-variance distributions.

d_a (3). Measure of sensitivity for SDT, assuming unequal-variance underlying distributions and using the root-mean-square standard deviation.

d'_e (3). Measure of sensitivity for SDT, assuming unequal-variance underlying distributions and using the average standard deviation.

D_{YN} (3). In ROC space on z coordinates, the minimum distance from the origin to the ROC.

decision space (1). The underlying distributions in an experiment, together with the observer's decision rule for making responses.

decision boundary (6). Multidimensional generalization of the criterion: the locus of points in a decision space that divides one response from another.

decisional separability (6). In a multidimensional representation, a decision rule that depends on only one dimension.

density function (A1). Function representing the likelihoods of possible values of a continuous random variable.

detection (I). Discrimination experiment in which one stimulus is the Null stimulus, or noise.

detection theory (1). A theory relating choice behavior to a psychological decision space. An observer's choices are determined by the distances between distributions due to different stimuli in this space (sensitivities) and by the manner in which the space is partitioned into regions corresponding to the possible responses.

detection with uncertainty (8). An experiment in which the stimulus to be detected varies from trial to trial.

deviation limit (11). In PEST and other adaptive procedures, the extent to which observed $p(c)$ must differ from $p(T)$ before the stimulus is changed.

difference threshold (1, 5). Empirical threshold in a discrimination experiment not involving the Null stimulus.

differencing models (7, 9). Models in which the observer uses the difference between observations from multiple intervals or dimensions as the basis for decision.

discrete random variable (A1). A random variable that takes on only a finite or countable number of values.

discrimination (I). The ability to distinguish between two stimulus classes, one of which may or may not be the Null stimulus, or noise. Also an experiment to measure this ability.

distance measure (1). A measure that has the characteristics of a distance. The sensitivity statistics d' and $\ln(\alpha)$ are distance measures.

distribution discrimination (12). Task in which stimuli are drawn from explicit distributions and the observer must determine which of them is the source of the stimulus presented.

distribution function (A1). Function giving the probability that a random variable is less than or equal to some value. For continuous variables, the integral of the density function up to that value; for discrete ones, the sum.

divided attention (8). Task in which attention to more than one dimension or channel is required for an optimal decision.

double high-threshold theory (4). A theory with three internal states and two high thresholds.

efficiency, relative (11, 12, 13). (a) In adaptive procedures, the ratio between the sweat factors of two statistics (or of the variances, if the number of trials is equal); a measure of relative precision, or repeatability over estimates. (b) In model comparisons, the square of the d' ratio. (c) In statistics, the ratio of the variances of two estimators of a parameter.

elementary event (A1). One of a finite number of equally probable events in a sample space.

empirical ROC (3). See ROC.

equivalent measures (1). Measures that are related by a monotonic transformation. Equivalent sensitivity measures have the same implied ROC, and equivalent bias measures have the same implied isobias curve.

error ratio (4). In a yes-no experiment, the ratio of misses to false alarms. A measure of response bias.

estimated probability (A1). A proportion used to estimate a true probability.

estimation (13). Process of approximating a population parameter from data.

Euclidean distance (1). Distance measured by the Pythagorean theorem.

expectation (A1). The mean of a random variable.

external noise (12). Variability limiting performance that arises from the stimulus itself rather than within the observer.

extrinsic uncertainty (8). Decline in performance due to uncertainty because of inherent limitation in the stimulus array.

F (1). The false-alarm rate.

false alarm (1). In a yes-no experiment, a response of "yes" to S_1 (the stimulus class for which "no" is correct).

false-alarm/hit pair (1). The false-alarm and hit rates considered as an ordered pair; graphically, a point in ROC space.

false-alarm rate (1). The proportion of false alarms on S_1 trials.

feature-complete factorial design (10). Identification design in which each value of one variable is combined with each value of the others.

feedback (3, 5). Information provided at the end of a trial about whether the response was correct.

fixed discrimination (5). A discrimination task in which only two stimulus classes can occur in a block of trials, so that only one sensitivity parameter is estimated.

forced choice (7, 10). A discrimination experiment in which m stimuli are presented on each trial, one containing a sample of S_2, the others samples of S_1.

General Recognition Theory (GRT) (6). Formulation of multidimensional SDT.

H (1). The hit rate.

high-threshold theory (4). A threshold theory with a finite number of internal states, one or more of which can only be activated by a specific corresponding stimulus. See single high-threshold theory and double high-threshold theory.

hit (1). In a yes-no experiment, a response of "yes" to S_2 (the stimulus class for which "yes" is correct).

hit rate (1). The proportion of hits on S_2 trials.

hypothesis testing (13). Statistical evaluation of statements about population parameters.

ideal observer (12). Decision strategy that uses all available information and thus maximizes performance.

identification (I, 5, 10). (a) absolute identification. (b) classification.

identification operating characteristic (10). In a simultaneous detection and identification experiment, the function relating the probability of both a correct detection and a correct identification to the probability of a false alarm.

implied ROC (1). See ROC.

incomplete correspondence experiments (Introduction). See correspondence experiment.

independent channels (10). Channels whose outputs are independent random variables.

independent-observation rule (8, 9). Rule by which the observer independently combines the observations in multiple intervals or channels to reach a decision.

independent random variables (A1). Two or more variables whose joint distribution is such that the value of one variable does not affect the value of another.

index (1). Same as statistic.

integrality (8). Dependence between dimensions, as measured operationally in the Garner paradigm.

integration rule (6). Rule for combining information by adding or subtracting values of multiple dimensions.

internal noise (12). Variability limiting performance that arises within the observer rather than from the stimulus itself.

internal representation (1). Same as decision space.

intrinsic uncertainty (8). Decline in performance due to uncertainty because of nonoptimal processing by the observer.

IOC (10). See identification operating characteristic.

isobias curve (2). A curve in ROC space connecting points with the same response bias but different sensitivities. An isobias curve may be theoretical (implied by a theory or sensitivity parameter) or empirical (observed in an experiment).

isosensitivity curve (1). Same as ROC.

joint distribution (6). Distribution of more than one variable.

just-noticeable-difference (*jnd*) (1, 5). See difference threshold.

least-squares fit (A1). Method of approximating data by a model so that the sum of the squared deviations between the model and data is as small as possible.

likelihood ratio (2). The odds that an event arose from one distribution rather than another. When the distributions are underlying ones due to two possible stimulus classes, a measure of response bias.

log odds transformation (1). A transformation that converts a proportion p to the natural log of $p/(1 - p)$.

logistic distribution (1). The form of underlying distribution assumed by Choice Theory for the one-interval design.

logistic regression (13). A statistical technique that can be used to test hypotheses about detection theory parameters.

logit (4). Unit proportional to the standard deviation of the logistic distribution, and equal to the natural log of $p/(1 - p)$.

low-threshold theory (4). A threshold theory with two internal states, each of which can be activated by either of the possible stimuli.

m-alternative forced choice (*m*AFC) (10). An m-interval experiment in which the observer must determine which interval contains a sample of S_2 (all others containing samples of S_1).

matching to sample (9). Same as ABX.

maximum-likelihood estimation (11, 13). Estimation of a parameter by finding the value for which the observed data are most likely.

maximum rule (6). A decision rule in which observations on all dimensions must exceed the respective criteria for a positive response to be made.

maximum $p(c)$ (6). The highest value of $p(c)$ that could be obtained by an observer with a given value of sensitivity (e.g., value of d').

mean category scale (5). A scale constructed from category scaling data, assigning to each stimulus the average of the categories used in responding to it.

mean sensitivity (13). Estimate of a parameter obtained by averaging estimates based on individual stimuli, sessions, or subjects.

mean (shift) integrality (8). Type of perceptual integrality in which the dependence between two dimensions is reflected by distribution means.

measure (1). Same as statistic.

method of constant stimuli (5, 11). A classification design in which a standard stimulus is followed by one of a set of comparison stimuli.

minimum rule (6). A decision rule in which an observation above criterion on any dimension is sufficient for a positive response to be made.

miss (1). In a yes-no experiment, a response of "no" to S_2 (the stimulus class for which "yes" is correct).

miss rate (1). The proportion of misses on S_2 trials.

Multidimensional Signal Detection Analysis (MSDA) (10). Method for assessing various types of independence in a feature-complete identification design.

multiple-look experiment (8). Design in which a sample of one stimulus class or the other is presented in each of several intervals.

Neyman–Pearson objective (2). Maximizing the hit rate while keeping the false-alarm rate at some fixed low level.

nonparametric measure (4). A measure making no distributional assumptions.

normal distribution (1). The form of underlying distribution assumed by SDT.

oddity (9). A design in which three (or more) stimuli are presented on each trial, one from one stimulus class, the rest from the other. The observer must choose the "odd" interval.

one-interval design (I). A paradigm in which one stimulus is presented on each trial.

optimal decision rule (3). A decision rule that serves to maximize some performance criterion, such as payoffs.

parameter (13). A characteristic of some population, according to a theory.

Parameter Estimation by Sequential Testing (PEST) (11). An adaptive procedure in which the decision to change stimuli is based on a Wald test, and the amount by which the stimulus is changed depends on the past history of the experimental run.

payoff function or matrix (3). The rewards associated with each stimulus–response outcome in a correspondence experiment.

$p(c)$ (1). Proportion correct.

$p(c)_{max}$ (6). Maximum possible value of $p(c)$.

perceptual dimensionality (6). Number of dimensions needed to describe sensitivities to all pairs of stimuli in the stimulus set.

perceptual independence (6). Property of a joint distribution in a decision space that is equal to the product of its marginal distributions.

perceptual integrality (8). Dependence between two dimensions of an underlying representation, measured across the entire stimulus set.

perceptual separability (8). Dependence between two dimensions of an underlying representation, measured within a single stimulus class.

point of subjective equality (5). In a two-response classification experiment, the stimulus for which each response is equally likely.

pooled estimate (13). Estimate of a parameter obtained by averaging response frequencies before other calculations.

positivity (1). The property of being always positive or zero.

presentation probability (3). The probability of presenting one of the possible stimulus classes.

probability function (A1). For a discrete random variable, the function giving the probability of each value of the variable.

probability summation (6). Advantage in performance due to multiple chances at success.

probit analysis (11). A procedure for fitting the normal distribution function to psychometric function data.

product rule (6). In a two-dimensional representation, the probability that $X < a$ and $Y < b$ equals the probability that $X < a$ times the probability that $Y < b$.

projection (6). A technique for reducing a two-dimensional representation to one dimension.

proportion correct (1). Either (a) the number of correct responses (hits and correct rejections) divided by the number of trials or (b) the average of the hit and correct-rejection rates. A measure of sensitivity.

PSE (5). Point of subjective equality.

pseudo-d' (5). Sensitivity between two stimuli that correspond to the same response.

psychoacoustics (12). Study of the relations between (auditory) stimulus characteristics and psychological measures.

psychometric function (5, 11). In a discrimination task, function relating probability of response, or sensitivity, to stimulus value.

psychophysics (12). Study of the relations between methodological characteristics and psychological measures.

$p(T)$ (11). See target proportion.

q (4). The hit rate "corrected" by the single high-threshold correction for guessing.

q_{2AFC}, q_{mAFC} (7, 10). The "corrected" hit rates in 2AFC and mAFC, found from the single high-threshold correction for guessing.

QUEST (11). An adaptive procedure incorporating an a priori distribution, and based on Bayesian principles.

random variable (A1). A function defined on a sample space, taking on different values probabilistically.

range-frequency model (5). A theory of category scaling according to which responses depend on the range of stimuli and the frequencies with which they are presented.

rating experiment (3). A one-interval experiment in which the set of (more than two) possible responses is used to express confidence that S_1 or S_2 was presented.

ratio scale (1). A scale that has a nonarbitrary zero, and that can be used to make meaningful statements about ratios. Distance measures are ratio scaled.

receiver operating characteristic (1). See ROC.

recognition (I, 10). (a) A discrimination experiment in which neither stimulus is Null. (b) An identification experiment.

regularity (1). A characteristic of ROCs. Regular ROCs increase from (0,0) to (1,1). The hit rate cannot be 1 unless the false-alarm rate is also 1, and the false-alarm rate cannot be 0 unless the hit rate is also 0.

relative context variance (5). Context variance divided by sensory variance.

relative criterion (2). The criterion location relative to a sensitivity measure.

reminder design (7). Paradigm in which a fixed "reminder" stimulus is presented on every trial.

response bias (1). The tendency to use a response with some frequency irrespective of the stimulus presented.

reversal (11). In an adaptive procedure, a change in stimulus value in the opposite direction of the previous change.

rms (3). Root-mean-square.

ROC (curve) (1). Receiver operating characteristic. A curve in ROC space connecting points with the same sensitivity but different response biases. An ROC may be theoretical (implied by a theory or sensitivity parameter) or empirical (observed in an experiment).

ROC space (1). The unit square, with false-alarm rate on the x-axis and hit rate on the y-axis.

root-mean-square (3). The square root of the average of the squares; a kind of average.

roving discrimination (7). A discrimination task in which more than two stimulus classes can occur in a block of trials, so that more than one sensitivity parameter is estimated.

s (slope of ROC) (3). Slope of the ROC, according to SDT, on z-transformed coordinates.

S' (4). Nonparametric measure of sensitivity in the two-response rating design.

same-different experiment (9). A two-interval experiment in which the observer must determine whether the two stimuli are the same or different.

sample space (A1). Set of all possible events that can, with some probability, occur.

sampling distribution (13, A1). Distribution of a statistic across repeated measurements.

saturated model (13). A model that fits the data perfectly and in which every possible effect is included.

sensitivity (1). (a) The ability to discriminate, that is, to capture the experimenter-defined correspondence by appropriate responding. (b) A measure of discriminability that is not affected by response bias.

sensory variance (5). Variability in sensitivity contributed by sensory coding.

separability (8). Independence between dimensions, as measured operationally in the Garner paradigm.

Signal Detection Theory (SDT) (1). (a) A version of detection theory in which underlying distributions are assumed to be normal. (b) Same as detection theory. [In this book, definition (a) always applies.]

simultaneous detection and identification (10). A task in which the observer must detect the presence of a stimulus, and also report which of several possible stimuli occurred.

simultaneous simple and compound detection (8). A detection task in which simple, null, or compound stimuli may occur.

single high-threshold theory (4). A theory with two internal states and a high threshold.

staircase method (11). An adaptive procedure in which the stimulus levels are chosen from a fixed (often uniformly spaced) set.

standard stimulus (5). Stimulus presented on every trial as a reference point for judging the comparison stimulus.

state diagram (4). A representation of the decision space for threshold theories, giving the probabilities with which each state arises from each stimulus and leads to each response.

statistic (1). A function of data, usually calculated to estimate the parameter of a theory.

subliminal perception (4, 10). In a simultaneous detection and recognition experiment, above-chance recognition performance on trials on which the detection response is "no," or for which detection sensitivity is zero.

summation rule (8). See integration rule.

sweat factor (11). Variance of an estimator multiplied by the number of trials needed to obtain it.

symmetry (1). Property of distance measures: distance from A to B equals distance from B to A.

target proportion (11). Level of $p(c)$ on the psychometric function corresponding to the empirical threshold being estimated by an adaptive procedure.

threshold, empirical (5, 11). The stimulus or stimulus difference corresponding, on the psychometric function, to a specific level of performance.

threshold, theoretical (4). The point in a decision space that, according to a threshold theory, divides one internal state from another.

threshold theory (4). A theory of discrimination postulating a small number of internal sensory states.

time-order error (7). The tendency for performance in a two-interval task to depend on the order of the stimuli and the time between them.

total sensitivity (5). Sensitivity to the difference between two endpoint stimuli, sometimes inferred from sensitivities to adjacent stimulus pairs.

trace variance (7). Variability in discrimination due to time lapse between intervals.

trading relations (5). Changes in two variables that result in the same response rate or sensitivity.

transformation (1). A function that converts input values to (generally) different output values.

transformed ROC (1). An ROC plotted on the coordinates $f(H)$ and $f(F)$, where f is the transformation applied to H and F to compute sensitivity. In SDT, $f = z$.

triangle inequality (1). A characteristic of distance measures: The distance from A to C must be less than or equal to the distance from A to B plus that from B to C.

triangular method (9). Oddity design with three intervals.

true probability (A1). Value of a probability in the population.

two-alternative forced-choice (5). A two-interval discrimination experiment in which S_1 and S_2 are presented in either order. m-alternative forced-choice for $m = 2$.

type-2 ROC (3). Graph in ROC space relating confidence judgments on correct trials to confidence judgments on incorrect trials.

UDTR (11). Up–down transformed method.

unboundedness (1). The property of having no theoretical maximum or minimum magnitude.

uncertain detection (8). See detection with uncertainty.

underlying distributions (1). The distributions of internal events arising from a stimulus set.

unsaturated model (13). A model in which not every possible effect is included; used to evaluate the statistical significance of the omitted effects.

up–down transformed method (11). An adaptive procedure in which the sequence of correct and incorrect responses since the last stimulus change determines the next stimulus according to a staircase method.

variance of a random variable (A1). Average squared deviation of values from the expectation.

Wald test (11). A decision process in which the stimulus is changed whenever $p(c)$ deviates sufficiently from $p(T)$.

yes rate (4). In a yes-no experiment, the proportion of "yes" responses. A measure of response bias.

yes-no design (I). A one-interval experiment in which there are two possible responses, which may or may not be "yes" and "no."

zROC (1). ROC on z coordinates.

z transformation (1). A transformation that converts a proportion to the z score such that the proportion of the area under a normal distribution below that z score is p.

References

Aaronson, D., & Watts, B. (1987). Extensions of Grier's computational formulas for A' and B'' to below-chance performance. *Psychological Bulletin, 102*, 439–442.

Abbott, E. A. (1991/1884). *Flatland*. Princeton, NJ: Princeton University Press.

Anderson, N. H. (1974). Algebraic models in perception. In E. C. Carterette & M. P. Friedman (Eds.), *Handbook of perception: Vol. 2. Psychophysical judgment and measurement* (pp. 215–298). New York: Academic Press.

Ashby, F. G., & Gott, R. E. (1988). Decision rules in the perception and categorization of multidimensional stimuli. *Journal of Experimental Psychology: Learning, Memory, and Cognition, 14*, 33–53.

Ashby, F. G., & Lee, W. W. (1991). Predicting similarity and categorization from identification. *Journal of Experimental Psychology: General, 120*, 150–172.

Ashby, F. G., & Maddox, W. T. (1994). A response time theory of separability and integrality in speeded classification. *Journal of Mathematical Psychology, 38*, 423–466.

Ashby, F. G., & Townsend, J. T. (1986). Varieties of perceptual independence. *Psychological Review, 93*, 154–179.

Balakrishnan, J. D. (1998). Some more sensitive measures of sensitivity and response bias. *Psychological Methods, 3*, 68–90.

Balakrishnan, J. D. (1999). Decision processes in discrimination: Fundamental misconceptions of signal detection theory. *Journal of Experimental Psychology: Human Perception and Performance, 25*, 1189–1206.

Banks, W. P. (1970). Signal detection theory and human memory. *Psychological Bulletin, 74*, 81–99.

Banks, W. P. (2000). Recognition and source memory as multivariate decision processes. *Psychological Science, 11*, 267–273.

Benzschawel, T., & Cohn, T. E. (1985). Detection and recognition of visual targets. *Journal of the Optical Society of America, A, 2*, 1543–1550.

Berg, B. G. (1989). Analysis of weights in multiple observation tasks. *Journal of the Acoustical Society of America, 86*, 1743–1746.

Berg, B. G., Robinson, D. E., & Grantham, W. (1989). Multiple observation tasks: A partitioned variance model.

Berliner, J. E., & Durlach, N. I. (1973). Intensity perception: IV. Resolution in roving-level discrimination. *Journal of the Acoustical Society of America, 53*, 1270–1287.

Bi, J. (2002). Variance of d' for the *same-different* method. *Behavior Research Methods, Instruments, & Computers, 34*, 37–45.

Bi, J., Ennis, D. M., & O'Mahony, M. (1997). How to estimate and use the variance of d' from difference tests. *Journal of Sensory Studies, 12*, 87–104.

Blough, D. (1958). A method for obtaining psychophysical thresholds from the pigeon. *Journal of Experimental Analysis of Behavior, 1*, 31–43.

Bock, R. D., & Jones, L. V. (1968). *The measurement and prediction of judgment and choice.* San Francisco: Holden-Day.

Bonnel, A.-M., & Miller, J. (1994). Attentional effects on concurrent psychophysical discriminations: Investigations of a sample-size model. *Perception & Psychophysics, 55*, 162–179.

Braida, L. D., & Durlach, N. I. (1972). Intensity perception: II. Resolution in one-interval paradigms. *Journal of the Acoustical Society of America, 51*, 483–502.

Braida, L. D., & Durlach, N. I. (1988). Peripheral and central factors in intensity perception. In G. M. Edelman, W. E. Gall, & W. M. Cowan (Eds.), *Auditory Function* (pp. 559–583). New York: John Wiley and Sons.

Broadbent, D. (1958). *Perception and communication.* London: Pergamon.

Brophy, A. L. (1985). Approximation of the inverse normal distribution function. *Behavior Research Methods, Instrumentation, & Computers, 17*, 415–417.

Byers, A. J., & Abrams, D. (1953). A comparison of the triangular and two-sample taste test methods. *Food Technology, 7*, 185–187.

Carney, A. E., Widin, G. P., & Viemeister, N. F. (1977). Noncategorical perception of stop consonants differing in VOT. *Journal of the Acoustical Society of America, 62*, 961–970.

Chase, S., Bugnacki, P., Braida, L. D., & Durlach, N. I. (1983). Intensity perception: XII. Effect of presentation probability on absolute identification. *Journal of the Acoustical Society of America, 73*, 279–284.

Chen, H., & Macmillan, N. A. (1990,). *Sensitivity and bias in same-different and 2AFC discrimination.* Paper presented at the Eastern Psychological Association, Philadelphia.

Clarke, F. R. (1957). Constant-ratio rule for confusion matrices in speech communication. *Journal of the Acoustical Society of America, 29*, 515–520.

Clarke, F. R., Birdsall, T. G., & Tanner, W. P., Jr. (1959). Two types of ROC curves and definitions of parameters. *Journal of the Acoustical Society of America, 31*, 629–630.

Cornsweet, T. N. (1962). The staircase method in psychophysics. *American Journal of Psychology, 75*, 485–491.

Cornsweet, T. N. (1970). *Visual perception.* New York: Academic Press.

Craven, B. J. (1992). A table of d' for M-alternative odd-man-out forced-choice procedures. *Perception & Psychophysics, 51*, 379–385.

Creelman, C. D. (1960). Detection of signals of uncertain frequency. *Journal of the Acoustical Society of America, 32*, 805–810.

Creelman, C. D. (1963a). Auditory sensitivity and pressure at the tympanum. *Journal of the Acoustical Society of America, 35*, 777 (Abstract).

Creelman, C. D. (1963b). Detection, discrimination, and the loudness of short tones. *Journal of the Acoustical Society of America, 35*, 1201–1205.

Creelman, C. D. (1965). Discriminability and scaling of linear extent. *Journal of Experimental Psychology, 70*, 192–200.

Creelman, C. D., & Macmillan, N. A. (1979). Auditory phase and frequency discrimination: A comparison of nine paradigms. *Journal of Experimental Psychology: Human Perception and Performance, 5*, 146–156.

Creelman, M. B. (1966). *The experimental investigation of meaning.* New York: Springer.

Dai, H., Versfeld, N. J., & Green, D. M. (1996). The optimum decision rules in the *same-different* paradigm. *Perception & Psychophysics, 58*, 1–9.

Darlington, R. B., & Carlson, P. M. (1987). *Behavioral statistics.* New York: The Free Press.

Davies, D. R., & Parasuraman, R. (1982). *The psychology of vigilance.* London: Academic.

Davis, E. T., & Graham, N. (1981). Spatial frequency uncertainty effects in the detection of sinusoidal gratings. *Vision Research, 21*, 705–712.

DeCarlo, L. T. (1998). Signal detection theory and generalized linear models. *Psychological Methods, 3*, 186–205.

Deese, J. (1959). On the prediction of occurrence of particular verbal intrusions in immediate recall. *Journal of Experimental Psychology, 58*, 17–22.

Delboef, J. R. L. (1883). *Examen critique de la loi psychophysique: Sa base et sa signification.* Paris: Baillere.

Diehl, R. L. (1981). Feature detectors for speech: A critical reappraisal. *Psychological Bulletin, 89*, 1–18.

Dixon, W. J., & Mood, A. M. (1948). A method for obtaining and analyzing sensitivity data. *Journal of the American Statistical Association, 43*, 109–126.

Donaldson, W. (1992). Measuring recognition memory. *Journal of Experimental Psychology: General, 121*, 275–277.

Donaldson, W. (1996). The role of decision processes in remembering and knowing. *Memory & Cognition, 24*, 523–533.

Donaldson, W., & Good, C. (1996). A′r: An estimate of area under isosensitivity curves. *Behavior Research Methods, Instruments, & Computers, 28*, 590–597.

Dorfman, D. D., & Alf, E., Jr. (1969). Maximum likelihood estimation of parameters of signal detection theory and determination of confidence intervals—Rating method data. *Journal of Mathematical Psychology, 6*, 487–496.

Dorfman, D. D., & Bernbaum, K. S. (1986). RSCORE-J: Pooled rating-method data: A computer program for analyzing pooled ROC curves. *Behavior Research Methods, Instruments, & Computers, 18*, 452–462.

Dosher, B. A. (1984). Discriminating preexperimental (semantic) from learned (episodic) associations: A speed-accuracy study. *Cognitive Psychology, 16*, 519–584.

Durlach, N. I., & Braida, L. D. (1969). Intensity perception: I. Preliminary theory of intensity resolution. *Journal of the Acoustical Society of America, 46*, 372–383.

Dusoir, A. E. (1975). Treatments of bias in detection and recognition models: A review. *Perception & Psychophysics, 17*, 167–178.

Dusoir, T. (1983). Isobias curves in some detection tasks. *Perception & Psychophysics, 33*, 403–412.

Egan, J. P. (1958). *Recognition memory and the operating characteristic* (Technical Note AFCRC-TN-58-51). Bloomington, IN: Indiana University Hearing and Communication Laboratory.

Egan, J. P. (1975). *Signal detection theory and ROC analysis.* New York: Academic Press.

Egan, J. P., & Clarke, F. R. (1956). Source and behavior theory in the use of a criterion. *Journal of the Acoustical Society of America, 28*, 1267–1269.

Egan, J. P., Schulman, A. I., & Greenberg, G. Z. (1959). Operating characteristics determined by binary decisions and by ratings. *Journal of the Acoustical Society of America, 31*, 768–773.

Ekman, P., O'Sullivan, M., & Frank, M. G. (1999). A few can catch a liar. *Psychological Science, 10*, 263–266.

Elliott, P. B. (1964). Tables of d′. In J. A. Swets (Ed.), *Signal detection and recognition by human observers* (pp. 651–684). New York: Wiley.

Elman, J. L. (1979). Perceptual origins of the phoneme boundary effect and selective adaptation in speech. *Journal of the Acoustical Society of America, 65*, 190–207.

Emerson, P. L. (1986a). Observations on maximum-likelihood and Bayesian methods of forced-choice sequential threshold estimation. *Perception & Psychophysics, 39*, 151–153.

Emerson, P. L. (1986b). A quadrature method for Bayesian sequential threshold estimation. *Perception & Psychophysics, 39*, 381–383.

Estes, W. K. (2002). Traps in the route to models of memory and decision. *Psychonomic Bulletin & Review, 9*, 3–25.

Fechner, G. T. (1860). *Elemente der Psychophysik* (Reissued 1964 by Bonset, Amsterdam ed.). Leipzig: Breitkopf & Hartel.

Fidell, S. (1982). Comments on Mulligan and Shaw's "Multimodal signal detection: Independent decisions vs. integration." *Perception & Psychophysics, 31*, 90.

Finney, D. J. (1971). *Probit analysis* (3rd ed.). New York: Cambridge University Press.

Foley, J. M., & Legge, G. E. (1981). Contrast detection and near-threshold discrimination in human vision. *Vision Research, 21*, 1041–1053.

Francis, M. A., & Irwin, R. J. (1995). Decision strategies and visual-field asymmetries in *same-different* judgments of word meaning. *Memory & Cognition, 23*, 301–312.

Frijters, J. E. R. (1979a). The paradox of discriminatory nondiscriminators resolved. *Chemical Senses and Flavor, 4*, 355–358.

Frijters, J. E. R. (1979b). Variations of the triangular method and the relationship of its unidimensional probabilistic models to three alternative forced-choice signal detection theory models. *British Journal of Mathematical and Statistical Psychology, 32*, 229–241.

Frijters, J. E. R., Kooistra, A., & Vereijken, P. F. G. (1980). Tables of d' for the triangular method and the 3-AFC signal detection procedure. *Perception & Psychophysics, 27*, 176–178.

Fullerton, G. S., & Cattell, J. M. (1892). *On the perception of small differences* (Publications of the University of Pennsylvania, 2). Philadelphia: University of Pennsylvania Press.

Garner, W. R. (1962). *Uncertainty and structure as psychological concepts*. New York: Wiley.

Garner, W. R. (1974). *The processing of information and structure*. Potomac, MD: Lawrence Erlbaum Associates.

Garner, W. R., Hake, H. W., & Eriksen, C. W. (1956). Operationism and the concept of perception. *Psychological Review, 63*, 149–159.

Geisler, W. S., & Chou, K. L. (1995). Separation of low-level and high-level factors in complex tasks: Visual search. *Psychological Review, 102*, 356–378.

Gerrits, E., & Schouten, M. E. H. (2004). Categorical perception depends on the discrimination task. *Perception & Psychophysics, 66*, 363–376.

Getty, D. J., Pickett, R. M., D'Orsi, C. J., & Swets, J. A. (1988). Enhanced interpretation of diagnostic images. *Investigative Radiology, 23*, 240–252.

Glanzer, M., & Adams, J. K. (1985). The mirror effect in recognition memory. *Memory & Cognition, 13*, 8–20.

Glanzer, M., & Bowles, N. (1976). Analysis of the word-frequency effect in recognition memory. *Journal of Experimental Psychology: Human Perception and Performance, 2*, 21–31.

Gourevitch, V., & Galanter, E. (1967). A significance test for one-parameter isosensitivity functions. *Psychometrika, 32*, 25–33.

Graham, C. H. (1950). Behavior, perception, and the psychophysical methods. *Psychological Review, 57*, 108–120.

Graham, N., Kramer, P., & Haber, N. (1985). Attending to the spatial frequency and spatial position of near-threshold visual patterns. In M. I. Posner & O. S. M. Marin (Eds.), *Attention and Performance XI* (pp. 269–284). Hillsdale, NJ: Lawrence Erlbaum Associates.

Graham, N., Kramer, P., & Yager, D. (1987). Signal-detection models for multidimensional stimuli: Probability distributions and combination rules. *Journal of Mathematical Psychology, 31*, 366–409.

Graham, N., & Nachmias, J. (1971). Detection of grating patterns containing two spatial frequencies: A test of single-channel and multiple-channel models. *Vision Research, 11*, 251–259.

Graham, N. V. (1989). *Visual pattern analyzers*. New York: Oxford University Press.

Green, D. M. (1960). Auditory detection of a noise signal. *Journal of the Acoustical Society of America, 32*, 121–131.

Green, D. M. (1961). Detection of auditory sinusoids of uncertain frequency. *Journal of the Acoustical Society of America, 33*, 897–903.

Green, D. M. (1988). *Profile analysis: Auditory intensity discrimination*. New York: Oxford University Press.

Green, D. M. (1990). Stimulus selection in adaptive psychophysical procedures. *Journal of the Acoustical Society of America, 87*, 2662–2674.

Green, D. M., & Luce, R. D. (1975). Parallel psychometric functions from a set of independent detectors. *Psychological Review, 82*, 483–486.

Green, D. M., & Swets, J. A. (1966). *Signal detection theory and psychophysics.* New York: Wiley.

Green, D. M., Weber, D. L., & Duncan, J. E. (1977). Detection and recognition of pure tones in noise. *Journal of the Acoustical Society of America, 62*, 948–954.

Grey, D. R., & Morgan, B. J. T. (1972). Some aspects of ROC curve fitting: Normal and logistic models. *Journal of Mathematical Psychology, 9*, 128–139.

Grier, J. B. (1971). Nonparametric indexes for sensitivity and bias: Computing formulas. *Psychological Bulletin, 75*, 424–429.

Hacker, M. J., & Ratcliff, R. (1979). A revised table of d' for M-alternative forced-choice. *Perception & Psychophysics, 26*, 168–170.

Hall, J. L. (1974). PEST: Note on the reduction of variance of threshold estimates. *Journal of the Acoustical Society of America, 55*, 1090–1091.

Hall, J. L. (1981). Hybrid adaptive procedure for estimation of threshold functions. *Journal of the Acoustical Society of America, 69*, 1763–1769.

Hall, J. L. (1983). A procedure for detecting variability of psychophysical thresholds. *Journal of the Acoustical Society of America, 73*, 663–667.

Hautus, M. J. (1995). Corrections for extreme proportions and their biasing effects on estimated values of d'. *Behavior Research Methods, Instruments, & Computers, 27*, 46–51.

Hautus, M. J. (1997). Calculating estimates of sensitivity from group data: Pooled versus averaged estimators. *Behavior Research Methods, Instruments, & Computers, 29*, 556–562.

Hautus, M. J., & Collins, S. (2003). An assessment of response bias for the *same-different* task: Implications for the single-interval task. *Perception & Psychophysics, 65*, 844–860.

Hautus, M. J., & Meng, X. (2002). Decision strategies in the ABX (matching-to-sample) psychophysical task. *Perception & Psychophysics, 64*, 89–106.

Hays, W. L. (1994). *Statistics* (5th ed.). New York: Holt, Rinehart, & Winston.

Hecht, S., Schlaer, S., & Pirenne, M. H. (1942). Energy, quanta, and vision. *Journal of General Physiology, 25*, 819–840.

Heller, L. M., & Trahiotis, C. (1995). The discrimination of samples of noise in monotic, diotic and dichotic conditions. *Journal of the Acoustical Society of America, 97*, 3775–3781.

Helson, H. (1964). *Adaptation level theory.* New York: Harper & Row.

Hintzman, D. L., & Curran, T. (1994). Retrieval dynamics of recognition and frequency judgments: Evidence for separate processes of familiarity and recall. *Journal of Memory and Language, 33*, 1–18.

Hodge, M. H. (1967). Some further tests of the constant-ratio rule. *Perception & Psychophysics, 2*, 429–437.

Hodge, M. H., & Pollack, I. (1962). Confusion matrix analysis of single and multidimensional auditory displays. *Journal of Experimental Psychology, 63*, 129–142.

Hodos, W. (1970). Nonparametric index of response bias for use in detection and recognition experiments. *Psychological Bulletin, 74*, 351–354.

Holender, D. (1986). Semantic activation without conscious identification. *Behavioral and Brain Sciences, 9*, 1–66.

Houtsma, A. J. M., Durlach, N. I., & Braida, L. D. (1980). Intensity perception: XI. Experimental results on the relation of intensity resolution to loudness matching. *Journal of the Acoustical Society of America, 68*, 807–813.

Ingham, J. G. (1970). Individual differences in signal detection. *Acta Psychologica, 34*, 39–50.

Irwin, R. J., & Francis, M. A. (1995a). Perception of simple and complex visual stimuli: Decision strategies and hemispheric differences in same-different judgments. *Perception, 24*, 787–809.

Irwin, R. J., & Francis, M. A. (1995b). *Psychophysical analysis of same-different judgments of letter parity*. Paper presented at the Fechner Day 95, Cassis, France.

Irwin, R. J., & Hautus, M. J. (1997). Likelihood-ratio decision strategy for independent observations in the *same-different* task: An approximation to the detection-theoretic model. *Perception & Psychophysics, 59,* 313–316.

Irwin, R. J., Hautus, M. J., & Francis, M. A. (2001). Indices of response bias in the same-different experiment. *Perception & Psychophysics, 63,* 1091–1100.

Jacoby, L. L., Woloshyn, V., & Kelley, C. (1989). Becoming famous without being recognized: Unconscious influences of memory produced by dividing attention. *Journal of Experimental Psychology: General, 118,* 115–125.

Jesteadt, W. (1980). An adaptive procedure for subjective judgments. *Perception & Psychophysics, 28,* 85–88.

Jesteadt, W. (in press). The variance of d' estimates obtained in yes-no and two-interval-forced-choice procedures. *Perception & Psychophysics.*

Jesteadt, W., & Bilger, R. C. (1974). Intensity and frequency discrimination in one- and two-interval paradigms. *Journal of the Acoustical Society of America, 55,* 1266–1276.

Jesteadt, W., & Sims, S. L. (1975). Decision processes in frequency discrimination. *Journal of the Acoustical Society of America, 57,* 1161–1168.

Johnson, D. M., Watson, C. S., & Kelly, W. J. (1984). Performance differences among intervals in forced-choice tasks. *Perception & Psychophysics, 35,* 553–557.

Jones, F. N. (1974). History of psychophysics and judgment. In E. C. Carterette & M. P. Friedman (Eds.), *Handbook of Perception: Vol. 2. Psychophysical judgment and measurement* (pp. 1–22). New York: Academic Press.

Kadlec, H. (1995). Multidimensional signal detection analyses (MSDA) for testing separability and independence: A Pascal program. *Behavior Research Methods, Instruments, & Computers, 27,* 442–458.

Kadlec, H. (1999). MSDA_2: Updated version of software for multidimensional signal detection analyses. *Behavior Research Methods, Instruments, & Computers, 31,* 384–385.

Kadlec, H., & Townsend, J. T. (1992a). Implications of marginal and conditional detection parameters for the separabilities and independence of perceptual dimensions. *Journal of Mathematical Psychology, 36,* 325–374.

Kadlec, H., & Townsend, J. T. (1992b). Signal detection analyses of dimensional interactions. In F. G. Ashby (Ed.), *Multidimensional probabilistic models of perception and cognition* (pp. 181–227). Hillsdale, NJ: Lawrence Erlbaum Associates.

Kaernbach, C. (1990). A single-interval adjustment-matrix (SIAM) procedure for unbiased adaptive testing. *Journal of the Acoustical Society of America, 88,* 2645–2655.

Kaernbach, C. (1991). Simple adaptive testing with the weighted up-down method. *Perception & Psychophysics, 49,* 227–229.

Kahneman, D., Solvic, P., & Tversky, A. (1982). *Judgment under uncertainty: Heuristics and biases.* Cambridge: Cambridge University Press.

Kaplan, H. L. (1975). The five distractors experiment: Exploring the critical band with contaminated white noise. *Journal of the Acoustical Society of America, 58,* 404–411.

Kaplan, H. L., Macmillan, N. A., & Creelman, C. D. (1978). Tables of d' for variable-standard discrimination paradigms. *Behavior Research Methods & Instrumentation, 10,* 796–813.

Kershaw, C. D. (1985). Statistical properties of staircase estimates from two interval forced choice experiments. *British Journal of Mathematical & Statistical Psychology, 38,* 35–43.

Kinchla, R., & Smyzer, F. (1967). A diffusion model of perceptual memory. *Perception & Psychophysics, 2,* 219–229.

King-Smith, P. E., & Rose, D. (1997). Principles of an adaptive method for measuring the slope of the psychometric function. *Vision Research, 37,* 1595–1604.

Kingston, J., & Macmillan, N. A. (1995). Integrality of nasalization and F_1 in vowels in isolation and before oral and nasal consonants: A detection-theoretic application of the Garner paradigm. *Journal of the Acoustical Society of America, 97*, 1261–1285.

Klatzky, R. L., & Erdelyi, M. H. (1985). The response criterion problem in tests of hypnosis and memory. *International Journal of Clinical and Experimental Hypnosis, 33*, 246–257.

Klein, S. A. (1985). Double-judgment psychophysics: Problems and solutions. *Journal of the Optical Society of America, A, 2*, 1560–1585.

Klein, S. A. (2001). Measuring, estimating, and understanding the psychometric function: A commentary. *Perception & Psychophysics, 63*, 1421–1455.

Köhler, W. (1923). Zur Theorie des Sukzessivvergleichs und der Zeitfehler. *Psychologische Forschung, 4*, 115–175.

Kollmeier, B., Gilkey, R. H., & Sieben, U. K. (1988). Adaptive staircase techniques in psychoacoustics: A comparison of human data and a mathematical model. *Journal of the Acoustical Society of America, 83*, 1852–1862.

Kontsevich, L. L., & Tyler, C. W. (1999). Bayesian adaptive estimation of psychometric function slope. *Vision Research, 39*, 2729–2737.

Kornbrot, D. E. (1978). Theoretical and empirical comparison of Luce's choice model and logistic Thurstone model of categorical judgment. *Perception & Psychophysics, 24*, 193–208.

Kotel'nikov, V. A. (1960). *The theory of optimum noise immunity* (R. A. Silverman, Trans.). New York: McGraw-Hill.

Krantz, D. H. (1969). Threshold theories of signal detection. *Psychological Review, 76*, 308–324.

Kubovy, M., Rapoport, A., & Tversky, A. (1971). Deterministic vs. probabilistic strategies in detection. *Perception & Psychophysics, 9*, 427–429.

Laming, D. (1986). *Sensory analysis.* New York: Academic Press.

Lee, W. (1963). Choosing among confusably distributed stimuli with specified likelihood ratios. *Perceptual & Motor Skills, 16*, 445–467.

Lee, W., & Janke, M. (1964). Categorizing externally distributed stimulus samples for three continua. *Journal of Experimental Psychology, 68*, 376–382.

Lee, W., & Janke, M. (1965). Categorizing externally distributed stimulus samples for unequal molar probabilities. *Psychological Reports, 17*, 79–90.

Leek, M. R. (2001). Adaptive procedures in psychophysical research. *Perception & Psychophysics, 63*, 1279–1292.

Leek, M. R., Hanna, T. E., & Marshall, L. (1991). An interleaved tracking procedure to monitor unstable psychometric functions. *Journal of the Acoustical Society of America, 90*, 1385–1397.

Leek, M. R., Hanna, T. E., & Marshall, L. (1992). Estimation of psychometric functions from adaptive tracking procedures. *Perception & Psychophysics, 51*, 247–256.

Levitt, H. L. (1971). Transformed up-down methods in psychophysics. *Journal of the Acoustical Society of America, 49*, 467–477.

Liberman, A. M., Harris, K. S., Hoffman, H. S., & Griffith, B. C. (1957). The discrimination of speech sounds within and across phoneme boundaries. *Journal of Experimental Psychology, 54*, 358–368.

Licklider, J. C. R. (1959). Three auditory theories. In S. Koch (Ed.), *Psychology: A study of a science* (Vol. 1, pp. 41–144). New York: McGraw-Hill.

Lieberman, H. R., & Pentland, A. P. (1982). Computer technology: Microcomputer-based estimation of psychophysical thresholds: The best PEST. *Behavior Research Methods & Instrumentation, 14*, 21–25.

Lim, J. S., Rabinowitz, W. M., Braida, L. D., & Durlach, N. I. (1977). Intensity perception: VIII. Loudness comparisons between different types of stimuli. *Journal of the Acoustical Society of America, 62*, 1256–1267.

Lindner, W. A. (1968). Recognition performance as a function of detection criterion in a simultaneous detection-recognition task. *Journal of the Acoustical Society of America, 44,* 204–211.

Lindsay, P. H., Taylor, M. M., & Forbes, S. S. (1968). Attention and multidimensional discrimination. *Perception & Psychophysics, 4,* 113–117.

Lisker, L. (1975). Is it VOT or a first-formant transition detector? *Journal of the Acoustical Society of America, 57,* 1547–1551.

Long, G. R. (1973). The role of the standard in auditory amplitude discrimination. *Perception & Psychophysics, 13,* 49–59.

Luce, R. D. (1959). *Individual choice behavior.* New York: Wiley.

Luce, R. D. (1963a). Detection and recognition. In R. D. Luce, R. R. Bush, & E. Galanter (Eds.), *Handbook of mathematical psychology* (Vol. 1, pp. 103–189). New York: Wiley.

Luce, R. D. (1963b). A threshold theory for simple detection experiments. *Psychological Review, 70,* 61–79.

Luce, R. D. (1986). *Response times: Their role in inferring elementary mental organization.* New York: Oxford University Press.

Luce, R. D., & Galanter, E. (1963). Psychophysical scaling. In R. D. Luce, R. R. Bush, & E. Galanter (Eds.), *Handbook of mathematical psychology* (Vol. 1, pp. 245–307). New York: Wiley.

Luce, R. D., & Krumhansl, C. L. (1988). Measurement, scaling, and psychophysics. In R. C. Atkinson, R. J. Herrnstein, G. Lindzey, & R. D. Luce (Eds.), *Stevens' handbook of experimental psychology* (2nd ed., pp. 3–74). New York: Wiley.

Macmillan, N. A. (1971). Detection and recognition of increments and decrements in auditory intensity. *Perception & Psychophysics, 10,* 233–238.

Macmillan, N. A. (1986). The psychophysics of subliminal perception. *Behavioral and Brain Sciences, 9,* 38–39.

Macmillan, N. A. (1987). Beyond the categorical/continuous distinction: A psychophysical approach to processing modes. In S. Harnad (Ed.), *Categorical perception: The groundwork of cognition* (pp. 53–83). New York: Cambridge University Press.

Macmillan, N. A., & Braida, L. D. (1985). Toward a psychophysics of the speech mode. *Bulletin of the Psychonomic Society, 23,* 278. (Abstract)

Macmillan, N. A., Braida, L. D., & Goldberg, R. F. (1987). Central and peripheral processes in the perception of speech and nonspeech sounds. In M. E. H. Schouten (Ed.), *The psychophysics of speech perception* (pp. 28–45). The Hague: Martinus Nijhoff Publishers.

Macmillan, N. A., & Creelman, C. D. (1990). Response bias: Characteristics of detection theory, threshold theory, and "nonparametric" measures. *Psychological Bulletin, 107,* 401–413.

Macmillan, N. A., & Creelman, C. D. (1996). Triangles in ROC space: History and theory of "nonparametric" measures of sensitivity and response bias. *Psychonomic Bulletin & Review, 3,* 164–170.

Macmillan, N. A., & Creelman, C. D. (1997). *d′plus*: A program to calculate accuracy and bias measures from detection and discrimination data. *Spatial Vision, 11,* 141–143.

Macmillan, N. A., Goldberg, R. F., & Braida, L. D. (1988). Resolution for speech sounds: Basic sensitivity and context memory on vowel and consonant continua. *Journal of the Acoustical Society of America, 84,* 1262–1280.

Macmillan, N. A., & Kaplan, H. L. (1985). Detection theory analysis of group data: Estimating sensitivity from average hit and false-alarm rates. *Psychological Bulletin, 98,* 185–199.

Macmillan, N. A., Kaplan, H. L., & Creelman, C. D. (1977). The psychophysics of categorical perception. *Psychological Review, 84,* 452–471.

Macmillan, N. A., Rotello, C. M., & Miller, J. O. (2004). The sampling distributions of Gaussian ROC statistics. *Perception & Psychophysics, 66,* 406–421.

Maddox, W. T. (1992). Perceptual and decisional separability. In F. G. Ashby (Ed.), *Multidimensional models of perception and cognition* (pp. 147–180). Hillsdale, NJ: Lawrence Erlbaum Associates.

Maddox, W. T., & Estes, W. K. (1997). Direct and indirect stimulus-frequency effects in recognition. *Journal of Experimental Psychology: Learning, Memory, and Cognition, 23,* 539–559.

Madigan, R., & Williams, D. (1987). Maximum likelihood procedures in two-alternative forced-choice: Evaluation and recommendations. *Perception & Psychophysics, 42,* 240–249.

Marascuilo, L. A. (1970). Extensions of the significance test for one-parameter signal detection hypotheses. *Psychometrika, 35,* 237–243.

Markowitz, J., & Swets, J. A. (1967). Factors affecting the slope of empirical ROC curves: Comparison of binary and rating responses. *Perception & Psychophysics, 2,* 91–100.

Marsh, R. L., & Hicks, J. L. (1998). Test formats change source-monitoring decision processes. *Journal of Experimental Psychology: Learning, Memory, and Cognition, 24,* 1137–1151.

McFadden, D., & Callaway, N. (1999). Better discrimination of small changes in commonly encountered than in less commonly encountered auditory stimuli. *Journal of Experimental Psychology: Human Perception and Performance, 25,* 543–560.

McKee, S. P., Klein, S. A., & Teller, D. Y. (1985). Statistical properties of forced-choice psychometric functions: Implications of probit analysis. *Perception & Psychophysics, 37,* 286–298.

Metz, C. E., & Kronman, H. B. (1980). Statistical significance tests for binormal ROC curves. *Journal of Mathematical Psychology, 22,* 218–243.

Miller, G. A. (1953). What is information measurement? *American Psychologist, 8,* 3–11.

Miller, G. A. (1956). The magical number seven, plus or minus two: Some limits on our capacity for processing information. *Psychological Review, 63,* 81–96.

Miller, J., & Ulrich, R. (2001). On the analysis of psychometric functions: The Spearman-Kärber method. *Perception & Psychophysics, 63,* 1399–1420.

Miller, J. O. (1996). The sampling distribution of d'. *Perception & Psychophysics, 58,* 65–72.

Miller, M. B., & Wolford, G. L. (1999). Theoretical commentary: The ro e of criterion shift in false memory. *Psychological Review, 106,* 398–405.

Moore, B. C. J. (2003). *An introduction to the psychology of hearing* (5th ed.). San Diego, CA: Academic Press.

Mulligan, R., & Shaw, M. (1980). Multimodal signal detection: Independent decisions vs. integration. *Perception & Psychophysics, 28,* 471–478.

Nachmias, J. (1981). On the psychometric function for contrast detection. *Vision Research, 21,* 215–223.

Neisser, U. (1967). *Cognitive psychology.* New York: Appleton-Century-Crofts.

Nisbett, R. E., & Wilson, T. C. (1977). Telling more than we can know: Verbal reports on mental processes. *Psychological Review, 84,* 231–259.

Noreen, D. L. (1981). Optimal decision rules for some common psychophysical paradigms. In S. Grossberg (Ed.), *Mathematical psychology and psychophysiology* (pp. 237–280). Providence, RI: American Mathematical Society.

Nosofsky, R. (1985). Luce's choice model and Thurstone's categorical judgment model compared: Kornbrot's data revisited. *Perception & Psychophysics, 37,* 89–91.

Nosofsky, R. M. (1984). Choice, similarity, and the context theory of classification. *Journal of Experimental Psychology: Learning, Memory, and Cognition, 10,* 104–114.

Ogilvie, J. C., & Creelman, C. D. (1968). Maximum-likelihood estimation of receiver operating characteristic curve parameters. *Journal of Mathematical Psychology, 5,* 377–391.

Osgood, C. E. (1958). *Method and theory in experimental psychology.* Oxford: Oxford University Press.

Parducci, A. (1974). Contextual effects: A range-frequency analysis. In E. C. Carterette & M. P. Friedman (Eds.), *Handbook of perception: Vol. 2. Psychophysical judgment and measurement* (pp. 127–141). New York: Academic Press.

Park, J., & Banaji, M. R. (2000). Mood and heuristics: The influence of happy and sad states on sensitivity and bias in stereotyping. *Journal of Personality and Social Psychology, 78,* 1005–1023.

Pastore, R. E., Friedman, C. J., & Baffuto, K. J. (1976). A comparative evaluation of the AX and two ABX procedures. *Journal of the Acoustical Society of America, 60,* S120 (Abstract).

Patalano, A. L., Smith, E. E., & Jonides, J. (2001). PET evidence for multiple strategies of categorization. *Cognitive, Affective & Behavioral Neuroscience, 1,* 360–370.

Pelli, D. G. (1985). Uncertainty explains many aspects of visual contrast detection and discrimination. *Journal of the Optical Society of America A, 2,* 1508–1532.

Pentland, A. (1980). Maximum likelihood estimation: The best PEST. *Perception & Psychophysics, 28,* 377–379.

Peterson, W. W., Birdsall, T. G., & Fox, W. C. (1954). The theory of signal detectability. *Transactions of the IRE Professional Group on Information Theory, PGIT-4,* 171–212.

Pierce, J. R., & Gilbert, E. N. (1958). On AX and ABX limens. *Journal of the Acoustical Society of America, 30,* 593–595.

Pollack, I., & Hsieh, R. (1969). Sampling variability of the area under the ROC-curve and of d'_e. *Psychological Bulletin, 71,* 161–173.

Pollack, I., & Norman, D. A. (1964). A nonparametric analysis of recognition experiments. *Psychonomic Science, 1,* 125–126.

Pollack, I., & Pisoni, D. B. (1971). On the comparison between identification and discrimination tests in speech perception. *Psychonomic Science, 24,* 299–300.

Press, W. H., Flannery, B. P., Teukolsky, S. A., & Vetterling, W. T. (1986). *Numerical recipes: The art of scientific computing.* New York: Cambridge University Press.

Pynn, C. T., Braida, L. D., & Durlach, N. I. (1972). Intensity perception: III. Resolution in small-range identification. *Journal of the Acoustical Society of America, 51,* 559–566.

Quick, R. F. (1974). A vector magnitude model of contrast detection. *Kybernetic, 16,* 65–67.

Rabin, M. D., & Cain, W. S. (1984). Odor recognition: Familiarity, identifiability, and encoding consistency. *Journal of Experimental Psychology: Learning, Memory, & Cognition, 10,* 316–325.

Rammsayer, T. H. (1992). An experimental comparison of the weighted up-down method and the transformed up-down method. *Bulletin of the Psychonomic Society, 30,* 425–427.

Ratcliff, R., McKoon, G., & Tindall, M. (1994). Empirical generality of data from recognition memory receiver-operating characteristic functions and implications for the global memory models. *Journal of Experimental Psychology: Learning, Memory, and Cognition, 20,* 763–785.

Ratcliff, R., Sheu, C.-F., & Gronlund, S. D. (1992). Testing global memory models using ROC curves. *Psychological Review, 99,* 518–535.

Reingold, E. M., & Merikle, P. M. (1988). Using direct and indirect measures to study perception without awareness. *Perception & Psychophysics, 44,* 563–575.

Repp, B. H. (1982). Phonetic trading relations and context effects: New experimental evidence for a speech mode of perception. *Psychological Bulletin, 92,* 81–110.

Robbins, H., & Monro, S. (1951). A stochastic approximation method. *Annals of Mathematical Statistics, 29,* 400–407.

Roediger, H. L., III, & McDermott, K. B. (1995). Creating false memories: Remembering words not presented in lists. *Journal of Experimental Psychology: Learning, Memory, and Cognition, 21,* 803–814.

Rosner, B. S. (1984). Perception of voice-onset-time continua: A signal detection analysis. *Journal of the Acoustical Society of America, 75,* 1231–1242.

Rotello, C. M., Macmillan, N. A., & Reeder, J. A. (2004). Sum-difference theory of remembering and knowing: A two-dimensional signal detection model. *Psychological Review, 111,* 588–616.

Sawusch, J. R., Nusbaum, H. C., & Schwab, E. (1980). Contextual effects in vowel perception: II. Evidence for two processing mechanisms. *Perception & Psychophysics, 27*, 421–434.

Schlauch, R. S., & Rose, R. M. (1990). Two-, three-, and four-interval forced-choice staircase procedures: Estimator bias and efficiency. *Journal of the Acoustical Society of America, 88*, 732–740.

Schönemann, P. H., & Tucker, L. R. (1967). A maximum likelihood solution for the method of successive intervals allowing for unequal stimulus dispersions. *Psychometrika, 32*, 403–417.

Schulman, A. I., & Mitchell, R. R. (1966). Operating characteristics from yes-no and forced-choice procedures. *Journal of the Acoustical Society of America, 40*, 473–477.

See, J. E., Warm, J. S., Dember, W. N., & Howe, S. R. (1997). Vigilance and signal detection theory: An empirical evaluation of five measures of response bias. *Human Factors, 39*, 14–29.

Shapiro, P. N., & Penrod, S. (1986). Meta-analysis of facial identification studies. *Psychological Bulletin, 100*, 139–156.

Shaw, M. L. (1980). Identifying attentional and decision-making components in information processing. In R. Nickerson (Ed.), *Attention and Performance VIII* (pp. 277–296). Hillsdale, NJ: Lawrence Erlbaum Associates.

Shaw, M. L. (1982). Attending to multiple sources of information: I. The integration of information in decision-making. *Cognitive Psychology, 14*, 353–409.

Shaw, M. L., & Mulligan, R. M. (1982). Models for bimodal signal detection: A reply to Fidell. *Perception & Psychophysics, 31*, 91–92.

Shelton, B. R., & Scarrow, I. (1984). Two-alternative versus three-alternative procedures for threshold estimation. *Perception & Psychophysics, 35*, 385–392.

Shepard, R. N. (1964). Attention and the metric structure of the stimulus space. *Journal of Mathematical Psychology, 1*, 54–87.

Shipley, E. F. (1960). A model for signal detection and recognition with uncertainty. *Psychometrika, 25*, 273–289.

Simpson, A. J., & Fitter, M. J. (1973). What is the best index of detectability? *Psychological Bulletin, 80*, 481–488.

Smith, J. E. K. (1982a). Recognition models evaluated: A commentary on Keren and Baggen. *Perception & Psychophysics, 31*, 183–189.

Smith, J. E. K. (1982b). Simple algorithms for M-alternative forced-choice calculations. *Perception & Psychophysics, 31*, 95–96.

Smith, W. D. (1995). Clarification of sensitivity measure A'. *Journal of Mathematical Psychology, 39*, 82–89.

Snodgrass, J. G., & Corwin, J. (1988). Pragmatics of measuring recognition memory: Applications to dementia and amnesia. *Journal of Experimental Psychology: General, 117*, 34–50.

Sorkin, R. D. (1962). Extensions of the theory of signal detectability to matching procedures in psychoacoustics. *Journal of the Acoustical Society of America, 34*, 1745–1751.

Sorkin, R. D. (1999). Spreadsheet signal detection. *Behavior Research Methods, Instruments, & Computers, 31*, 46–54.

Sorkin, R. D., & Dai, H. (1994). Signal detection analysis of the ideal group. *Organizational Behavior and Human Decision Processes, 60*, 1–13.

Sorkin, R. D., Hays, C. J., & West, R. (2001). Signal-detection analysis of group decision making. *Psychological Review, 108*, 183–203.

Spence, K. W. (1937). Differential responding in animals to stimuli varying within a single dimension. *Psychological Review, 44*, 430–444.

Sperling, G. A., & Dosher, B. A. (1986). Strategy and optimization in human information processing. In K. Boff, L. Kaufman, & J. Thomas (Eds.), *Handbook of perception and performance* (Vol. 1, pp. 2-1–2-65). New York: Wiley.

Starr, S. J., Metz, C. E., Lusted, L. B., & Goodenough, D. J. (1975). Visual detection and localization of radiographic images. *Radiology, 116*, 533–538.

Stevens, S. S. (1975). *Psychophysics: Introduction to its perceptual, neural, and social prospects.* New York: Wiley.

Stretch, V., & Wixted, J. T. (1998). Decision rules for recognition memory confidence judgments. *Journal of Experimental Psychology: Learning, Memory, and Cognition, 24,* 1397–1410.

Swets, J. A. (1959). Indices of signal detectability obtained with various psychophysical procedures. *Journal of the Acoustical Society of America, 31,* 511–513.

Swets, J. A. (1973). The relative operating characteristic in psychology. *Science, 182,* 990–1000.

Swets, J. A. (1986a). Form of empirical ROCs in discrimination and diagnostic tasks. *Psychological Bulletin, 99,* 181–198.

Swets, J. A. (1986b). Indices of discrimination or diagnostic accuracy: Their ROCs and implied models. *Psychological Bulletin, 99,* 100–117.

Swets, J. A., & Pickett, R. M. (1982). *Evaluation of diagnostic systems: Methods from signal detection theory.* New York: Academic Press.

Swets, J. A., Shipley, E. F., McKee, J. M., & Green, D. M. (1959). Multiple observations of signals in noise. *Journal of the Acoustical Society of America, 31,* 514–521.

Tanner, T. A., Haller, R. W., & Atkinson, R. C. (1967). Signal recognition as influenced by presentation schedules. *Perception & Psychophysics, 2,* 349–358.

Tanner, T. A., Rauk, J. A., & Atkinson, R. C. (1970). Signal recognition as influenced by information feedback. *Journal of Mathematical Psychology, 7,* 259–274.

Tanner, W. P., Jr. (1956). Theory of recognition. *Journal of the Acoustical Society of America, 28,* 882–888.

Tanner, W. P., Jr. (1961). Physiological implications of psychophysical data. *Annals of the New York Academy of Sciences, 89,* 752–765.

Tanner, W. P., Jr., & Swets, J. A. (1954). A decision-making theory of visual detection. *Psychological Review, 61,* 401–409.

Taylor, M. M., & Creelman, C. D. (1967). PEST: Efficient estimates on probability functions. *Journal of the Acoustical Society of America, 41,* 782–787.

Taylor, M. M., Forbes, S. M., & Creelman, C. D. (1983). PEST reduces bias in forced-choice psychophysics. *Journal of the Acoustical Society of America, 74,* 1367–1374.

Thurstone, L. L. (1927a). A law of comparative judgment. *Psychological Review, 34,* 273–286.

Thurstone, L. L. (1927b). Psychophysical analysis. *American Journal of Psychology, 38,* 369–389.

Torgerson, W. S. (1958). *Theory and methods of scaling.* New York: Wiley.

Trahiotis, C., & Bernstein, L. R. (1990). Detectability of interaural delays over select spectral regions: Effects of flanking noise. *Journal of the Acoustical Society of America, 87,* 810–813.

Treisman, A. M., & Gelade, G. (1980). A feature-integration theory of attention. *Cognitive Psychology, 12,* 97–136.

Treisman, M. (1998). Combining information: Probability summation and probability averaging in detection and discrimination. *Psychological Methods, 3,* 252–265.

Treutwein, B. (1995). Adaptive psychophysical procedures. *Vision Research, 35,* 2503–2522.

Tulving, E. (1985). Memory and consciousness. *Canadian Journal of Psychology, 26,* 1–12.

Tulving, E., & Lindsay, P. H. (1967). Identification of simultaneously presented simple visual and auditory stimuli. *Acta Psychologica, 27,* 101–109.

Tversky, A., & Kahneman, D. (1971). Belief in the law of small numbers. *Psychological Bulletin, 76,* 105–110.

Uchanski, R. M., Braida, L. D., & Durlach, N. I. (1981). Variability of interaural loudness comparisons. *Journal of the Acoustical Society of America [Suppl. 1], 69,* S104 (Abstract).

Urban, F. M. (1908). *The application of statistical methods to the problems of psychophysics.* Philadelphia: Psychological Clinic Press.

van Meter, D., & Middleton, D. (1954). Modern statistical approaches to reception in communication theory. *Transactions of the IRE Professional Group on Information Theory, PGIT-4,* 119–141.

Versfeld, N. J., Dai, H., & Green, D. M. (1996). The optimum decision rules for the oddity task. *Perception & Psychophysics, 58,* 10–21.

Vogels, R., & Orban, G. A. (1986). Decision processes in visual discrimination of line orientation. *Journal of Experimental Psychology: Human Perception and Performance, 12,* 115–143.

von Békésy, G. (1947). A new audiometer. *Acta Otolaryngology, 35,* 411–422.

Wald, A. (1947). *Sequential analysis.* New York: Wiley.

Waldmann, M. R., & Göttert, R. (1989). Response-bias in below-chance performance: Computation of the parametric measure β. *Psychological Bulletin, 106,* 338–340.

Watson, A. B., & Fitzhugh, A. (1990). The method of constant stimuli is inefficient. *Perception & Psychophysics, 47,* 87–91.

Watson, A. B., & Pelli, D. G. (1983). QUEST: A Bayesian adaptive psychometric method. *Perception & Psychophysics, 33,* 113–120.

Watt, R. J., & Andrews, D. P. (1981). APE: Adaptive probit estimation of psychometric functions. *Current Psychological Reviews, 1,* 205–214.

Wetherill, G. B., & Levitt, H. (1965). Sequential estimation of points on a psychometric function. *British Journal of Mathematical and Statistical Psychology, 18,* 1–10.

Wickelgren, W. A. (1968). Unidimensional strength theory and component analysis of noise in absolute and comparative judgments. *Journal of Mathematical Psychology, 5,* 102–122.

Wickelgren, W. A. (1969). Associative strength theory of recognition memory for pitch. *Journal of Mathematical Psychology, 6,* 13–61.

Wier, C. C., Jesteadt, W., & Green, D. M. (1976). A comparison of method-of-adjustment and forced-choice procedures in frequency discrimination. *Perception & Psychophysics, 19,* 75–79.

Wixted, J. T., & Stretch, V. (2000). The case against a criterion-shift account of false memory. *Psychological Review, 107,* 368–376.

Woodworth, R. S. (1938). *Experimental psychology.* New York: Holt.

Yasuhara, M., & Kuklinski, T. T. (1978). Category boundary effect for grapheme perception. *Perception & Psychophysics, 23,* 97–104.

Yellott, J. I., Jr. (1977). Relationship between Luce's choice axiom, Thurstone's theory of comparative judgment, and the double exponential distribution. *Journal of Mathematical Psychology, 15,* 109–144.

Yonelinas, A. P. (1997). Recognition memory ROCs for items and associative information: The contribution of recollection and familiarity. *Memory & Cognition, 25,* 747–763.

Zhang, J., & Mueller, S. T. (in press). A note on ROC analysis and non-parametric estimation of sensitivity. *Psychometrika.*

Author Index

Subject Index

A

A' (area under the one-point ROC),
  100–103
Absolute judgment, 113
  see also Classification, one-dimen-
    sional; Classification, multi-
    dimensional; Identification,
    absolute
ABX, 229–235
  decision space, 231–234
  differencing model, 233–234
  hit and false-alarm rates, 230
  independent-observation model,
    230–233
  vs other designs, 234–235, 253–255
  response bias, 232–233
  ROC, 232–233
  sensitivity, 232–233, 380–400
  threshold model, 234–235
Accuracy, *see* Sensitivity
Adaptation level, 129–130
Adaptive methods, 267–296
  components of, 277
  decision rules, 277–280
  evaluation of, 289–292
  vs nonadaptive methods, 269–270, 276
  for slope estimation, 293–294
  stepping rules, 281–285
  stopping rules, 285–286
  target proportion, 277, 280–281
  see also Adaptive Probit Estimation;
    Békésy audiometry; Hall's

adaptive method;
  Kaernbach's adaptive
  method; Maximum-likeli-
  hood estimation of thresholds; PEST; QUEST; UDTR;
  Wald rule
Adaptive Probit Estimation (APE), 294
A_g (area under the multipoint ROC), 64,
  330
α (sensitivity measure in Choice Theory),
  95–96
  and d', 95–96
Animal experiments, 229–233, 271–272
AOC, 206–209
a priori probabilities, *see* Presentation
  probabilities
Area theorem, 170–171
Area under the ROC, 170–175
  for multipoint ROC, *see* A_g
  for one-point ROC, *see* A'
  in SDT, *see* A_z
Attention
  capacity model, 207–209
  divided, 188, 204–205
  incomplete, 46–47
  operating characteristic, *see* AOC
  selective, 188, 203–204
  see also Uncertain detection; Uncer-
    tainty
Audiology, 269–270
AX design, *see* Same-different
A_z (area under the SDT ROC), 63, 172,
  330–331

"For the last many years I have been suggesting Macmillan and Creelman to those who ask me for a reference to detection theory. It is an excellent book and has proved useful to a wide variety of behavioral scientists who need detection theory as a tool. I am delighted to have this new edition to recommend, an edition which includes material that should make it of use to still more investigators. The new information about multidimensional signal-detection theory allows analysis of more complex experimental designs and, even more importantly from my persective, analysis of situations where there are multiple detectors, or channels, or pathways."

—*Norma Graham, Columbia University*

"Rarely, I believe, has a book so fine in its first edition been as enhanced in its second. It continues to serve handsomely as a handbook, neatly laying out practically everything an experimenter needs in order to select from and apply a wide range of methods and measures. Its purpose as a textbook has been notably advanced: for example, early chapters on basic detection theory and alternatives are reorganized to make fundamental ideas more accessible and the later material on complex stimuli and methods is integrated by a tutorial treatment of recent developments in multidimensional detection theory. This volume's friendliness to the reader, and its broad coverage and considerable sophistication (do see the "essays"), make it highly suitable for the student and very likely informative even for the experienced investigator."

—*John A. Swets, BBN Technologies*

Detection Theory, Second Edition is an introduction to one of the most important tools for analysis of data where choices must be made and performance is not perfect. Originally developed for evaluation of electronic detection, detection theory was adopted by psychologists as a way to understand sensory decision making, then embraced by students of human memory. It has since been utilized in areas as diverse as animal behavior and X-ray diagnosis.

This book covers the basic principles of detection theory, with separate initial chapters on measuring detection and evaluating decision criteria. Some other features include:

- complete tools for application, including flowcharts, tables, pointers, and software;
- student-friendly language;
- complete coverage of content area, including both one-dimensional and multidimensional models;
- separate, systematic coverage of sensitivity and response bias measurement;
- integrated treatment of threshold and nonparametric approaches;
- an organized, tutorial level introduction to multidimensional detection theory;
- popular discrimination paradigms presented as applications of multidimensional detection theory; and
- a new chapter on ideal observers and an updated chapter on adaptive threshold measurement.

ISBN 0-8058-4231-4

90000

9 780805 842319